AF352980

Genomics and Models of Nerve Sheath Tumors

Genomics and Models of Nerve Sheath Tumors

Editors

Angela C. Hirbe
Christine A. Pratilas
Rebecca D. Dodd

MDPI • Basel • Beijing • Wuhan • Barcelona • Belgrade • Manchester • Tokyo • Cluj • Tianjin

Editors

Angela C. Hirbe
Washington University
School of Medicine
USA

Christine A. Pratilas
Johns Hopkins University
School of Medicine
USA

Rebecca D. Dodd
Holden Comprehensive Cancer
Center at University of Iowa
USA

Editorial Office
MDPI
St. Alban-Anlage 66
4052 Basel, Switzerland

This is a reprint of articles from the Special Issue published online in the open access journal *Genes* (ISSN 2073-4425) (available at: https://www.mdpi.com/journal/animals/special_issues/ Farm_Animal_Transport).

For citation purposes, cite each article independently as indicated on the article page online and as indicated below:

LastName, A.A.; LastName, B.B.; LastName, C.C. Article Title. *Journal Name* **Year**, *Article Number*, Page Range.

ISBN 978-3-03943-489-3 (Hbk)
ISBN 978-3-03943-490-9 (PDF)

Contents

About the Editors

Angela C. Hirbe is a graduate of the Washington University M.D. Ph.D. program and completed her residency in Internal Medicine and fellowship in Oncology as part of the Physician Scientist Training Program. She has had a longstanding interest in neurofibromatosis research; her post-doctoral work was performed in the laboratory of Dr. David Gutmann, where she used next-generation sequencing technologies to identify beta-III spectrin as a protein involved in malignant peripheral nerve sheath tumor (MPNST) pathogenesis and developed a mouse model for this deadly type of sarcoma. This work transitioned into her own lab when she joined the faculty at Washington University.

Dr. Hirbe is currently an Assistant Professor in the Division of Medical Oncology in the Departments of Medicine and Pediatrics at Washington University in St. Louis. Her laboratory continues to use genomics to identify drivers in MPNST pathogenesis that can be exploited as diagnostic biomarkers or therapeutic targets.

Clinically, Dr. Hirbe is a member of the sarcoma section and part of the Adolescent Young Adult Cancer Program and the Neurofibromatosis Center at Washington University. Her clinical practice is geared at caring for patients with cancer predisposition syndromes such as neurofibromatosis type 1 and Li–Fraumeni syndrome as well as treating patients with any type of sarcoma. She has particular expertise in the care of patients with MPNST.

Christine A. Pratilas is Director of the Pediatric Sarcoma Program at the Sidney Kimmel Comprehensive Cancer Center and Associate Professor of Oncology and Pediatrics at the Johns Hopkins University School of Medicine. She joined the Hopkins pediatric sarcoma and solid tumor team in 2014, after completing her fellowship and spending the early years of her career at Memorial Sloan Kettering Cancer Center.

During her years at MSKCC, Dr. Pratilas focused her research on signal transduction, the molecular events that both activate and negatively regulate cancer cell signaling pathways. This knowledge helps to determine how to best deploy novel targeted therapies and to predict how resistance to these agents may emerge over time. One of her most important contributions to date has been advancing our understanding of the molecular alterations in a protein called BRAF, how RAF inhibitors work to inhibit cancer cell growth, and how cancer cellular networks adapt to RAF and MEK inhibitors.

Her current laboratory research focuses on deregulated ERK signaling in solid tumors, including pediatric sarcomas. ERK signaling output is activated in many human tumors, including those with BRAF and RAS mutations and those with loss of NF1. Several pediatric solid tumors express these mutations, including a subset of rhabdomyosarcoma, neuroblastoma, and MPNST. Her lab is investigating whether MEK inhibition can effectively inhibit ERK output in pediatric sarcomas with activation of ERK signaling. A primary focus is to determine the biochemical and adaptive signaling response to MEK inhibition and other targeted therapies, to identify mechanisms by which tumors with loss of NF1 or mutations in RAS can evade MEK inhibition, and to identify targets for more effective and combination therapy.

Dr. Pratilas' clinical expertise is in the management of children, adolescents, and young adults with all sarcomas, and specifically rhabdomyosarcoma and MPNST. In addition, she sees children with melanocytic neoplasms and melanoma and children with NF1-associated non-CNS neoplasms and other rare pediatric cancers. She has a strong interest in pediatric cancer genetics and, together with the Clinical Genetics Service and the Comprehensive Neurofibromatosis Center at Johns Hopkins, provides care for children with Li–Fraumeni syndrome, neurofibromatosis type 1 (NF1), RAS-opathies, and other cancer predisposition syndromes.

Rebecca D. Dodd is an Assistant Professor of Internal Medicine in Hematology/Oncology and Bone Marrow Transplantation at the University of Iowa Holden Comprehensive Cancer Center. Dr. Dodd completed her Ph.D. and post-doctoral fellowship at Duke University, where she focused on sarcoma metastasis and MPNST biology.

Dr. Dodd's lab uses powerful in vivo model systems to address complex questions in cancer biology. Her research program uses CRISPR/Cap and Cre-loxP technology for translational oncology research. Over the past decade, her group has built new mouse models of MPNST to investigate events that are difficult to study in patient populations. These models include new approaches for in vivo modeling of cancer, including novel somatic CRISPR/Cas9-based tumorigenesis tools to generate MPNSTs in wild-type adult mice. Other areas of interest include (1) epigenetically targeted therapies, (2) novel gene-editing tools, and (3) the tumor microenvironment.

Editorial

Special Issue: "Genomics and Models of Nerve Sheath Tumors"

Angela C. Hirbe [1,*], Rebecca D. Dodd [2,*] and Christine A. Pratilas [3,*]

1   Siteman Cancer Center, Washington University School of Medicine, 660 South Euclid Avenue, Campus Box 8076, St. Louis, MO 63110, USA
2   Holden Comprehensive Cancer Center, Carver College of Medicine, University of Iowa, 285 Newton Road, Iowa City, IA 52242, USA
3   The Sidney Kimmel Comprehensive Cancer Center, Johns Hopkins University School of Medicine, Baltimore, MD 21287, USA
*   Correspondence: hirbea@wustl.edu (A.C.H.); rebecca-dodd@uiowa.edu (R.D.D.); cpratil1@jhmi.edu (C.A.P.)

Received: 21 August 2020; Accepted: 28 August 2020; Published: 1 September 2020

Keywords: genomics; mouse models; NF1; nerve sheath tumors

Nerve sheath tumors arising in the context of neurofibromatosis type 1 (NF1) include benign tumors such as cutaneous, diffuse and plexiform neurofibromas; atypical neurofibromas or atypical neurofibromatosis neoplasms of uncertain biological potential (ANNUBP); and the aggressive soft tissue sarcoma, the malignant peripheral nerve sheath tumor (MPNST). Even benign tumors often represent a significant cause of morbidity for many patients, due to disfigurement, disability, or organ dysfunction. MPNST are aggressive, often metastasize, and are often lethal. An expanding body of literature related to genomic alterations common to MPNST, signaling events that regulate tumorigenesis, and novel models that recapitulate the human tumor, has informed novel therapeutic approaches. Despite numerous clinical trials, curative responses to treatment remain limited for patients with this malignancy. Here, we have compiled a series of articles that focus on the genomics of MPNST and the latest models generated to study these tumors.

Included in this Special Edition are six manuscripts that present original research highlighting novel therapeutic strategies, models, and genomic findings, as well as a whitepaper describing consortium efforts to genomically characterize MPNST. Staedke et al. [1] present a chemoprevention strategy repurposing two drugs already in clinical use for other indications (mebendazole and cyclooxygenase-2 inhibitors), utilizing one of the most commonly used preclinical models for preclinical testing of MPNST, the *cis Nf1+/−;Tp53+/− (NPcis)* mouse model [2,3]. In these studies, they report that mebendazole reduces levels of RAS-GTP, delays the formation of solid malignancy in at-risk mice, and increases survival. Further clinical studies are needed to validate the potential of this strategy in humans, but the study demonstrates the feasibility of a prevention strategy for NF1-associated malignancy. The article by Scherer et al. highlights newer mouse models of MPNST that use somatic CRISPR/Cas9 tumorigenesis to generate genomically-matched tumors in different background strains of wild-type mice [4]. This is the first study to systematically evaluate the impact of host strain on CRISPR/Cas9-generated mouse models and identifies several key strain-dependent phenotypes, including impacts on tumor onset and the tumor immune landscape. Moon and Tompkins et al. performed a comprehensive genomic analysis of multiple areas from within a single large MPNST. These authors identify varied genomic profiles within each area, highlighting the need for further studies on intra-tumoral heterogeneity in order to truly understand the genomic composition of any given tumor [5]. Such studies are critical to aid in our understanding of tumor and patient responsiveness and non-responsiveness to a range of therapies. Miller et al., on behalf of the Genomics of MPNST (GeM) consortium, present a whitepaper describing the composition, design, and analysis plan of this consortium, founded by the NF Research Initiative at Boston Children's Hospital. These authors have aimed to perform the most comprehensive genomic analysis of the largest cohort of MPNST to date, data from which will be shared on an outward-facing web-based interface made available to other investigators, in order to accelerate collaborative and therapy-directed research [6]. Grit et al. describe their experiments using reverse phase phospho-proteome array (RPPA) analysis of murine MPNST models to determine mechanisms of resistance to commonly-used therapies, including DNA damaging agents (doxorubicin) and kinase inhibitors (MET and MEK inhibitors). These authors observed profound signaling plasticity in treated tumors, with key activation of the AXL and NFkB pathways that were associated with the development of resistance [7]. Banerjee et al. set out to design an integrative approach that combined multiple transcriptomic and genomic datasets, the analysis of which would be poised to identify new therapeutic avenues in MPNST. Gene expression data from four independent studies were integrated and analyzed using a transfer learning-inspired approach to identify latent variables (LV)—groups of genes derived from larger repositories of gene expression datasets that exhibit common transcriptomic patterns relevant to a specific subset of samples—and thereby uncover previously unknown biology. To assess the biological underpinnings of uncharacterized LVs, a tumor immune cell deconvolution analysis was used, which indicated the presence of activated mast cells and M2 macrophages in all tumor types, as well as CD4 memory T-cells [8]. The findings uncovered using these computational approaches suggest potential biological signatures rich for experimental and clinical investigation.

The Special Edition also includes three review articles. Lemberg et al. have compiled a collated summary of sequencing efforts in MPNST published in the past two decades, using a total of 12 studies to summarize the range of incidences of the most common mutations in *NF1, CDKN2A, TP53, EED* and *SUZ12*. In this article, the authors review the initial findings of *NF1* as the gene responsible for neurofibromatosis type 1, its function as a RAS-GTPase-activating protein (RAS-GAP), and the spectrum of alterations in *NF1* found in human disease. They then further summarize 16 additional genomic studies, covering 10 other recurrently altered genes, including *BRAF, MET, EGFR, TYK2, ATRX* and others [9]. Williams and Largaespada review the range of published MPNST model systems, including genetically-engineered mouse models (GEMM), the genes involved, and the limitations of these models. They elaborate on the commonly used *NPCis* mouse, its genetic design, and the tumors that develop in these mice, as well as human-derived cell lines and xenografts. The use of synthetic lethality screens to identify combination drug therapies is explored, as are dysregulated signaling pathways that represent targets for molecularly based therapies [10]. The review article by Zhang et al. discusses the current biological understanding of polycomb repressive complex 2 (PRC2) loss in MPNST, which is a frequently-mutated pathway in these tumors. This article also highlights PRC2 function in normal Schwann cell development and nerve injury repair, in addition to discussing potential therapies that target PCR2 deficiency in tumor cells [11].

In conclusion, the articles that we have assembled in this Special Edition on Genomics and Models of Nerve Sheath Tumors highlight the most recent scientific advances on the genomic composition of malignant peripheral nerve sheath tumors and review novel efforts to model and study these tumors. While a wide range of benign, borderline and malignant nerve sheath tumors affect individuals with neurofibromatosis type 1, our collection of articles here focuses primarily on malignant nerve sheath tumors and underscores the pressing need for novel therapies. As genomic and transcriptomic capabilities continue to advance at an impressive pace, the hope is that an improved understanding of the genetics, and therefore the pathobiology, of these tumors, will ultimately lead to effective therapies that result in deeper and more durable responses, and therefore improved survival rates for these patients.

Author Contributions: A.C.H., R.D.D., and C.A.P. contributed to conceptualization, writing, and editing. All authors have read and agreed to the published version of the manuscript.

Funding: This research required no outside funding.

Acknowledgments: The Special Issue editors would like to thank all of the authors and reviewers who contributed to this Special Edition.

Conflicts of Interest: The authors declare no conflict of interest.

References

1. Staedtke, V.; Gray-Bethke, T.; Riggins, G.J.; Bai, R.Y. Preventative Effect of Mebendazole against Malignancies in Neurofibromatosis 1. *Genes* **2020**, *11*, 762. [CrossRef] [PubMed]
2. Cichowski, K.; Shih, T.S.; Schmitt, E.; Santiago, S.; Reilly, K.; McLaughlin, M.E.; Bronson, R.T.; Jacks, T. Mouse models of tumor development in neurofibromatosis type 1. *Science* **1999**, *286*, 2172–2176. [CrossRef] [PubMed]
3. Vogel, K.S.; Klesse, L.J.; Velasco-Miguel, S.; Meyers, K.; Rushing, E.J.; Parada, L.F. Mouse tumor model for neurofibromatosis type 1. *Science* **1999**, *286*, 2176–2179. [CrossRef] [PubMed]
4. Scherer, A.; Stephens, V.R.; McGivney, G.R.; Gutierrez, W.R.; Laverty, E.A.; Knepper-Adrian, V.; Dodd, R.D. Distinct Tumor Microenvironments Are a Defining Feature of Strain-Specific CRISPR/Cas9-Induced MPNSTs. *Genes* **2020**, *11*, 583. [CrossRef] [PubMed]
5. Moon, C.I.; Tompkins, W.; Wang, Y.; Godec, A.; Zhang, X.; Pipkorn, P.; Miller, C.A.; Dehner, C.; Dahiya, S.; Hirbe, A.C. Unmasking Intra-tumoral Heterogeneity and Clonal Evolution in NF1-MPNST. *Genes* **2020**, *11*, 499. [CrossRef] [PubMed]
6. Miller, D.T.; Cortes-Ciriano, I.; Pillay, N.; Hirbe, A.C.; Snuderl, M.; Bui, M.M.; Piculell, K.; Al-Ibraheemi, A.; Dickson, B.C.; Hart, J.; et al. Genomics of MPNST (GeM) Consortium: Rationale and Study Design for Multi-Omic Characterization of NF1-Associated and Sporadic MPNSTs. *Genes* **2020**, *11*, 387. [CrossRef] [PubMed]
7. Grit, J.L.; Pridgeon, M.G.; Essenburg, C.J.; Wolfrum, E.; Madaj, Z.B.; Turner, L.; Wulfkuhle, J.; Petricoin, E.F., 3rd; Graveel, C.R.; Steensma, M.R. Kinome Profiling of NF1-Related MPNSTs in Response to Kinase Inhibition and Doxorubicin Reveals Therapeutic Vulnerabilities. *Genes* **2020**, *11*, 331. [CrossRef] [PubMed]
8. Banerjee, J.; Allaway, R.J.; Taroni, J.N.; Baker, A.; Zhang, X.; Moon, C.I.; Pratilas, C.A.; Blakeley, J.O.; Guinney, J.; Hirbe, A.; et al. Integrative Analysis Identifies Candidate Tumor Microenvironment and Intracellular Signaling Pathways that Define Tumor Heterogeneity in NF1. *Genes* **2020**, *11*, 226. [CrossRef] [PubMed]
9. Lemberg, K.M.; Wang, J.; Pratilas, C.A. From Genes to -Omics: The Evolving Molecular Landscape of Malignant Peripheral Nerve Sheath Tumor. *Genes* **2020**, *11*, 691. [CrossRef] [PubMed]
10. Williams, K.B.; Largaespada, D.A. New Model Systems and the Development of Targeted Therapies for the Treatment of Neurofibromatosis Type 1-Associated Malignant Peripheral Nerve Sheath Tumors. *Genes* **2020**, *11*, 477. [CrossRef] [PubMed]
11. Zhang, X.; Murray, B.; Mo, G.; Shern, J.F. The Role of Polycomb Repressive Complex in Malignant Peripheral Nerve Sheath Tumor. *Genes* **2020**, *11*, 287. [CrossRef] [PubMed]

 *genes* 

Article

Preventative Effect of Mebendazole against Malignancies in Neurofibromatosis 1

Verena Staedtke [1,*], Tyler Gray-Bethke [1], Gregory J. Riggins [2] and Ren-Yuan Bai [2,*]

[1] Department of Neurology, Johns Hopkins University School of Medicine, Baltimore, MD 21231, USA; tgraybe1@jhmi.edu
[2] Department of Neurosurgery, Johns Hopkins University School of Medicine, Baltimore, MD 21231, USA; griggin1@jhmi.edu
* Correspondence: vstaedt1@jhmi.edu (V.S.); rbai1@jhmi.edu (R.-Y.B.)

Received: 26 May 2020; Accepted: 30 June 2020; Published: 8 July 2020

Abstract: Patients with RASopathy Neurofibromatosis 1 (NF1) are at a markedly increased risk of the development of benign and malignant tumors. Malignant tumors are often recalcitrant to treatments and associated with poor survival; however, no chemopreventative strategies currently exist. We thus evaluated the effect of mebendazole, alone or in combination with cyclooxygenase-2 (COX-2) inhibitors, on the prevention of NF1-related malignancies in a *cis Nf1+/−;Tp53+/−* (NPcis) mouse model of NF1. Our in vitro findings showed that mebendazole (MBZ) inhibits the growth of NF1-related malignant peripheral nerve sheath tumors (MPNSTs) through a reduction in activated guanosine triphosphate (GTP)-bound Ras. The daily MBZ treatment of NPcis mice dosed at 195 mg/kg daily, initiated 60 days after birth, substantially delayed the formation of solid malignancies and increased median survival ($p < 0.0001$). Compared to placebo-treated mice, phosphorylated extracellular signal-regulated kinase (pERK) levels were decreased in the malignancies of MBZ-treated mice. The combination of MBZ with COX-2 inhibitor celecoxib (CXB) further enhanced the chemopreventative effect in female mice beyond each drug alone. These findings demonstrate the feasibility of a prevention strategy for malignancy development in high-risk NF1 individuals.

Keywords: neurofibromatosis 1 (NF1); mebendazole (MBZ); COX-2 inhibitor; MPNST; malignancy; sarcoma; chemoprevention

1. Introduction

RASopathy Neurofibromatosis 1 (NF1) is an autosomal dominant hereditary cancer predisposition syndrome that affects ~1:3000 individuals [1]. It is caused by mutations in the *neurofibromin 1 (Nf1)* tumor suppressor gene, which encodes the GTPase-activating protein-related domain (GRD) that catalyzes the inactivation of Ras by accelerating guanosine triphosphate (GTP) hydrolysis to guanosine diphosphate (GDP) [2]. In NF1 individuals, loss of *Nf1* results in high levels of activated Ras, leading to the formation of multiple benign and malignant tumors via multiple effector pathways, including the Ras–MAPK pathway, with subsequent activation of the RAF–MEK–ERK cascade.

Patients with NF1 have an increased cancer risk and mortality, and lower survival compared with the general population [3,4]. Based on the Finnish NF1 Registry, the estimated lifetime cancer risk in patients with NF1 is 59.6%, with an estimated cumulative cancer risk of ~25% and ~39% by age 30 and 50 years, whereas the respective percentages in the general Finnish population are much lower, at 30.8%, 0.8% and 3.9% [3]. The most common malignancies are of nervous system origin, such as malignant peripheral nerve sheath tumors (MPNSTs) and astrocytomas, which comprise 63% of all malignancies [3]. Other malignancies include breast cancer, rhabdomyosarcomas, pheochromocytoma, gastrointestinal stromal tumor (GIST), malignant fibrous histiocytoma, and thyroid cancer [3].

MPNST is a very aggressive spindle cell sarcoma which accounts for the majority of cancer deaths in all NF1 patients and is a hallmark complication of this condition [3–6]. MPNST may arise from any of the pre-existing plexiform neurofibromas distributed throughout a patient's body. Unfortunately, there is no way of knowing which individual and, more specifically, which lesions within any one individual are likely to behave in a malignant fashion and thus many patients require regular screening with standard radiographic techniques such as MRI and PET/CT. Patients with *Nf1* microdeletion, i.e., a large deletion of the *Nf1* gene and its flanking regions, are especially susceptible to MPNSTs [7,8].

NF1-specific malignancies, including MPNSTs, typically manifest early in life and are responsible for the relative excess in cancer incidence and mortality observed in children and young adults [4]. Those malignancies are typically very difficult to treat and current therapies have shown little long-term benefit despite extensive research efforts [9]; however, early chemoprevention to delay cancer occurrence and reduce cancer risk remains largely unexplored. The success of chemoprevention has been impressively demonstrated in epithelial malignancies, particularly breast, prostate and colorectal cancers, with the use of selective estrogen receptor modulators (SERM) (e.g., tamoxifen), 5α-reductase inhibitors (e.g., finasteride) and cyclooxygenase-2 (COX-2) inhibitors, a type of non-steroidal anti-inflammatory drug (NSAID, e.g., sulindac, aspirin, celecoxib) that inhibited the appearance of colorectal polyps in various familial colorectal cancer predisposing syndromes [10].

The development of new chemical agents for chemoprevention is a long, difficult and expensive process. A potential strategy to circumvent these challenges is to discover new uses for compounds with an established track record of safe and long-term use in humans, alone or in combination with already known cancer prevention agents, such as widely used cyclooxygenase-2 (COX-2) inhibitors, whose anti-neoplastic effects are mediated through the inhibition of angiogenesis via decreasing COX-2-induced vascular endothelial growth factor (VEGF) production [11] and apoptosis via altered caspase signaling [12,13]. Notably, COX-2 overexpression has been found in a variety of sarcomas and has been associated with poor prognosis [14–16], thus suggesting that COX-2 inhibitors could play a role in NF1 cancer prevention.

We previously identified that mebendazole (MBZ), an FDA-approved low molecular weight benzimidazole derivative with a lengthy track record of safe long-term human use, significantly reduced tumor growth and improved survival in the animal models of glioblastoma multiforme (GBM) and medulloblastoma (Sonic Hedgehog (SHH) Group and c-Myc/OTX2 amplified Group 3) and also reduced tumor formation in a Familial Adenomatous Polyposis (FAP) colon cancer model [17–20]. A number of mechanisms for MBZ's anti-neoplastic activity have been proposed by us and others, including microtubule disruption, pro-apoptosis, and the inhibition of growth factor signaling through the blockage of various tyrosine kinases, particularly VEGFR2 [17,18].

The current study evaluates the feasibility of a cancer prevention strategy using non-toxic MBZ alone and in combination with COX-2 inhibitors in a *cis Nf1+/–;Tp53+/–* (NPcis) mouse model of NF1 [21]. Like NF1 patients, NPcis mice spontaneously develop predominantly soft tissue sarcomas including MPNSTs (genetically engineered murine (GEM) PNSTs) and malignant Triton tumors, as well as rhabdomyosarcomas and astrocytomas that severely limit their life expectancy to ~5 months [21–24]. The addition of heterozygous *Tp53* knock-out (KO) accelerates the cancer development, which mimics the secondary mutations required for the transformation to malignancies such as MPNST, where the second copy of *Nf1* is also lost due to the loss of heterozygosity (LOH) [21,22].

2. Material and Methods

2.1. Tissue Culture and Cell Lines

The human NF1-associated MPNST cell line NF90.8 was provided by Dr. Michael Tainsky (Wayne University, Detroit, MI) and sNF96.2 was purchased from the American Type Culture Collection (ATCC; Manassas, VA, USA). Cells were cultured in DMEM (ATCC) supplemented with 10% fetal bovine serum (FBS) (Sigma, St. Louis, MO, USA) and penicillin/streptomycin (Thermo Fisher,

Waltham, MA, USA). These cell lines were not authenticated. All cells were tested and found free of mycoplasma contamination.

2.2. Reagents and Antibodies

Rabbit anti-Nf1 antibody (A300-140A, Lot 3) was purchased from Bethyl Laboratories and anti-βActin horseradish peroxidase (HRP) antibody (C-11, SC-1615HRP, Lot G3015) was purchased from Santa Cruz Biotech. An Active Ras Detection Kit (#8821, antibody Lot 7), including the anti-Ras antibody, was purchased from Cell Signaling Technology.

2.3. Assays

A Ras activity assay was performed according to the manufacturer's instructions for the Active Ras Detection Kit (Cell Signaling Technology, Danvers, MA, USA). Briefly, cells were lysed with the Lysis/Binding/Wash buffer and pelleted, then the supernatant was used as the cell lysate. In the positive control, 5 µL of 10 mM GTPγS was added to 500 µL of lysates and incubated at 30 °C for 15 min. Cell lysates were incubated with glutathione resin, together with the purified GST-Raf1-RBD protein at 4 °C for 1 h in a spin cup. The resin was washed and the bound proteins were eluted by incubating with dithiothreitol (DTT)-containing sample buffer at RT for 2 min. Eluted samples were heated and analyzed by anti-Ras Western blotting.

A cell proliferation assay was performed using Cell Counting Kit-8 from Dojindo Molecular Technologies. Cells in 100 µL media in a 96-well plate were incubated with 10 µL of WST-8, a tetrazolium salt, at 37 °C in a tissue culture incubator. Absorbance was measured at 450 nm in a PerkinElmer Victor^3 plate reader. Half maximal inhibitory concentrations (IC_{50}s) were determined by incubating cells at a range of concentrations for 72 h and were calculated by GraphPad Prism 5.0 using the log (inhibitor) vs. response function and non-linear fit.

2.4. Chemoprevention in NPcis Mice

NPcis (*cis Nf1+/−;Tp53+/−*) mice in C57BL/6 background (B6;129S2-Trp53tm1Tyj Nf1tm1Tyj/J, Stock No: 008191, Jackson Laboratory) were bred by pairing male heterozygous NPcis mice with the female wildtype mice to better generate MPNST animals [21,23]. Since homozygous *Nf1/Tp53* KO mice are embryonically lethal, only heterozygous and wildtype pups were born [21,25]. Mice were genotyped via qPCR by Transnetyx using the following primer pairs: Nf1 wildtype (WT) (5′-GGTATTGAATTGAAGCACCTTTGTTTGG-3′, 5′-CGTTTGGCATCATCATTATGCTTACA-3′, reporter: 5′-AATATATGACCCCATGGCTGTC-3′), Nf1 KO (5′-TGGAGAGGCTTTTTGCTTCCT-3′, 5′-CGTTTGGCATCATCATTATGCTTACA-3′, reporter: 5′-CTGCTCGACATGGCTG-3′), Tp53 WT (5′-GTGAGGTAGGGAGCGACTTC-3′, 5′-TTGTAGTGGATGGTGGTATACTCAGA-3′, Reporter: 5′-CCTGGATCCTGTGTCTTC-3′) and Tp53 KO (5′-TGTTTTGCCAAGTTCTAATTCCATCAGA-3′, 5′-TTGTAGTGGATGGTGGTATACTCAGA-3′, reporter: 5′-ACAGGATCCTCTAGAGTCAG-3′). At day 60 after birth, heterozygous mice were started on the medicated feed or water. The mouse diet consisting of 45 kcal% fat containing soybean oil and lard for fat (D12451, Research Diets) was used as the control feed. Diets with 175, 195, 215 or 250 mg/kg of MBZ polymorph C (Aurochem Laboratories Ltd., Mumbai, India) or 1000 ppm (mg/kg) celecoxib (Sigma) were manufactured with the D12451 formulation in color codes. Sulindac (Sigma) was added to drinking water at 160 ppm (0.5 mg/day) in 4 mM sodium phosphate buffer as previously described [20]. Animals were palpated weekly for tumors and survival and cause of death, as detailed in the Results section, were recorded. All animal experiments were performed under an approved protocol and in accordance with Johns Hopkins Animal Care and Use guidelines.

2.5. Immunohistochemistry

Mouse tumors were first fixed by formalin and embedded in paraffin. For hematoxylin & eosin (H&E) staining, the section was de-paraffinized and stained by the standard hematoxylin and eosin procedure to visualize tissue structures. For immunostaining, rabbit anti-Erk1/2 (Cell Signaling, Cat. No. 9102) and anti-pErk1/2 (Thermo Fisher, Waltham, MA, USA, Cat. No. 36-8800) antibodies were used. Sections were de-paraffinized using a standard procedure and blocked using 1.9% H_2O_2 in methanol at room temperature for 10 min. Sections were heated at 100 °C for 20 min in the antigen retrieval citra solution (BioGenex, San Ramon, CA) and blocked by the serum-free protein blocker (Dako, Glostrup, Denmark, Cat. No. X0909) for 5 min at room temperature. After incubation with the rabbit anti-Erk1/2 or anti-pErk1/2 antibody diluted at 1:50 overnight at 4 °C, biotin-conjugated anti-rabbit IgG (Jackson ImmunoResearch, West Grove, PA, USA, Cat. No. 111-066-144) was applied for 20 min at room temperature, followed by washing and incubation with streptavidin peroxidase (Biogenex, Fremont, CA, USA, Cat. No. HK330-9KT) for 15 min at room temperature. Antibody binding was visualized by the 3,3'-Diaminobenzidine (DAB) chromogen system (Dako, Glostrup, Denmark). Subsequently, sections were counterstained by hematoxylin. Immunohistochemistry (IHC) quantification of representative tumor tissue sections was carried out with open source software Fiji ImageJ (NIH, Bethesda, MD, USA) using JPEG files. Mean optical density (OD) was calculated as the log average (maximal intensity/mean intensity) after image processing with color deconvolution and background subtraction.

2.6. Statistical Analysis

The results are presented as a mean value plus or minus the standard deviation. Data were analyzed by GraphPad Prism 5.0. The *p*-values were determined by a Mantel–Cox test. A *p*-value under 0.05 was accepted as statistically significant.

3. Results

3.1. MBZ Inhibited NF1-Derived MPNST Cell Lines through Ras Inhibition

Human MPNST cells NF90-8 and sNF96.2, both derived from NF1 patients, were treated with MBZ for 72 h at indicated concentrations, revealing favorable IC_{50} levels at 0.18 and 0.32 µM, respectively (Figure 1A). Because NF1-associated tumors are mainly driven by Ras hyperactivation, we studied MBZ's ability to inhibit Ras activity in the NF90-8 cell line by exposing NF90-8 cells to different concentrations of MBZ (0.2 and 1 µM) for 24 h. The activated form of GTP-bound Ras, detected by GST-Raf1-RBD fusion protein binding, was reduced in MBZ-treated NF90-8 cells in a concentration-dependent manner (Figure 1B). This confirmed the Ras inhibitory effect of MBZ in vitro.

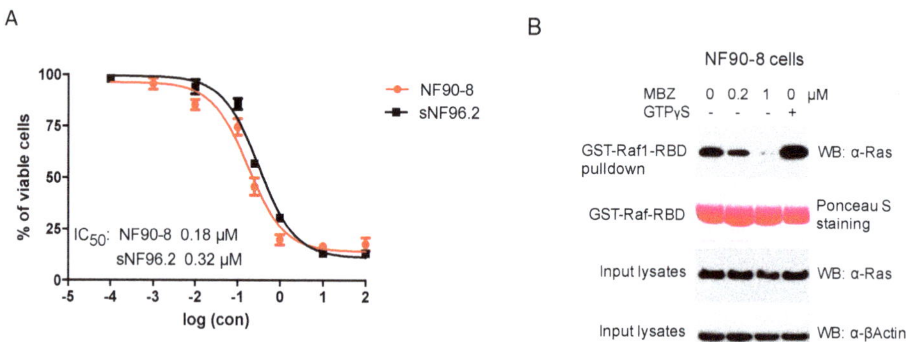

Figure 1. Mebendazole (MBZ) inhibits malignant peripheral nerve sheath tumor (MPNST) cells and Ras activity. (**A**) IC_{50s} of MBZ with NF90-8 and sNF96.2 cells were measured at 0.18 and 0.32 µM, respectively. Cells were incubated with MBZ or DMSO for 72 h and viable cells were determined with WST-8 and calculated as percentage of the control. Data are presented as mean ± s.d. (**B**) RASopathy

Neurofibromatosis 1 (NF1)-deficient NF90-8 cells were treated with MBZ at 0.2 and 1 µM for 24 h and cell lysates were incubated with GST-Raf1-RBD (the Ras-binding domain) coupled with glutathione resin. The pulldown products were analyzed by anti-Ras western blot, showing the activated GTP-bound Ras protein. Lysates incubated with GTPγS were used as positive controls.

3.2. MBZ Delayed Tumor Formation and Improves Survival in NPcis Mice

As reported before, *cis Nf1+/−;Tp53 +/−* (NPcis) mice are naturally predisposed to a number of solid malignancies, which typically form ~3–5 months after birth: 77% will develop soft tissue sarcomas—of which 60–65% are MPNSTs, 20% malignant Triton tumors, 10% rhabdomyosarcomas, 10% leiomyosarcomas and fibrohistiocytomas, 14% lymphomas, 8% carcinomas, and 1% neuroblastomas [21–23]; astrocytomas have also been reported [24,26].

To determine the most effective and tolerable long-term MBZ dose in vivo, 60-day old male and female NPcis mice were separated into groups and provided with control feed or continuous medicated feed containing 175, 195, 215 or 250 mg/kg MBZ. This range was calculated based on our previously established maximal dose of 50 mg/kg MBZ via oral gavage and the estimated daily food intake of a mouse [17]. Mice were weighed weekly and examined for signs of toxicity over 4 weeks. In the higher MBZ dosing groups of 250 and 215 mg/kg diets, nearly all mice showed evidence of excessive toxicity, including ruffled fur and significant weight loss between 10–15% thereby precluding the long-term use of those doses and establishing 195 mg/kg MBZ feed as the most suitable diet for long-term chemoprevention in these mice (Figure 2A,B).

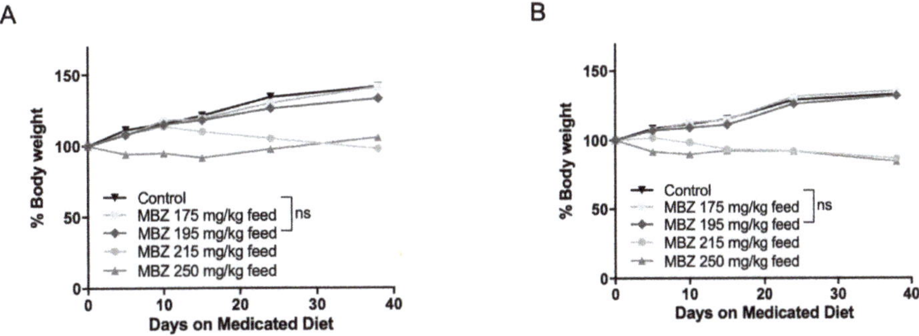

Figure 2. Dose-dependent MBZ toxicity in *cis Nf1+/−;Tp53+/−* (NPcis) mice. The 60-day old NPcis mice were provided with MBZ feed at indicated concentrations. Shown is the 30-day weight of (**A**) male and (**B**) female mice on the MBZ diet with the indicated doses. $n = 5$ mice per each MBZ dosing group.

In order to investigate the tumor-preventative effects of MBZ, continuous oral administration of MBZ via 195 mg/kg feed was initiated at 60 days after birth, before the formation of any malignancies. Mice were palpated weekly for the presence of any tumors. For the purpose of this study, 'Solid Malignancies' were defined as any type of sarcoma and astrocytoma, in addition to neuroblastomas and carcinomas, while 'Others' included non-solid malignancies such as lymphomas, leukemias and unknown causes of death.

MBZ treatment started at the age of 60 days significantly increased the overall median survival for male, female and combined cohorts (Figure 3A). In MBZ-treated mice, the time to tumor occurrence was significantly delayed compared to untreated control animals: 50% of all control mice had developed tumors and succumbed to disease by the age of 160 days, whereas in the MBZ-treated cohort, the tumor occurrence and median mortality was delayed by 32 days to 192 days (Figure 3B). Although observed in male and female NPcis mice alike, MBZ's cancer preventative effect appeared to be more pronounced in males, with an increase in median survival by 34.5 days compared to 14 days in female mice (Figure 3B). Figure 3C demonstrates that MBZ's chemopreventative effect was specific to mice with solid malignancies and did not affect the median survival of other, i.e., non-solid malignancy-related and unknown, causes of death both in male and female mice (Figure 3C). Lastly, MBZ treatment resulted in a ~25% reduction in solid cancer-related causes of death, thus demonstrating the feasibility of such a cancer prevention strategy in these NPcis mice (Figure 3D)

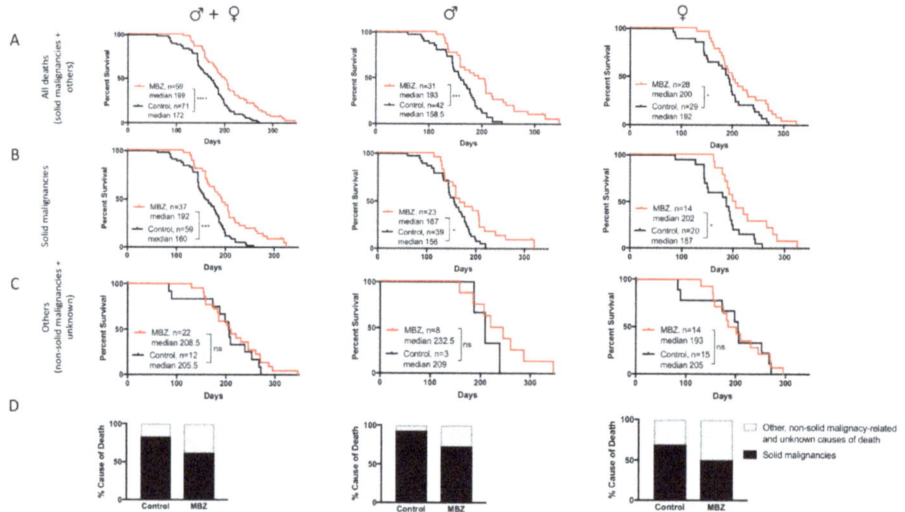

Figure 3. MBZ delays cancer onset and improves survival in NPcis mice. Shown are the Kaplan–Meier curves for MBZ-treated NPcis mice, initiated 60 days after birth, in comparison to controls for (**A**) overall survival, (**B**) solid malignancy-related mortality and (**C**) others, i.e., non-solid malignancy-related and unknown causes of mortality, analyzed as combined (males and females, left) male (middle) and female (right) cohorts. Animal numbers are provided for the specific groups in each graph and were analyzed with a two-sided log-rank test. * $p \leq 0.05$; *** $p \leq 0.001$, **** $p \leq 0.0001$; ns = not significant. (**D**) Percentage distribution of malignancy-related cause of death of MBZ-treated NPcis mice compared to controls, analyzed as combined (males and females, left) male (middle) and female (right) cohorts.

3.3. MBZ Reduced pERK Activity in Tumors In Vivo

In NPcis mice, the loss of *Nf1* leads to the hyperactivation of Ras, with the subsequent activation of the downstream effector ERK that is reflected by elevated levels of phosphorylated ERK (pERK) in MPNSTs and other related tumors. Immunohistochemistry showed that continuous MBZ treatment with a 195 mg/kg diet reduced pERK levels in sarcomas of NPcis mice compared to untreated mice (Figure 4). An analysis of the DAB staining intensity in three independent MBZ-treated tumor samples confirmed these results, with a reduced mean optical density (OD) of 0.02 in MBZ-treated samples compared to 0.05 in controls, while ERK staining was similar between both groups, with mean intensities of 0.07 and 0.08 for MBZ-treated and untreated tumors, respectively (Figure 4).

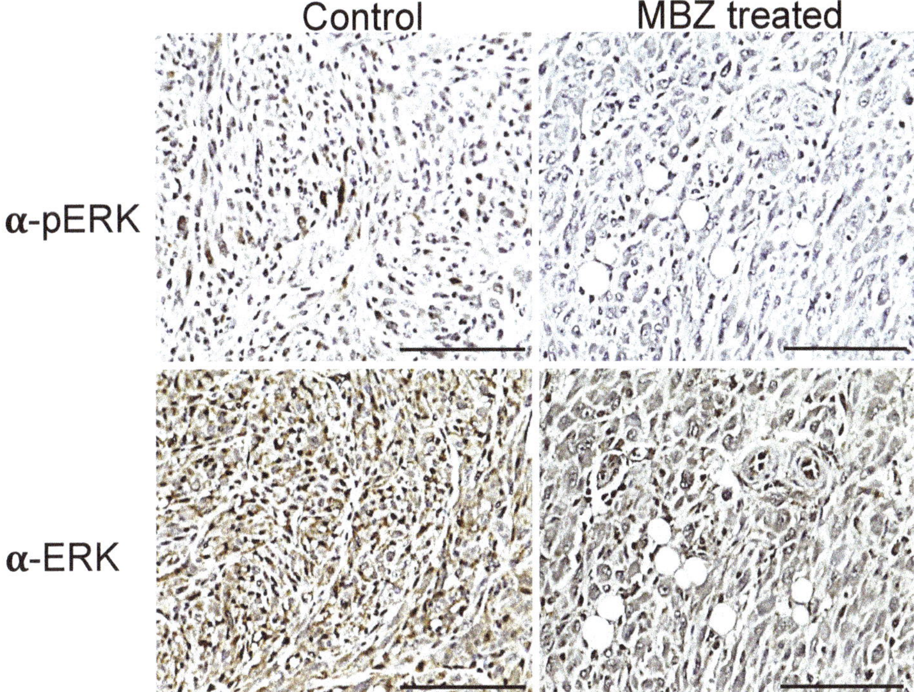

Figure 4. MBZ reduces ERK (pERK) in treated NPcis mice. Representative images of tumors from untreated controls (left) and MBZ-treated NPcis mice (left) were stained for pERK1/2 (upper row) and ERK1/2 (lower row). pERK staining was visualized in brown in untreated controls but reduced in tumors of MBZ-treated mice. Each scale bar represents 100 μm.

3.4. Cancer-Preventative Effects of CXB and MBZ Are Similar in NPcis Mice

The antitumor effect of selective COX-2 inhibitors, such as sulindac (SUL) and celecoxib (CXB), has been shown in several malignancies and cancer predisposition syndromes. In the NPcis mouse model, we found that MBZ-treated mice had a longer overall median survival of 199 days compared to CXB, with 193 days; however, this difference was not statistically significant (Figure 5A). When compared to untreated controls, CXB's effect on median survival was statistically increased in male NPcis mice with solid malignancies, while female mice showed a notable, but statistically insignificant, increase in survival compared to controls. Furthermore, CXB was substantially more effective in delaying the onset of malignancies than SUL, which showed a median survival of 171.5 days and failed to demonstrate any effect in male or female mice compared to controls (Figure 5A,B). Like MBZ, neither SUL nor CXB had an effect on the survival of non-cancer related causes (Figure 5C). Consistent with our findings, we also noticed a ~25% decline in cancer-related cause of death in CXB-treated mice (Figure 5D).

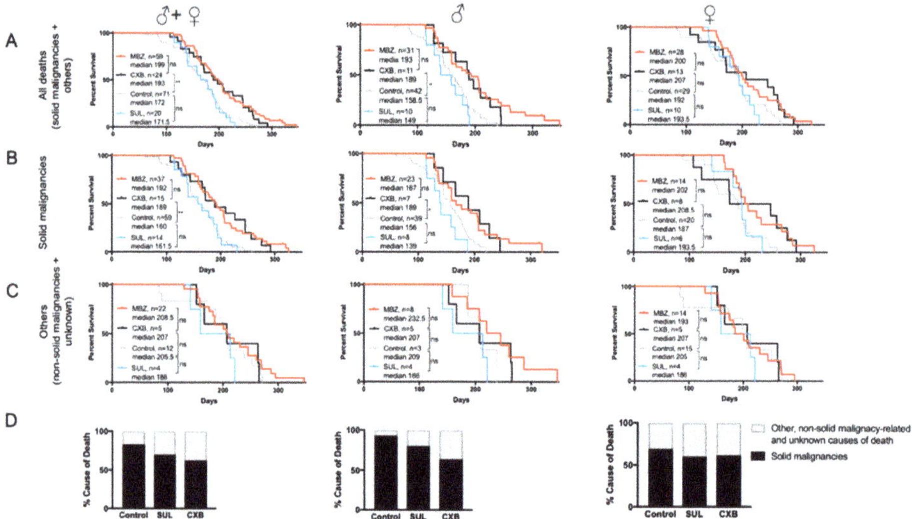

Figure 5. MBZ and cyclooxygenase-2 (COX-2) inhibitor celecoxib (CXB) are similarly effective in preventing cancer in NPcis mice. Shown are the Kaplan–Meier curves for CXB and sulindac (SUL)-treated NPcis mice, initiated 60 days after birth, in comparison to MBZ-treated mice and untreated controls for (**A**) overall survival, (**B**) solid malignancy-related mortality and (**C**) other, non-solid malignancy-related and unknown causes of mortality, analyzed as combined (males and females, left) male (middle) and female (right) cohorts. Animal numbers are provided for the specific groups in each graph and analyzed with two-sided log-rank test. * $p \leq 0.05$; ** $p \leq 0.01$; ns = not significant. (**D**) Percentage distribution of solid malignancy-related cause of death by CXB and SUL treated NPcis mice compared to controls, analyzed as combined (males and females, left) male (middle) and female (right) cohorts.

3.5. MBZ Is More Effective than Combined MBZ with CXB

Combined treatment with MBZ and CXB significantly increased median survival in NPcis mice compared to controls. However, the observed overall survival benefit appeared inferior to the effect achieved by MBZ or CXB alone, however, the difference is not statistically significant (Figure 6A). When investigating gender-specific effects, we found that dual use of MBZ and CXB in female NPcis mice was successful in delaying solid cancer occurrence and substantially enhancing the median survival beyond what was achieved by each agent alone and untreated controls (Figure 6B). This stands in contrast to male mice with solid malignancies, who did not experience any additional survival benefits from the combination treatment in comparison to single agent MBZ or CXB (Figure 6B). Figure 6C demonstrated that the combination therapy of MBZ and CXB resulted in an earlier mortality from non-solid cancer-related causes, particularly for male mice, indicating possibly the presence of toxicity, which we had assessed beforehand for each agent separately but not in combination (Figure 6C). However, the number of mice who died in the MBZ/CXB cohort due to other, non-solid malignancy-related and unknown causes, were small and thus, limiting our ability to conclusively interpret these results. When analyzing cause of death in MBZ/CXB-treated mice, we observed a reduction in solid cancer-related causes in comparison to the controls, as expected, which was largely comparable with what was seen with single agent use (Figure 6D).

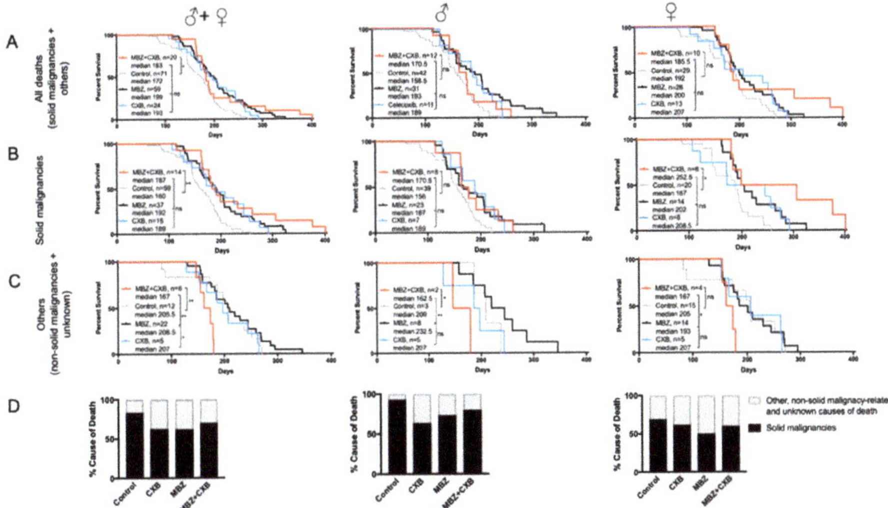

Figure 6. Combination of MBZ and CXB enhances survival in female NPcis mice. Shown are the Kaplan–Meier curves for NPcis mice treated with combined MBZ and CXB, initiated 60 days after birth, in comparison to CXB and MBZ alone for (**A**) overall survival, (**B**) solid malignancy-related mortality and (**C**) other, non-solid malignancy-related and unknown causes of mortality, analyzed as combined (males and females, left) male (middle) and female (right) cohorts. Animal numbers are provided for the specific groups in each graph and analyzed with two-sided log-rank test. * $p \leq 0.05$; ** $p \leq 0.01$; ns = not significant. (**D**) Percentage distribution of solid malignancy-related cause of death by combined MBZ/CXB treated NPcis mice compared to MBZ, CXB and controls, analyzed as combined (males and females, left) male (middle) and female (right) cohorts.

4. Discussion

Our previous work showed that MBZ's anti-tumor effect in glioblastomas and medulloblastomas is caused by multiple different mechanisms, such as the inhibition of microtubule formation and VEGFR2 autophosphorylation [17,18], which was corroborated by other investigators, applied to various preclinical cancer models and ultimately translated into clinical trials for adult and pediatric patients with cancer (NCT03925662, NCT03628079, NCT02644291, NCT01729260, NCT01837862). In the current study, we expanded MBZ's scope of application to chemoprevention, i.e., the use of drugs to reduce the risk of cancer development, in high-risk patients of NF1. NF1 is the most common tumor predisposition syndrome in which the loss of tumor suppressor neurofibromin leads to the activation of the Ras proto-oncogene and the development of dozens of benign and malignant tumors. MPNSTs and gliomas are the most common NF1-specific cancers, accounting for 63% of malignancies and a substantial mortality burden in adults younger than 40 years of age; other sarcomas (e.g., rhabdomyosarcomas), gastrointestinal stromal tumors, pheochromocytomas and breast cancers may also occur at a higher frequency compared to the non-affected population [3,4]. MPNSTs in NF1 patients have been particularly recalcitrant to treatment, with overall survival times that are shorter than those of patients with spontaneous MPNSTs. Surgical removal of a high-risk, pre-cancerous lesion is the only prophylactic modality that may reduce mortality, but has unfortunately been associated with morbidity.

In this study, we found that MBZ inhibited the growth of NF1-related MPNST cells in vitro and substantially delayed tumor formation in NPcis mice when initiated 60 days after birth, without overt disease. Interestingly, the effect was different between genders, with male mice experiencing a more substantial protective effect than female mice, who tend to develop tumors later than their male counterparts and have a longer median survival. A similar observation was made in mice treated with combined CXB with MBZ, which resulted in the largest delay in tumor occurrence and superior median survival in female mice, while males did not experience any benefit from the combination therapy compared to single agent MBZ or CXB. These potential gender-linked distinctions of cancer preventatives are important to realize, as this could impact the clinical applicability of such agents and patient management. Notably, the male bias of NPcis mice in developing MPNST has been reported before [27]. One could also envision intrinsic factors such as the tumor microenvironment, inflammation and differences in the sex hormones as potential causes of this phenomenon [26,28,29]; however, the underlying mechanism is unclear and should be investigated further in animals and humans. Our data further suggest that MBZ's cytotoxic effect on NF1-related malignancies may result from a reduction in activated GTP-bound Ras and a subsequent decrease in pERK in MBZ-treated malignancies in vivo, thus directly targeting the molecular underpinnings for tumor development in this condition. The potential significant impact of such chemoprevention on the mortality rate of cancer in the NF1 patient population can be envisioned from the success of NSAIDs and other agents on reducing the risk of colorectal, prostate and breast cancer. We therefore hope that, by demonstrating the feasibility of a chemopreventative approach for NF1, this will stimulate a rational approach to interrogate already existing databases for drugs that appear to decrease Ras activity and/or increase NF1 expression as a preventative drug discovery pipeline in these patients in order to reduce cancer occurrence and mortality.

Chemoprevention may involve the perturbation of a variety of steps in tumor initiation, promotion and progression. As such, COX-2 overexpression leads to cancer cell proliferation, neovascularization, and suppression of apoptosis and thus is associated with a worse prognosis in various malignancies, especially sarcomas [14–16]. It is therefore not surprising that overexpression of COX-2 has also been observed in NF1-associated MPNSTs and that selective COX-2 inhibition had an antitumor effect on these cells [30]. Our study confirmed these results and showed that the selective COX-2 inhibitor CXB, but not the non-selective COX inhibitor SUL, delayed cancer occurrence and increased median survival in both male and female NPcis mice.

Effective chemoprevention requires the need to identify a high-risk patient population and compounds or drug combinations with very low toxicity to allow long-term use in humans. When initiated 60 days after birth, long-term daily continuous MBZ administration was well tolerated in male and female NPcis mice, with stable weights using 195 mg/kg MBZ feed. This is in line with human data, which demonstrate a >40-year history of safe and continuous use for parasitic infections and cystic echinococcosis. This, along with the observed Ras inhibitory effect, could make MBZ an attractive candidate for long-term chemoprevention in the NF1 patient population. It should be noted that rigorous monitoring for adverse reactions would be required, as unexpected and expected toxicities could develop from long-term use of cancer preventative agents, particularly when multiple agents are used, and the benefits should clearly outweigh any potential risks. Given the heterogeneity of clinical symptoms among NF1 patients, it is doubtful that all NF1 patients would experience the same benefits and patient groups at high or low risk would have to be defined. For example, the low-risk NF1 population would include individuals with NF1 Arg1809, NF1 Arg1038Gly, NF1 Met992del, and NF1 Met1149 mutations, all of which are known not to develop any tumors or malignancies [31–34]. In contrast, the largest benefit would likely be observed in patients with a severe phenotype characterized by a higher tumor burden and a higher risk of malignancies. This group of patients would include individuals with large *Nf1* microdeletions, in which the lifetime risk for MPNST is increased to 16–26% [7,35]; patients with an NF1 p.844–848 missense mutation, who have a higher predisposition for symptomatic neurofibromas, optic pathway gliomas and malignancies compared with the general NF1-affected population [36] and NF1 patients with Arg1276 variants, who are also at a higher risk of developing symptomatic tumors and MPNSTs [31].

In summary, this study lays an important foundation for the effective and feasible chemoprevention of malignancies in patients with NF1, which has the potential to delay or perhaps even prevent the malignant transformation of MPNST and other NF1-related malignancies, decrease the need for surgical intervention and reduce the use of antineoplastic therapies in this patient population. Further research is necessary to evaluate these findings in a larger animal studies, such as the NF1 pig model, and to determine whether the observed effects will result in improved clinical outcomes.

Author Contributions: R.-Y.B. and V.S. designed and performed experiments, analyzed data and wrote the manuscript. T.G.-B. provided technical assistance. G.J.R. participated in scientific discussion. All authors have read and agreed to the published version of the manuscript.

Funding: V.S. and R.-Y.B. were supported by the Children's Tumor Foundation Drug Discovery Initiative (DDI) Award and the Department of Defense (DOD) W81XWH1810236. V.S. was also supported by NCI 5K08CA230179.

Conflicts of Interest: A patent on MBZ has been filed by JHU, with R-Y.B, V.S. and G.J.R. listed as inventors. The terms of these arrangements are managed by the Johns Hopkins University. In addition, V.S. serves as scientific advisor of the Gilbert Family Foundation—Gene Therapy Initiative outside the submitted work. T.G-B declares no conflict of interest.

References

1. Carey, J.C.; Baty, B.J.; Johnson, J.P.; Morrison, T.; Skolnick, M.; Kivlin, J. The Genetic Aspects of Neurofibromatosis. *Ann. N. Y. Acad. Sci.* **1986**, *486*, 45–56. [CrossRef] [PubMed]
2. Ratner, N.; Miller, S.J. A RASopathy gene commonly mutated in cancer: The neurofibromatosis type 1 tumour suppressor. *Nat. Rev. Cancer* **2015**, *15*, 290–301. [CrossRef] [PubMed]
3. Uusitalo, E.; Rantanen, M.; Kallionpaa, R.A.; Poyhonen, M.; Leppavirta, J.; Yla-Outinen, H.; Riccardi, V.M.; Pukkala, E.; Pitkaniemi, J.; Peltonen, S.; et al. Distinctive Cancer Associations in Patients With Neurofibromatosis Type 1. *J. Clin. Oncol.* **2016**, *34*, 1978–1986. [CrossRef] [PubMed]
4. Peltonen, S.; Kallionpaa, R.A.; Rantanen, M.; Uusitalo, E.; Lahteenmaki, P.; Poyhonen, M.; Pitkaniemi, J.; Peltonen, J. Pediatric malignancies in neurofibromatosis type 1: A population-based cohort study. *Int. J. Cancer* **2019**, *145*, 2926–2932. [CrossRef]
5. Evans, D.G.; Baser, M.; McGaughran, J.; Sharif, S.; Howard, E.; Moran, A. Malignant peripheral nerve sheath tumours in neurofibromatosis 1. *J. Med. Genet.* **2002**, *39*, 311–314. [CrossRef]
6. Mautner, V.-F.; Kluwe, L.; Friedrich, R.; Roehl, A.C.; Bammert, S.; Hogel, J.; Cooper, D.N.; Kehrer-Sawatzki, H.; Spori, H. Clinical characterisation of 29 neurofibromatosis type-1 patients with molecularly ascertained 1.4 Mb type-1 NF1 deletions. *J. Med. Genet.* **2010**, *47*, 623–630. [CrossRef]
7. Kehrer-Sawatzki, H.; Mautner, V.-F.; Cooper, D.N. Emerging genotype-phenotype relationships in patients with large NF1 deletions. *Qual. Life Res.* **2017**, *136*, 349–376. [CrossRef]
8. Pasmant, E.; Sabbagh, A.; Spurlock, G.; Laurendeau, I.; Grillo, E.; Hamel, M.-J.; Martin, L.; Barbarot, S.; Leheup, B.; Rodriguez, D.; et al. NF1 microdeletions in neurofibromatosis type 1: From genotype to phenotype. *Hum. Mutat.* **2010**, *31*, E1506–E1518. [CrossRef]
9. Staedtke, V.; Bai, R.-Y.; Blakeley, J. Cancer of the Peripheral Nerve in Neurofibromatosis Type 1. *Neurotherapeutics* **2017**, *14*, 298–306. [CrossRef]
10. Cuzick, J. Preventive therapy for cancer. *Lancet Oncol.* **2017**, *18*, e472–e482. [CrossRef]
11. Gallo, O.; Franchi, A.; Magnelli, L.; Sardi, I.; Vannacci, A.; Boddit, V.; Chiarugi, V.; Masini, E. Cyclooxygenase-2 Pathway Correlates with VEGF Expression in Head and Neck Cancer. Implications for Tumor Angiogenesis and Metastasis. *Neoplasia* **2001**, *3*, 53–61. [CrossRef]
12. Hsu, A.-L.; Ching, T.-T.; Wang, D.-S.; Song, X.; Rangnekar, V.M.; Chen, C.-S. The Cyclooxygenase-2 Inhibitor Celecoxib Induces Apoptosis by Blocking Akt Activation in Human Prostate Cancer Cells Independently of Bcl-2. *J. Biol. Chem.* **2000**, *275*, 11397–11403. [CrossRef] [PubMed]
13. Jendrossek, V. Targeting apoptosis pathways by Celecoxib in cancer. *Cancer Lett.* **2013**, *332*, 313–324. [CrossRef] [PubMed]
14. Endo, M.; Matsumura, T.; Yamaguchi, T.; Yamaguchi, U.; Morimoto, Y.; Nakatani, F.; Kawai, A.; Chuman, H.; Beppu, Y.; Shimoda, T.; et al. Cyclooxygenase-2 overexpression associated with a poor prognosis in chondrosarcomas. *Hum. Pathol.* **2006**, *37*, 471–476. [CrossRef] [PubMed]
15. Lee, C.H.; Roh, J.-W.; Choi, J.-S.; Kang, S.; Park, I.-A.; Chung, H.H.; Jeon, Y.-T.; Kim, B.-G.; Park, N.-H.; Kang, S.-B.; et al. Cyclooxygenase-2 Is an Independent Predictor of Poor Prognosis in Uterine Leiomyosarcomas. *Int. J. Gynecol. Cancer* **2011**, *21*, 668–672. [CrossRef] [PubMed]

16. Wang, S.; Gao, H.; Zuo, J.; Gao, Z. Cyclooxygenase-2 expression correlates with development, progression, metastasis, and prognosis of osteosarcoma: A meta-analysis and trial sequential analysis. *FEBS Open Bio* **2019**, *9*, 226–240. [CrossRef]

17. Bai, R.-Y.; Staedtke, V.; Aprhys, C.M.; Gallia, G.L.; Riggins, G.J. Antiparasitic mebendazole shows survival benefit in 2 preclinical models of glioblastoma multiforme. *Neuro-Oncology* **2011**, *13*, 974–982. [CrossRef]

18. Bai, R.-Y.; Staedtke, V.; Rudin, C.M.; Bunz, F.; Riggins, G.J. Effective treatment of diverse medulloblastoma models with mebendazole and its impact on tumor angiogenesis. *Neuro-Oncology* **2014**, *17*, 545–554. [CrossRef]

19. Bai, R.-Y.; Staedtke, V.; Wanjiku, T.; Rudek, M.A.; Joshi, A.; Gallia, G.L.; Riggins, G.J. Brain Penetration and Efficacy of Different Mebendazole Polymorphs in a Mouse Brain Tumor Model. *Clin. Cancer Res.* **2015**, *21*, 3462–3470. [CrossRef]

20. Williamson, T.; Bai, R.-Y.; Staedtke, V.; Huso, D.; Riggins, G.J. Mebendazole and a non-steroidal anti-inflammatory combine to reduce tumor initiation in a colon cancer preclinical model. *Oncotarget* **2016**, *7*, 68571–68584. [CrossRef]

21. Cichowski, K.; Shih, T.S.; Schmitt, E.; Santiago, S.; Reilly, K.; McLaughlin, M.E.; Bronson, R.T.; Jacks, T. Mouse Models of Tumor Development in Neurofibromatosis Type 1. *Science* **1999**, *286*, 2172–2176. [CrossRef] [PubMed]

22. Vogel, K.S.; Klesse, L.J.; Velasco-Miguel, S.; Meyers, K.; Rushing, E.J.; Parada, L.F. Mouse Tumor Model for Neurofibromatosis Type 1. *Science* **1999**, *286*, 2176–2179. [CrossRef] [PubMed]

23. Reilly, K.M.; Broman, K.W.; Bronson, R.T.; Tsang, S.; Loisel, D.A.; Christy, E.S.; Sun, Z.; Diehl, J.; Munroe, D.J.; Tuskan, R.G. An imprinted locus epistatically influences Nstr1 and Nstr2 to control resistance to nerve sheath tumors in a neurofibromatosis type 1 mouse model. *Cancer Res.* **2006**, *66*, 62–68. [CrossRef] [PubMed]

24. Reilly, K.M.; Tuskan, R.G.; Christy, E.; Loisel, D.A.; Ledger, J.; Bronson, R.T.; Smith, C.D.; Tsang, S.; Munroe, D.J.; Jacks, T. Susceptibility to astrocytoma in mice mutant for Nf1 and Trp53 is linked to chromosome 11 and subject to epigenetic effects. *Proc. Natl. Acad. Sci. USA* **2004**, *101*, 13008–13013. [CrossRef] [PubMed]

25. Jacks, T.; Shih, T.S.; Schmitt, E.M.; Bronson, R.T.; Bernards, A.; Weinberg, R.A. Tumour predisposition in mice heterozygous for a targeted mutation in Nf1. *Nat. Genet.* **1994**, *7*, 353–361. [CrossRef]

26. Reilly, K.M.; Loisel, D.A.; Bronson, R.T.; McLaughlin, M.E.; Jacks, T. Nf1; Trp53 mutant mice develop glioblastoma with evidence of strain-specific effects. *Nat. Genet.* **2000**, *26*, 109–113. [CrossRef]

27. Walrath, J.C.; Fox, K.; Truffer, E.; Alvord, W.G.; Quinones, O.A.; Reilly, K.M. Chr 19A/J modifies tumor resistance in a sex- and parent-of-origin-specific manner. *Mamm. Genome* **2009**, *20*, 214–223. [CrossRef]

28. Dorak, M.T.; Karpuzoglu, E. Gender Differences in Cancer Susceptibility: An Inadequately Addressed Issue. *Front. Genet.* **2012**, *3*, 268. [CrossRef]

29. Rubin, J.B.; Lagas, J.S.; Broestl, L.; Sponagel, J.; Rockwell, N.; Rhee, G.; Rosen, S.F.; Chen, S.; Klein, R.S.; Imoukhuede, P.; et al. Sex differences in cancer mechanisms. *Biol. Sex Differ.* **2020**, *11*, 1–29. [CrossRef]

30. Hakozaki, M.; Tajino, T.; Konno, S.; Kikuchi, S.; Yamada, H.; Yanagisawa, M.; Nishida, J.; Nagasawa, H.; Tsuchiya, T.; Ogose, A.; et al. Overexpression of Cyclooxygenase-2 in Malignant Peripheral Nerve Sheath Tumor and Selective Cyclooxygenase-2 Inhibitor-Induced Apoptosis by Activating Caspases in Human Malignant Peripheral Nerve Sheath Tumor Cells. *PLoS ONE* **2014**, *9*, e88035. [CrossRef]

31. Koczkowska, M.; Callens, T.; Chen, Y.; Gomes, A.; Hicks, A.D.; Sharp, A.; Johns, E.; Uhas, K.A.; Armstrong, L.; Bosanko, K.A.; et al. Clinical spectrum of individuals with pathogenic NF1 missense variants affecting p.Met1149, p.Arg1276, and p.Lys1423: Genotype–phenotype study in neurofibromatosis type 1. *Hum. Mutat.* **2019**, *41*, 299–315. [CrossRef] [PubMed]

32. Koczkowska, M.; Callens, T.; Gomes, A.; Sharp, A.; Chen, Y.; Hicks, A.D.; Aylsworth, A.S.; Azizi, A.A.; Basel, D.G.; Bellus, G.; et al. Expanding the clinical phenotype of individuals with a 3-bp in-frame deletion of the NF1 gene (c.2970_2972del): An update of genotype-phenotype correlation. *Genet. Med.* **2018**, *21*, 867–876. [CrossRef] [PubMed]

33. Rojnueangnit, K.; Xie, J.; Gomes, A.; Sharp, A.; Callens, T.; Chen, Y.; Liu, Y.; Cochran, M.; Abbott, M.-A.; Atkin, J.; et al. High Incidence of Noonan Syndrome Features Including Short Stature and Pulmonic Stenosis in Patients carrying NF1 Missense Mutations Affecting p.Arg1809: Genotype–Phenotype Correlation. *Hum. Mutat.* **2015**, *36*, 1052–1063. [CrossRef]

34. Upadhyaya, M.; Huson, S.M.; Davies, M.; Thomas, N.; Chuzhanova, N.; Giovannini, S.; Evans, D.G.; Howard, E.; Kerr, B.; Griffiths, S.; et al. An Absence of Cutaneous Neurofibromas Associated with a 3-bp Inframe Deletion in Exon 17 of the NF1 Gene (c.2970-2972 delAAT): Evidence of a Clinically Significant NF1 Genotype-Phenotype Correlation. *Am. J. Hum. Genet.* **2006**, *80*, 140–151. [CrossRef] [PubMed]
35. De Raedt, T.; Brems, H.; Wolkenstein, P.; Vidaud, D.; Pilotti, S.; Perrone, F.; Mautner, V.; Frahm, S.; Sciot, R.; Legius, E. Elevated Risk for MPNST in NF1 Microdeletion Patients. *Am. J. Hum. Genet.* **2003**, *72*, 1288–1292. [CrossRef]
36. Koczkowska, M.; Chen, Y.; Callens, T.; Gomes, A.; Sharp, A.; Johnson, S.; Hsiao, M.-C.; Chen, Z.; Balasubramanian, M.; Barnett, C.P.; et al. Genotype-Phenotype Correlation in NF1: Evidence for a More Severe Phenotype Associated with Missense Mutations Affecting NF1 Codons 844–848. *Am. J. Hum. Genet.* **2018**, *102*, 69–87. [CrossRef]

Article

Distinct Tumor Microenvironments Are a Defining Feature of Strain-Specific CRISPR/Cas9-Induced MPNSTs

Amanda Scherer [1,2], **Victoria R. Stephens** [1,3], **Gavin R. McGivney** [2,4], **Wade R. Gutierrez** [2,4,5], **Emily A. Laverty** [1,2], **Vickie Knepper-Adrian** [1,2] and **Rebecca D. Dodd** [1,2,*]

1   Department of Internal Medicine, University of Iowa, Iowa City, IA 52242, USA; amanda-scherer@uiowa.edu (A.S.); victoria.r.stephens@vanderbilt.edu (V.R.S.); emily-laverty@uiowa.edu (E.A.L.); vickie-knepper@uiowa.edu (V.K.-A.)
2   Holden Comprehensive Cancer Center, University of Iowa, Iowa City, IA 52242, USA; gavin-mcgivney@uiowa.edu (G.R.M.); wade-gutierrez@uiowa.edu (W.R.G.)
3   PREP program, University of Iowa, Iowa City, IA 52242, USA
4   Cancer Biology Graduate Program, University of Iowa, Iowa City, IA 52242, USA
5   Medical Scientist Training Program, University of Iowa, Iowa City, IA 52242, USA
*   Correspondence: rebecca-dodd@uiowa.edu; Tel.: +1-319-335-4962

Received: 5 May 2020; Accepted: 21 May 2020; Published: 23 May 2020

Abstract: The tumor microenvironment plays important roles in cancer biology, but genetic backgrounds of mouse models can complicate interpretation of tumor phenotypes. A deeper understanding of strain-dependent influences on the tumor microenvironment of genetically-identical tumors is critical to exploring genotype–phenotype relationships, but these interactions can be difficult to identify using traditional Cre/loxP approaches. Here, we use somatic CRISPR/Cas9 tumorigenesis approaches to determine the impact of mouse background on the biology of genetically-identical malignant peripheral nerve sheath tumors (MPNSTs) in four commonly-used inbred strains. To our knowledge, this is the first study to systematically evaluate the impact of host strain on CRISPR/Cas9-generated mouse models. Our data identify multiple strain-dependent phenotypes, including changes in tumor onset and the immune microenvironment. While BALB/c mice develop MPNSTs earlier than other strains, similar tumor onset is observed in C57BL/6, 129X1 and 129/SvJae mice. Indel pattern analysis demonstrates that indel frequency, type and size are similar across all genetic backgrounds. Gene expression and IHC analysis identify multiple strain-dependent differences in CD4+ T cell infiltration and myeloid cell populations, including M2 macrophages and mast cells. These data highlight important strain-specific phenotypes of genomically-matched MPNSTs that have implications for the design of future studies using similar *in vivo* gene editing approaches.

Keywords: CRISPR/Cas9; MPNST; mouse models; sarcoma; tumor microenvironment

---

1. Introduction

Mouse models are a cornerstone of cancer research and have produced a wealth of mechanistic insights into tumor biology. While mice from a wide variety of genetic backgrounds are used for *in vivo* cancer modeling, there is strong evidence that strain-dependent phenotypes can complicate interpretation of results. Within similar genetic contexts, mouse strain can impact tumor susceptibility, disease onset, metastatic potential, and the spectrum of cancer development [1–5]. Multiple strain-dependent cancer phenotypes can be attributed to background-specific modifying loci [6,7]. Classic examples include tumor development in $Nf1^{+/-}$; $p53^{+/-}$ mice (*NPcis*), which have high incidences of astrocytomas and malignant peripheral nerve sheath tumors (MPNST) on a C57BL/6 background but are less tumor prone on other genetic backgrounds. Extensive genetic mapping experiments determined that astrocytoma susceptibility is linked to an imprinted locus on chromosome 11, while MPNST formation is associated with polymorphisms in the nerve sheath tumor resistance (*Nstr*) genes [8–10]. The development of neurofibromas, benign nerve sheath tumors that are precursor lesions to MPSNTs, is also strain dependent. Schwann cell-specific overexpression of neuregulin in *p53*$^{+/-}$ mice (P$_0$-GGFβ3; p53$^{+/-}$) drives neurofibroma formation on a mixed background, but mice fail to develop tumors after backcrossing onto an inbred C57BL/6J background [11]. In addition to tumorigenesis events, metastatic phenotypes can also be dramatically influenced by genetic background, as observed in *Pten*-driven prostate cancer models [12,13] and MMTV-PyMT-driven mammary tumors [14].

Strain-dependent variations in the tumor microenvironment (TME) can also profoundly impact cancer phenotypes. The TME is comprised of a diverse array of extracellular matrix and stromal cells including cancer-associated fibroblasts, endothelial cells, and immune infiltrates. Variations in the immune systems of common inbred strains are well documented [15]. For example, C57BL/6 mice have elevated neutrophils and splenic macrophages, but decreased B cell and CD4+ T cell populations compared to BALB/c and 129/SvHsd mice [16,17]. Polarization of macrophage function is strain dependent, with enrichment of classically-activated, pro-inflammatory M1 macrophages in Th1-oriented mouse strains such as C57BL/6, while immunosuppressive M2 macrophages are predominant in Th2-oriented mouse strains such as BALB/c [18]. Functional activity of immune cells is also heavily influenced by mouse background, including the cytotoxic capacity of NK cells [19] and macrophage recruitment [20].

Multiple tumor phenotypes can be attributed to differences in host immune function, including metastatic potential and therapeutic response. Depletion of myeloid cell-derived MMP9 in MMTV-PyVT models slows metastatic progression in C57BL/6 mice, but had no impact on pulmonary metastases in an FVB/N background [21]. In syngeneic transplant models, antibody blocking experiments demonstrate that melanoma metastasis is dependent on strain-specific NK cell activity [19]. These differences in the strain-dependent immune landscape have implications for immunotherapy response in preclinical models [22–26]. Multiple groups have reported that while immunosuppressive cells predominate in poorly-responsive models, cytotoxic effector cells are prevalent in tumors of responsive models.

A deeper understanding of the impact of host strain background on the TME of genetically-identical tumors is necessary to help guide future experimental design and interpretation of preclinical cancer studies. The nature of genetically-engineered mouse models (GEMMs) and syngeneic cell transplant models have necessitated that data are obtained from tumors arising in a limited number of genetic contexts and tissues. Therefore, most basic and translational studies utilize only a single inbred mouse strain, and the majority of primary model studies have been conducted predominantly in C57BL/6 and 129/S mice. However, this current paradigm of using a small number of genetic backgrounds does not address the important role of TME variation as a determinant of cancer phenotype.

The development of somatic CRISPR/Cas9 tumorigenesis approaches allows for direct comparisons of host TME in genetically-identical tumors. We have recently published a CRISPR/Cas9-induced model of soft-tissue sarcoma in wild-type mice [27]. This approach delivers an adenovirus expressing Cas9 and guide RNAs targeting *Nf1* and *p53* into the sciatic nerve of adult mice to generate high-fidelity malignant peripheral nerve sheath tumors (MPNSTs), a high-grade sarcoma of the myelinating nerve sheath. This system allows for introduction of multiple somatic mutations into adult animals surrounded by native, non-mutant stroma and an intact immune system. By introducing somatic gene alterations into adult mice without the need for lengthy and costly backcrossing, CRISPR/Cas9 approaches can assess genetic events in different murine backgrounds. Because this approach uses exogenous delivery of Cas9, it can be applied to a mouse from any strain or pre-existing genetically-engineered model. This adaptability is important to facilitate studies that rely on specific strains for experimental models, such as in the fields of metabolic disease and immunology.

To our knowledge, a systematic study examining the impact of host strain on CRISPR/Cas9-generated mouse models has not been undertaken. Here, we use CRISPR/Cas9 approaches to determine the influence of mouse background on genetically-identical MPNSTs. We report variations in tumor onset, immune landscape, and TME-associated gene expression across MPNSTs generated in four classically inbred strains. These data highlight important strain-specific phenotypes of genomically-matched MPNSTs that have implications for the future design of studies using similar in vivo gene editing approaches. Ultimately, CRISPR/Cas9 tumorigenesis approaches may provide unique opportunities to explore TME-dependent events by leveraging the diversity of stromal landscapes across tumor models from distinct genetic backgrounds.

2. Materials and Methods

2.1. Animals

All animal experiments were performed in accordance with protocols approved by the University of Iowa Institutional Animal Care and Use Committee (IACUC) and adhere to the NIH Guide for the Care and Use of Laboratory Animals. C57BL/6 (stock #556) and BALB/c mice (stock #555) were purchased from Charles River Laboratories. 129X1 mice (stock #000691) were purchased from Jackson Laboratories. Wild-type 129Sv/Jae mice were bred and maintained at the University of Iowa.

2.2. CRISPR/Cas9 Generated MPNSTs and Growth Analysis

Adenovirus containing Cas9 and sgRNAs targeting *Nf1* and *p53* was purchased from ViraQuest (North Liberty, Iowa) [27]. Prior to injection, virus was mixed with DMEM and calcium phosphate as previously described [28–30]. Tumors were generated by injection of 25 uL of prepared virus into the left sciatic nerve of mice. When tumors reached a volume of 150 mm^3 (Day 1), they were measured by calipers 3 times weekly. Tumor volumes were calculated using the formula $V = (\pi \times L \times W \times H)/6$, with L, W, and H representing the length, width, and height of the tumor in mm, respectively. Tumors were harvested when they reached a volume of 1500 mm^3 or earlier if animals showed signs of distress, in accordance with IACUC guidelines at the University of Iowa. Tissue was collected for histology, RNA, and generation of cell lines.

2.3. Generation of Cell Lines from MPNSTs

Cell lines were derived from terminally-harvested MPNSTs. Tumors were finely minced and digested in dissociation buffer Collagenase Type IV (700 units/mL, Thermo, 17104-019, Thermo Fisher Scientific, Waltham, MA, USA) and dispase (2.4 units/mL, Thermo, 17105-041, Thermo Fisher Scientific, Waltham, MA, USA) in PBS for 1–1.5 h at 37 °C on an orbital shaker. Dissociated tissue was passed through a sterile 70 μM cell strainer (Fisherbrand, 22363548, Thermo Fisher Scientific, Waltham, MA, USA), washed once with PBS, and resuspended in DMEM (Gibco, 11965-092, Thermo Fisher Scientific, Waltham, MA, USA). Cells were cultured in DMEM containing 10% FBS, 1% penicillin-streptomycin (Gibco, 15140-122, Thermo Fisher Scientific, Waltham, MA, USA) and 1% sodium pyruvate (Gibco, 11360-070, Thermo Fisher Scientific, Waltham, MA, USA). After 10 passages, cells were used for indel analysis and subsequent studies.

2.4. Indel Analysis

Indel pattern analysis was previously described [31]. Genomic regions of *Nf1* and *p53* that spanned the gRNA target sites were amplified by PCR using Phusion high-fidelity DNA polymerase (NEB, M0530L). PCR primers for *Nf1* indels generate a 569 bp fragment in wild-type cells while those used to amplify *p53* indels result in a 520 bp fragment in wild-type cells. Primer sequences are listed in Supplementary Table S1. PCR amplicons were purified with the Monarch PCR and DNA Cleanup Kit (NEB T1030S). Sanger sequencing was performed by the Genomics Division of the Iowa Institute of Human Genetics at the University of Iowa. Indel frequencies were quantified from the chromatograms by sequence trace analysis using Synthego ICE [32]. Indels > 50 bp were determined by band size on a 2% agarose gel.

2.5. Histology and Immunohistochemistry

Upon harvest, a portion of tumor tissue was stored in 10% neutral buffered formalin for fixation and subsequent paraffin embedment. Formalin-fixed paraffin embedded tumors were sectioned and stained with hematoxylin (Vector Laboratories, H-3401, Burlingame, CA, USA) and eosin (Sigma-Aldrich, 586-X, St. Louis, MO, USA) to evaluate tissue morphology. All immunostaining was conducted with citrate-based antigen retrieval (Vector Laboratories, H-3300, Burlingame, CA, USA). The following antibodies were used: S100 (Abcam, ab4066, Cambridge, United Kingdom), Ki67 (BD Biosciences, 556003), CD4 (Abcam, ab183685), CD8a (Thermo Fischer Scientific, 14-0808-82, Waltham, MA, USA), Foxp3 (Thermo Fisher Scientific, 14-4777-82, Waltham, MA, USA), and F4/80 (Thermo Fisher Scientific 14-4801-82, Waltham, MA, USA). To visualize mast cells, slides were stained with toluidine blue solution (0.02% toluidine blue in 1% NaCl, pH 2.2) for 2 min, followed by two washes in distilled water and three washes in 100% ethanol. At least five tumors per group were analyzed, and quantification of cells staining positive was performed on 6 independent fields. The 20× fields were used for all analyses except for Ki67, which used 40× fields. Imaging was performed using an EVOS XL Core Imaging System (Thermo Fisher Scientific, AMEX1000, Waltham, MA, USA).

2.6. Quantitative RT-PCR

Upon harvest, tumor tissue was stored in RNA Later (AM7020, Thermo Fisher Scientific) at $-20\,^{\circ}\mathrm{C}$. Tumors ($n = 5$ per strain) were homogenized in liquid nitrogen and resuspended in Trizol (15596018, Thermo Fisher Scientific, Waltham, MA, USA). cDNA was synthesized from 1 ug of RNA using iScript (1708891, Bio-Rad). RT-qPCR was performed with Power-up Sybr Green 2x Master Mix (A25778, Thermo Fisher Scientific, Waltham, MA, USA) per the manufacturer's instructions on an Applied Biosystems 7900HT instrument using the $\Delta\Delta C_t$ relative to B2M expression (Genomics Division of the Iowa Institute of Human Genetics, University of Iowa). Primer sequences are listed in Supplementary Table S1 [24].

2.7. Statistical Analysis

Statistical analysis was performed using GraphPad Prism 8. Tumor growth kinetics, IHC quantification, and gene expression were analyzed using a one-way ANOVA with Tukey's multiple comparison test. Sample sizes for IHC and qRT-PCR analysis were 5 per group. Comparison of survival curves was performed using the log-rank (Mantel–Cox) test. For all studies, a p value of less than 0.05 was considered statistically significant.

3. Results

3.1. Host Strain Determines Tumor Onset for Genetically-Identical MPNSTs

To determine the impact of murine background strain on MPNST development, we generated somatic CRISPR/Cas9-induced tumors in four commonly-used laboratory strains: 129/SvJae, C57BL/6, 129X1, and BALB/c. Importantly, the 129/SvJae mice serve as reference controls, as this strain was used in our prior study [27]. We injected the sciatic nerve of 10–13 mice per background with adenovirus containing Cas9 and guide RNAs for *Nf1* and *p53* (Ad-Cas9 + gNF1 + gp53). This approach was previously shown to generate high-fidelity, *Nf1/p53*-null MPNSTs at the site of injection within 3–4 months. Similar to our prior data, 129/SvJae mice in the current study develop tumors at ~80 days post-injection (Figure 1A). Tumor onset is similar in C57BL/6 and 129X1 mice, arising at an average of 82 and 93 days, respectively. In contrast, BALB/c mice develop MPNSTs earlier than other strains, with tumors developing with an average onset of 61 days. After tumor detection, MPNSTs were measured 3x/weekly to obtain proliferative rates, which are calculated from a uniform initiating size of 150 mm^3. The average time for tumors to double in volume is 7–8 days, which is similar across all backgrounds (Figure 1B). Tumor proliferation was also examined by immunohistochemistry for Ki67 in terminally-harvested MPNSTs. Ki67 indices are similar in tumors from all strains, supporting the observation that host strain does not influence MPNST proliferation (Figure 1C). Histological analysis confirms MPNST morphology in all tumors, with S100 positivity noted in tumors from each background (Figure 1D). Taken together, these data show that somatic CRISPR/Cas9 tumorigenesis approaches can generate MPNSTs in a broad spectrum of wild-type mice, and that background strain can influence tumor initiation in genetically-matched tumors.

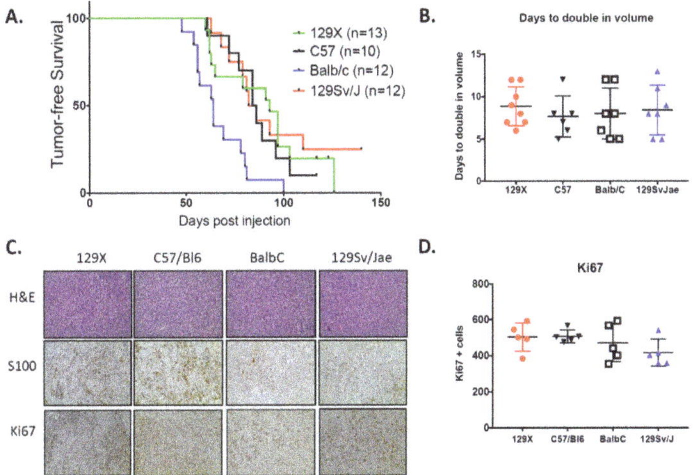

Figure 1. Host strain determines tumor onset but does not alter tumor growth kinetics. (**A**) Kaplan–Meyer curve of tumor-free survival. Formation of *Nf1/p53*-deleted malignant peripheral nerve sheath tumors (MPNSTs) is accelerated in BALB/c mice. Tumor initiation occurs within a similar timeframe in mice from 129/SvJae, C57BL/6, and 129X1 backgrounds. (**B**) Growth kinetics are similar across all background strains for genetically-identical MPNSTs (*n*= 6–8 tumors per strain). Growth rates are calculated as the number of days required for tumors to double from an initial volume of 150 mm^3. 129X1 (red circles), C57BL/6 (black triangles), BABL/c (white squares), and 129/SvJae (blue triangles). (**C**) Representative images of MPNSTs from different host strains stained for H&E (20×), S100 (20×), and Ki67 (20×). (**D**) Quantification of Ki67 confirms that background strain does not alter the rate of tumor proliferation (*n*= 5 tumors per strain). (**B,D**) analyzed by one-way ANOVA with Tukey's multiple comparison test.

3.2. Indel Analysis Reveals Unique Patterns of Gene Disruption

Indel signatures can determine the spectrum and frequency of CRISPR/Cas9-induced events in individual tumors. We generated tumor-derived cell lines to evaluate the unique indel patterns within each MPNST (Figure 2). Our analysis confirms the presence of *Nf1* and *p53* indels in all tumors. Additionally, no wild-type sequence is detectable in any cell line, suggesting complete disruption of the targeted regions. As CRISPR/Cas9 generates indels by random reassembly of DNA, we investigated the types of indels generated with each guide RNA. To focus this analysis, we evaluated indels that occur at > 5% frequency. Across 14 tumor-derived cell lines, we observe 24 indels in *Nf1* and 33 indels in *p53*. Several cell lines have a simple signature, containing predominantly one indel, while others have complex signatures comprised of up to five distinct variants per gene. The majority of cell lines contain multiple *p53* indels, as a single dominant indel of *p53* is detected in only 4/14 (29%) of cells. Single indels in *Nf1* are more frequent, with 7/14 (50%) of cell lines containing a solitary *Nf1* indel event. Insertions are less common than deletions, with only 1/14 (7%) of cell lines harboring *Nf1* insertions and 6/14 (43%) of cell lines harboring *p53* insertions. Indeed, only one cell line does not have a deletion event in *p53*, with a single predominant insertion being the only indel event detected within the sample. In our analysis, CRISPR-generated insertions are genetically small (1–2 bp), while deletions occur within a larger range (1 bp to > 20 bp). In *p53* indels, we observe a trend towards smaller deletions (<10 bp), which occur in 23/27 (85%) of deletion events. All of the indels detected in *Nf1* were either frameshift (FS) mutations (20/24) or indels ≥ 20 bp (4/24) that are the most likely to disrupt protein function by shifting the reading frame and inducing premature termination, nonsense mediated decay (NMD), or alterations in protein structure [33,34]. For indels detected in *p53*, 24/33 were FS mutations and 3/33 were deletions ≥ 20 bp. We did not identify any strain-specific trends in indel type, size, or frequency in this analysis, suggesting that in vivo CRISPR/Cas9 genomic editing occurs similarly across different murine backgrounds.

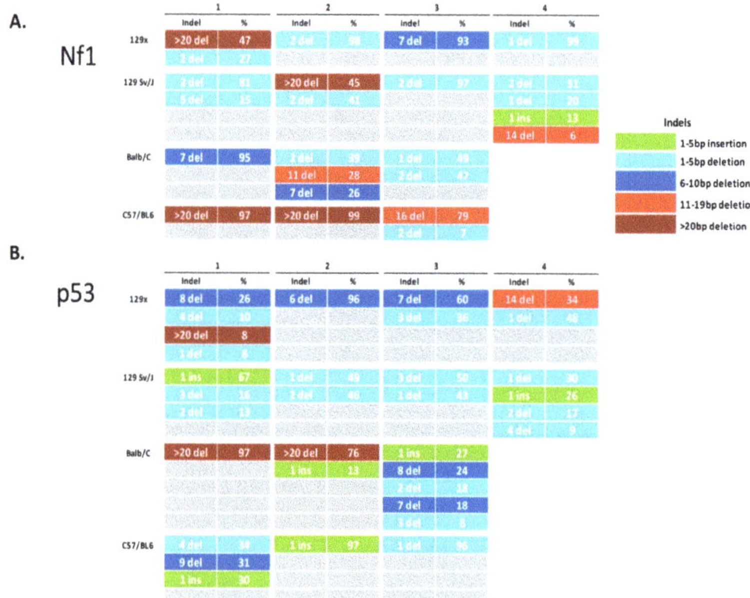

Figure 2. CRISPR/Cas9-induced insertions and deletions detected in *Nf1* and *p53* in MPNST-derived cell lines from different genetic backgrounds. Indel pattern analysis of the sgRNA-targeted regions of *Nf1* (**A**) and *p53* (**B**) demonstrates disruption of genomic targets in all tumors. The majority of indels detected in both *Nf1* and *p53* are frameshift mutations that result in inactivation of targeted proteins.

3.3. Immunological Diversity of MPNSTs Is a Hallmark of Genetic Background

Data from genetically-engineered mouse models strongly support a role for host strain in distinct patterns of immune cell activation [18,24]. Therefore, we hypothesized that there are strain-dependent differences in the composition of the immune landscape in our CRISPR/Cas9 generated MPNSTs. To examine the tumor microenvironment in genetically-identical tumors from different mouse strains, we performed histological analysis for populations of innate and adaptive immune cells that play key roles in MPNST biology, including CD4+ T cells, CD8+ T cells, regulatory T lymphocytes (Tregs), macrophages and mast cells in five tumors per genetic background (Supplementary Figure S1).

Levels of tumor-infiltrating cytotoxic CD8+ T lymphocytes are similar across all host strains (Figure 3A). In contrast, amounts of CD4+ T lymphocytes are highly dependent on background strain, with MPNSTs from C57BL/6 mice having lower CD4+ T infiltration than tumors on 129Sv/Jae, BALB/c, and 129X1 backgrounds (Figure 3B). MPNSTs from 129Sv/Jae mice display a heterogenous distribution of CD4+ T lymphocytes, with a wide variability of cell number across individual tumors. Regulatory T cells levels are highly variable across individual tumors, most likely due to the rare nature of these cells. In several tumors, we were unable to detect a single Treg in the sample. Analysis of multiple tumors determined that MPNSTs from 129X1 mice have higher levels of Tregs than MPNSTs from C57BL/6 or BALB/c mice (Figure 3C). Analysis of macrophage levels by F4/80 staining shows increased macrophage infiltration in MPNSTs from BALB/c mice (Figure 3D). Mast cells, histamine-rich myeloid cells with a strong role in MPNST biology [30,35], are enriched in MPNSTs from BALB/c mice (Figure 3E). The lowest levels of mast cells are observed in tumors from C57BL/6 mice. Taken together, these observations demonstrate the broad diversity of immune landscapes in MPNSTs from different background strains.

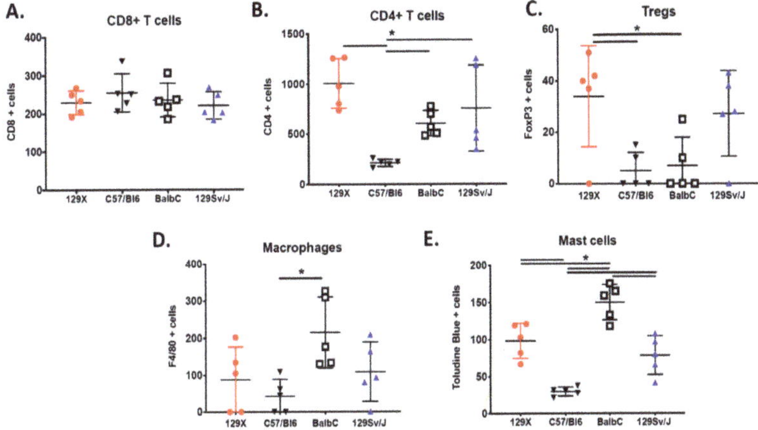

Figure 3. The MPNST immune landscape is determined by genetic background. (**A**) Levels of CD8+ T cells in terminally-harvested MPNSTs are similar across all host strains. (**B**) Infiltration of CD4+ T cells are significantly lower in tumors from C57BL/6 mice compared to MPNSTs in mice from 129X1, BALB/c, and 129/SvJae backgrounds. (**C**) Foxp3+ Tregs are detected at higher levels in tumors from 129X1 mice compared to C57BL/6 and BALB/c mice. (**D**) MPNSTs from BALB/c mice have significantly higher levels of infiltrating F4/80+ macrophages compared to C57BL/6 mice. (**E**) Mast cell infiltration is higher in tumors from BALB/c mice compared to 129/SvJae, C57BL/6, and 129X1 mice. Mast cell levels are lowest in MPNSTs from C57BL/6 mice. 129X1 (red circles), C57BL/6 (black triangles), BABL/c (white squares), and 129/SvJae (blue triangles). Analyzed by one-way ANOVA with Tukey's multiple comparison test. A *p*-value of less than 0.05 is considered statistically significant and is denoted by "*" (*n* = 5 tumors per strain).

3.4. Gene Expression of the MPNST Microenvironment

Given the broad variability of strain-dependent immune infiltration observed in our IHC data, we chose to perform extensive gene expression analysis of key tumor microenvironmental markers [24]. Using real-time qPCR analysis of whole tumor lysates from five tumors per background, we evaluated expression levels of pathways involved in innate immunity, adaptive immunity, angiogenesis, and cytokine signaling (Figure 4A and Supplementary Figure S2). These data provide insight into key tumor–stroma interactions and reveal extensive heterogeneity across host strains and individual tumors.

We first examined expression of tumor-associated macrophage (TAM) genes, since they are one of the most differentially-regulated immune cell populations between host strains. Expression of *Arg1* mRNA, a marker of immunosuppressive M2 macrophages, is elevated in MPNSTs from BALB/c mice (Figure 4B). Of note, *Arg1* is the only gene in our analysis that is statistically different between host backgrounds (*p* = 0.0156, one-way ANOVA). There were no differences in levels of the M1 macrophage marker *iNos1/Nos2* in tumors from different host strains (Figure 4C), suggesting that the influx of macrophages in MPNSTs from BALB/c mice consists of *Arg1*-expressing TAMs of the M2 subtype. This finding is consistent with data demonstrating that expression of the pro-immunogenic, M1 macrophage transcription factor *Stat3* is similar across backgrounds.

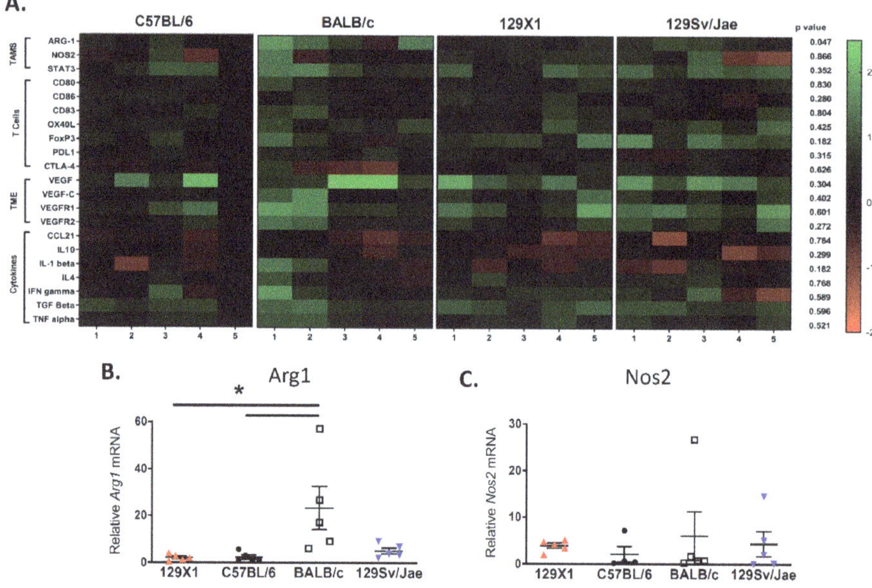

Figure 4. Expression of key genes in the MPNST microenvironment. (**A**) RT-qPCR analysis of markers for innate immunity, adaptive immunity, angiogenesis, and cytokine signaling in terminally-harvested tumors shows a large degree of heterogeneity between host strains and individual tumors. Samples are normalized to a single C57BL/6 tumor, shown as reference ($n = 5$ tumors per strain). (**B**) Expression analysis determines that MPNSTs from BALB/c mice express significantly higher levels of *Arg-1* mRNA, a marker of immunosuppressive M2 macrophages, when compared to tumors from 129/SvJae, C57BL/6, and 129X1 mice. (**C**) In contrast, levels of *Nos2* mRNA, a marker of M1 macrophages, is similar in tumors from all background strains. 129X1 (red circles), C57BL/6 (black triangles), BABL/c (white squares), and 129/SvJae (blue triangles). Analyzed by one-way ANOVA with Tukey's multiple comparison test. A *p*-value of less than 0.05 is considered statistically significant and is denoted by "*".

To further explore T lymphocyte populations, we examined genes involved in T cell activation and signaling. Expression of APC-resident co-stimulatory molecules—including *CD80*, *CD86*, *OX40L*, and *PDL1*—are similar across host strains. Similarly, expression of *CTLA-4*, an inhibitory receptor that negatively regulates T cell responses, and *CD83*, a marker of activated CD4+ T lymphocytes and dendritic cells, is not strain dependent. Levels of the regulatory T cell marker *FoxP3* are not statistically different across strains due to extensive heterogeneity between tumors, although trends are similar to IHC findings in Figure 3.

We next examined expression of angiogenesis genes, including *Vegf*, *Vegfr1*, and *Vegfr2*, in addition to the lymphangiogenic growth factor *Vegfc*. While several individual tumors display high expression of these growth factors, there are no statistically significant differences between host strains. Finally, we examined expression of key cytokines involved in immune activation, including proinflammatory molecules (*Tnfa*, *Ifng*, *IL4*, *IL1b*, and *Ccl21*) and immune-suppressive cytokines (*IL10* and *Tgfb*). Several cytokines have similar expression across all tumors, including *Tnfa*, *Ifng*, and *Ccl21*. Other cytokines (including *Tgfb*, *IL4*, *IL10*, and *IL1b*) display more variability across individual tumors, although this was not associated with specific background strains. Taken together, this gene expression analysis highlights key strain-dependent differences in the composition of the tumor microenvironment—most notably, the elevation of M2 macrophages in MPNSTs from BALB/c mice.

4. Discussion

The genetic background of murine cancer models can determine critical phenotypes such as disease onset, metastatic potential, immune response, and treatment outcome. To examine the impact of mouse strain on the biology of genetically-identical tumors, we used somatic CRISPR/Cas9 tumorigenesis approaches to generate MPNSTs in four commonly-used, classically inbred strains. We evaluated the influence of mouse strain on tumor growth, histology, indel pattern, immune cell infiltration, and expression of TME markers. Our data indicate that background strain impacts tumor latency, immune composition, and gene expression of genetically-identical MPNSTs. In particular, BALB/c mice exhibit multiple strain-dependent tumor phenotypes, including acceleration of tumor onset, elevated mast cell infiltration, and enrichment of M2 macrophages. In contrast, MPNSTs generated in C57BL/6 mice display decreased levels of T lymphocytes. Taken together, these data highlight the importance of considering host strain in the design and interpretation of tumor studies.

CRISPR/Cas9 approaches can facilitate the study of cancer-relevant questions that are difficult to address using conventional Cre/loxP methods. The requirement for complex backcrossing and the potential for persistent modifier loci with traditional GEMM approaches complicates data interpretation, and it has been challenging to examine the impact of background strain on the immune landscape of genetically-matched tumors. While multiple groups have reported broad immunological diversity in different syngeneic cell transplant models generated within the same background strain [22–26], our data identify multiple strain-specific differences in tumor infiltration by myeloid and adaptive immune cells in isogenic MPNSTs. Of note, tumors from C57BL/6 mice have the lowest levels of infiltrating CD4+ T lymphocytes. This observation is in line with published work examining the immune microenvironment in a series of cell transplant models from C57BL/6 and BALB/c mice. One study found that CD4+ T lymphocytes comprise only 1–4% of total CD45 cells in syngeneic C57BL/6 models—including MC38, LL/2, and B16F10 tumors—while populations of CD4+ T lymphocytes account for 6–10% of total immune cells in syngeneic BALB/c models such as CT26, RENCA, and 4T1 [22]. We also observed increased Tregs by IHC analysis in MPNSTs from 129X1 mice. However, it is difficult to compare our findings to the other 129-derived tumor models, as there are few published studies that include 129-based models in cross-strain analysis of immune infiltration.

Our data also found enrichment of mast cells in MPNSTs from BALB/c mice. Increased mast cell levels are associated with accelerated onset of MPNSTs in *Nf1* haploinsufficient mouse models [30]. In neurofibromas, *Nf1*$^{+/-}$ mast cells are essential to tumor formation due to critical SCF-mediated interactions with *Nf1*$^{+/-}$ Schwann cells [35]. Indeed, mast cells may play tumor promoting roles in multiple cancers—including colorectal and pancreatic—by supporting an immunosuppressive microenvironment or altering ECM homeostasis [36]. However, the prognostic significance of mast cells varies greatly across different cancer types. While a mechanistic role for mast cells in MPNST development has not been shown, a study in a small number of patient samples ($n = 34$) found that mast cell density did not correlate with patient survival [37]. Mast cell function is strain dependent, with bone marrow-derived mast cells (BMMCs) from BALB/c mice displaying more robust responses than BMMCs from other backgrounds. For example, in response to allergenic challenge, BMMCs from BALB/c mice degranulate more efficiently [38], produce higher amounts of newly-synthesized mediators [39], and infiltrate more rapidly into bronchial tissue than BMMCs from C57BL/6 mice [40]. This increased activity of mast cells in BALB/c mice, combined with elevated mast cell infiltration in BALB/c-derived MPNSTs, could partially explain the accelerated tumor onset phenotype in this strain.

One of the strongest strain-dependent immune phenotypes we observed was enrichment of macrophages in MPNSTs from BALB/c mice. In syngeneic tumor models, macrophage infiltration is highly variable and is more dependent upon cancer type than host strain [22,23]. For example, macrophages account for ~18% of total CD45+ immune cells in both RENCA (BALB/c hosts) and Lewis Lung carcinomas (C57BL/6 hosts), while macrophages make up only ~5% of immune cells in CT26 (BALB/c hosts) and B16 melanoma (C57BL/6 hosts) models [22]. Our data also identify a strong M2 polarization in TAMs from BALB/c-derived tumors by upregulation of *Arg1* expression.

This strain-specific enrichment of M1/M2 macrophages is a well-documented phenotype. As M2 macrophages predominantly promote wound healing and tissue homoeostasis, the M1/M2 polarization can have important phenotypic consequences. For example, in response to challenge with *Leishmania*, C57BL/6 mice can eliminate infection by activation of an M1/Th1 response, but BALB/c mice succumb to infection due to the inability of their M2 macrophages to mount an effective response [18].

It is important to note that the M1/M2 definition of macrophages represents a phenotypic spectrum, rather than a binary characterization. The strict definition of M1 vs. M2 has recently been broadened with the discoveries of in vivo populations that exist along a mixed M1/M2/monocyte spectrum that support plasticity among myeloid populations [41]. Indeed, macrophage diversity is widespread among mouse models, as demonstrated with data from the hybrid mouse diversity panel (HMDP) that was developed to examine immunological variation across different host backgrounds. By using a panel of 83 inbred mouse strains, this resource can perform gene association studies to better understand and map complex traits [42]. A genome-wide study of peritoneal macrophage transcriptomes from the HMDP identified a natural spectrum of macrophage activation phenotypes and confirmed that the M1/M2 axis is a major macrophage polarization phenotype in vivo [43]. Of particular importance to cancer biology, the M1 and M2 paradigm of macrophage polarization does not clearly apply to TAMs, which are strongly influenced by tumor location and external cues from the surrounding microenvironment [44]. TAM subsets can express both M1 and M2 markers simultaneously, suggesting that they display a more complex activation scenario than the simple M1/M2 activation status [41,44,45]. Nonetheless, an appreciation of strain-dependent macrophage polarity is important for interpretation and design of in vivo tumor models examining macrophage tumor biology.

One interesting observation from our study is the acceleration of tumor initiation in BALB/c mice. Several groups have reported accelerated tumor formation in $p53^{+/-}$ BALB/c mice in comparison to C57BL/6 mice [2–4]. However, these studies did not induce spatially-restricted tumors in adult mice. One possible explanation for earlier tumor onset of *Nf1/p53*-driven MPNSTs in BALB/c mice is their strain-specific mutation in *Ink4a* (also known as *p16*). *Ink4a* is a member of the *Cdkn2a* locus that is fundamental to cell cycle entry and progression [46]. The *Cdkn2a* (*Ink4a/Arf*) allele is a well-documented example of a strain-dependent genetic variant that can impact cancer progression [47,48]. Indeed, the increased susceptibility of BALB/c mice for various cancer types has been linked to the presence of a hypomorphic *Ink4a* allele caused by mutations in the promoter region [48]. Since disruptions in *Cdkn2a* are commonly observed in clinical MPNST samples, we postulate that acceleration of tumor onset in BALB/c mice may be partially due to disruption of this locus.

These studies underscore the need to use a diverse toolkit of mouse backgrounds in cancer biology, as the reliance on single strain studies can be a barrier to a robust understanding of cancer progression [49]. We believe there is immense strength in applying a broad diversity of in vivo models to better account for the large interindividual variation of immune systems across human populations [50,51]. Additionally, these data suggest that caution must be taken in interpretation of preclinical studies, with respect to potential influences of complex, strain-specific interactions between the TME and tumor cells. Further studies are necessary to determine whether strain-specific immune landscapes would alter therapeutic outcomes in preclinical MPNST models. It is plausible that enrichment of either T lymphocytes or macrophages could alternatively impact immunotherapy response. However, chemotherapy outcomes may be less dependent upon immune composition, as we reported that murine MPNSTs with distinct myeloid cell compositions respond similarly to doxorubicin/ifosfamide-containing regimens [30]. Taken together, our findings highlight how CRISPR/Cas9 tumorigenesis approaches can provide new experimental opportunities to leverage the immunological diversity of inbred mouse strains to reveal new features of the tumor microenvironment that drive MPNST progression.

Supplementary Materials: The following are available online at http://www.mdpi.com/2073-4425/11/5/583/s1, Table S1: Primer and guide RNA Sequences, Figure S1: IHC of innate and adaptive immune cells in CRISPR/Cas9-generated MPNSTs. Macrophages (F4/80 staining; 40×) and mast cells (toludine blue staining; 20×)

are enriched in MPNSTs from BABL/c mice. Cytotoxic T cells (CD8 staining; 20×) are similar across all strains. Helper T cells (CD4 staining; 20×) are enriched in 129X1 and 129Sv/Jae tumors. Regulatory T cells (FoxP3 staining, 40×) are enriched in 129X1 tumors, Figure S2: Quantitative RT-PCR data from heatmap. Expression levels of genes in the MPNST microenvironment examining macrophages (A), adaptive immunity (B–H), angiogenesis and lymphangiogenesis (I–L), and cytokines (M–S).

Author Contributions: R.D.D. and A.S. conceived and designed the study. A.S., V.R.S., W.R.G., G.R.M., E.A.L., and V.K.-A. performed the experiments. A.S. and V.R.S. collected and analyzed the data. R.D.D. and A.S. interpreted the data. R.D.D. and A.S. wrote the paper. R.D.D. acquired the funding and supervised the study. All authors reviewed and approved the manuscript.

Funding: This work was supported by an American Cancer Society Internal Review Grant IRG-15-176-40 [RDD], Department of Defense CDMRP Neurofibromatosis Research Program W81XWH-18-1-0174 [RDD], University of Iowa PREP R25 GM116686 [VS], T32 GM067795 [WRG], T32 GM007337 [WRG], T32 CA078586 [GRM], and an NCI Core Grant P30 CA086862 [University of Iowa Holden Comprehensive Cancer Center].

Acknowledgments: We are grateful to colleagues in the University of Iowa Sarcoma Research Group and the Henry, Dupuy, and Stipp labs for their critical feedback throughout this study. Sanger sequencing and qRT-PCR data were obtained at the Genomics Division of the Iowa Institute of Human Genetics, which is supported, in part, by the University of Iowa Carver College of Medicine and the Holden Comprehensive Cancer Center (National Cancer Institute of the National Institutes of Health under Award Number P30CA086862).

Conflicts of Interest: The authors declare no conflict of interest.

References

1.   Reilly, K.M. The Effects of Genetic Background of Mouse Models of Cancer: Friend or Foe? *Cold Spring Harb. Protoc.* **2016**, *2016*, pdb.top076273. [CrossRef] [PubMed]

2.   Kuperwasser, C.; Hurlbut, G.D.; Kittrell, F.S.; Dickinson, E.S.; Laucirica, R.; Medina, D.; Naber, S.P.; Jerry, D.J. Development of spontaneous mammary tumors in BALB/c p53 heterozygous mice. A model for Li-Fraumeni syndrome. *Am. J. Pathol.* **2000**, *157*, 2151–2159. [CrossRef]

3.   Koch, J.G.; Gu, X.; Han, Y.; El-Naggar, A.K.; Olson, M.V.; Medina, D.; Jerry, D.J.; Blackburn, A.C.; Peltz, G.; Amos, C.I.; et al. Mammary tumor modifiers in BALB/cJ mice heterozygous for p53. *Mamm. Genome Off. J. Int. Mamm. Genome Soc.* **2007**, *18*, 300–309. [CrossRef] [PubMed]

4.   Blackburn, A.C.; Hill, L.Z.; Roberts, A.L.; Wang, J.; Aud, D.; Jung, J.; Nikolcheva, T.; Allard, J.; Peltz, G.; Otis, C.N.; et al. Genetic mapping in mice identifies DMBT1 as a candidate modifier of mammary tumors and breast cancer risk. *Am. J. Pathol.* **2007**, *170*, 2030–2041. [CrossRef] [PubMed]

5.   Brandt, L.P.; Albers, J.; Hejhal, T.; Pfundstein, S.; Gonçalves, A.F.; Catalano, A.; Wild, P.J.; Frew, I.J. Mouse genetic background influences whether HrasG12V expression plus Cdkn2a knockdown causes angiosarcoma or undifferentiated pleomorphic sarcoma. *Oncotarget* **2018**, *9*, 19753–19766. [CrossRef] [PubMed]

6.   Dragani, T.A. 10 years of mouse cancer modifier loci: Human relevance. *Cancer Res.* **2003**, *63*, 3011–3018.

7.   Dietrich, W.F.; Lander, E.S.; Smith, J.S.; Moser, A.R.; Gould, K.A.; Luongo, C.; Borenstein, N.; Dove, W. Genetic identification of Mom-1, a major modifier locus affecting Min-induced intestinal neoplasia in the mouse. *Cell* **1993**, *75*, 631–639. [CrossRef]

8.   Reilly, K.M.; Loisel, D.A.; Bronson, R.T.; McLaughlin, M.E.; Jacks, T. Nf1;Trp53 mutant mice develop glioblastoma with evidence of strain-specific effects. *Nat. Genet.* **2000**, *26*, 109–113. [CrossRef]

9.   Reilly, K.M.; Tuskan, R.G.; Christy, E.; Loisel, D.A.; Ledger, J.; Bronson, R.T.; Smith, C.D.; Tsang, S.; Munroe, D.J.; Jacks, T. Susceptibility to astrocytoma in mice mutant for Nf1 and Trp53 is linked to chromosome 11 and subject to epigenetic effects. *Proc. Natl. Acad. Sci. USA* **2004**, *101*, 13008–13013. [CrossRef]

10.   Reilly, K.M.; Broman, K.W.; Bronson, R.T.; Tsang, S.; Loisel, D.A.; Christy, E.S.; Sun, Z.; Diehl, J.; Munroe, D.J.; Tuskan, R.G. An imprinted locus epistatically influences Nstr1 and Nstr2 to control resistance to nerve sheath tumors in a neurofibromatosis type 1 mouse model. *Cancer Res.* **2006**, *66*, 62–68. [CrossRef]

11.   Brosius, S.N.; Turk, A.N.; Byer, S.J.; Brossier, N.M.; Kohli, L.; Whitmire, A.; Mikhail, F.M.; Roth, K.A.; Carroll, S.L. Neuregulin-1 overexpression and Trp53 haploinsufficiency cooperatively promote de novo malignant peripheral nerve sheath tumor pathogenesis. *Acta Neuropathol. (Berl.)* **2014**, *127*, 573–591. [CrossRef] [PubMed]

12.   Chen, M.-L.; Xu, P.-Z.; Peng, X.; Chen, W.S.; Guzman, G.; Yang, X.; Di Cristofano, A.; Pandolfi, P.P.; Hay, N. The deficiency of Akt1 is sufficient to suppress tumor development in Pten+/- mice. *Genes Dev.* **2006**, *20*, 1569–1574. [CrossRef] [PubMed]

13. Wang, S.; Gao, J.; Lei, Q.; Rozengurt, N.; Pritchard, C.; Jiao, J.; Thomas, G.V.; Li, G.; Roy-Burman, P.; Nelson, P.S.; et al. Prostate-specific deletion of the murine Pten tumor suppressor gene leads to metastatic prostate cancer. *Cancer Cell* **2003**, *4*, 209–221. [CrossRef]

14. Lifsted, T.; Le Voyer, T.; Williams, M.; Muller, W.; Klein-Szanto, A.; Buetow, K.H.; Hunter, K.W. Identification of inbred mouse strains harboring genetic modifiers of mammary tumor age of onset and metastatic progression. *Int. J. Cancer* **1998**, *77*, 640–644. [CrossRef]

15. Sellers, R.S.; Clifford, C.B.; Treuting, P.M.; Brayton, C. Immunological variation between inbred laboratory mouse strains: Points to consider in phenotyping genetically immunomodified mice. *Vet. Pathol.* **2012**, *49*, 32–43. [CrossRef]

16. Hensel, J.A.; Khattar, V.; Ashton, R.; Ponnazhagan, S. Characterization of immune cell subtypes in three commonly used mouse strains reveals gender and strain-specific variations. *Lab. Investig. J. Technol. Methods Pathol.* **2019**, *99*, 93–106. [CrossRef]

17. Chen, J.; Harrison, D.E. Quantitative trait loci regulating relative lymphocyte proportions in mouse peripheral blood. *Blood* **2002**, *99*, 561–566. [CrossRef]

18. Mills, C.D.; Kincaid, K.; Alt, J.M.; Heilman, M.J.; Hill, A.M. M-1/M-2 macrophages and the Th1/Th2 paradigm. *J. Immunol. Baltim. Md 1950* **2000**, *164*, 6166–6173. [CrossRef]

19. Foerster, F.; Boegel, S.; Heck, R.; Pickert, G.; Rüssel, N.; Rosigkeit, S.; Bros, M.; Strobl, S.; Kaps, L.; Aslam, M.; et al. Enhanced protection of C57 BL/6 vs. Balb/c mice to melanoma liver metastasis is mediated by NK cells. *Oncoimmunology* **2018**, *7*, e1409929. [CrossRef]

20. White, P.; Liebhaber, S.A.; Cooke, N.E. 129 × 1/SvJ mouse strain has a novel defect in inflammatory cell recruitment. *J. Immunol. Baltim. Md 1950* **2002**, *168*, 869–874. [CrossRef]

21. Martin, M.D.; Carter, K.J.; Jean-Philippe, S.R.; Chang, M.; Mobashery, S.; Thiolloy, S.; Lynch, C.C.; Matrisian, L.M.; Fingleton, B. Effect of ablation or inhibition of stromal matrix metalloproteinase-9 on lung metastasis in a breast cancer model is dependent on genetic background. *Cancer Res.* **2008**, *68*, 6251–6259. [CrossRef]

22. Mosely, S.I.S.; Prime, J.E.; Sainson, R.C.A.; Koopmann, J.-O.; Wang, D.Y.Q.; Greenawalt, D.M.; Ahdesmaki, M.J.; Leyland, R.; Mullins, S.; Pacelli, L.; et al. Rational Selection of Syngeneic Preclinical Tumor Models for Immunotherapeutic Drug Discovery. *Cancer Immunol. Res.* **2017**, *5*, 29–41. [CrossRef]

23. Yu, J.W.; Bhattacharya, S.; Yanamandra, N.; Kilian, D.; Shi, H.; Yadavilli, S.; Katlinskaya, Y.; Kaczynski, H.; Conner, M.; Benson, W.; et al. Tumor-immune profiling of murine syngeneic tumor models as a framework to guide mechanistic studies and predict therapy response in distinct tumor microenvironments. *PLoS ONE* **2018**, *13*, e0206223. [CrossRef] [PubMed]

24. Lechner, M.G.; Karimi, S.S.; Barry-Holson, K.; Angell, T.E.; Murphy, K.A.; Church, C.H.; Ohlfest, J.R.; Hu, P.; Epstein, A.L. Immunogenicity of murine solid tumor models as a defining feature of in vivo behavior and response to immunotherapy. *J. Immunother. (Hagerstown Md.: 1997)* **2013**, *36*, 477–489. [CrossRef] [PubMed]

25. Grasselly, C.; Denis, M.; Bourguignon, A.; Talhi, N.; Mathe, D.; Tourette, A.; Serre, L.; Jordheim, L.P.; Matera, E.L.; Dumontet, C. The Antitumor Activity of Combinations of Cytotoxic Chemotherapy and Immune Checkpoint Inhibitors Is Model-Dependent. *Front. Immunol.* **2018**, *9*, 2100. [CrossRef] [PubMed]

26. De Luca, R.; Neri, D. Potentiation of PD-L1 blockade with a potency-matched dual cytokine-antibody fusion protein leads to cancer eradication in BALB/c-derived tumors but not in other mouse strains. *Cancer Immunol. Immunother. CII* **2018**, *67*, 1381–1391. [CrossRef] [PubMed]

27. Huang, J.; Chen, M.; Whitley, M.J.; Kuo, H.-C.; Xu, E.S.; Walens, A.; Mowery, Y.M.; Van Mater, D.; Eward, W.C.; Cardona, D.M.; et al. Generation and comparison of CRISPR-Cas9 and Cre-mediated genetically engineered mouse models of sarcoma. *Nat. Commun.* **2017**, *8*, 15999. [CrossRef] [PubMed]

28. Dodd, R.D.; Añó, L.; Blum, J.M.; Li, Z.; Van Mater, D.; Kirsch, D.G. Methods to generate genetically engineered mouse models of soft tissue sarcoma. *Methods Mol. Biol. Clifton NJ* **2015**, *1267*, 283–295. [CrossRef]

29. Dodd, R.D.; Mito, J.K.; Eward, W.C.; Chitalia, R.; Sachdeva, M.; Ma, Y.; Barretina, J.; Dodd, L.; Kirsch, D.G. NF1 deletion generates multiple subtypes of soft-tissue sarcoma that respond to MEK inhibition. *Mol. Cancer Ther.* **2013**, *12*, 1906–1917. [CrossRef]

30. Dodd, R.D.; Lee, C.-L.; Overton, T.; Huang, W.; Eward, W.C.; Luo, L.; Ma, Y.; Ingram, D.R.; Torres, K.E.; Cardona, D.M.; et al. NF1+/- Hematopoietic Cells Accelerate Malignant Peripheral Nerve Sheath Tumor Development without Altering Chemotherapy Response. *Cancer Res.* **2017**, *77*, 4486–4497. [CrossRef] [PubMed]

31. Maresch, R.; Mueller, S.; Veltkamp, C.; Öllinger, R.; Friedrich, M.; Heid, I.; Steiger, K.; Weber, J.; Engleitner, T.; Barenboim, M.; et al. Multiplexed pancreatic genome engineering and cancer induction by transfection-based CRISPR/Cas9 delivery in mice. *Nat. Commun.* **2016**, *7*, 10770. [CrossRef] [PubMed]

32. Synthego. ICE v2 CRISPR Analysis Tools. Available online: https://www.synthego.com/products/bioinformatics/crispr-analysis (accessed on 3 January 2020).

33. Lindeboom, R.G.H.; Supek, F.; Lehner, B. The rules and impact of nonsense-mediated mRNA decay in human cancers. *Nat. Genet.* **2016**, *48*, 1112–1118. [CrossRef] [PubMed]

34. You, K.T.; Li, L.S.; Kim, N.-G.; Kang, H.J.; Koh, K.H.; Chwae, Y.-J.; Kim, K.M.; Kim, Y.K.; Park, S.M.; Jang, S.K.; et al. Selective Translational Repression of Truncated Proteins from Frameshift Mutation-Derived mRNAs in Tumors. *PLoS Biol.* **2007**, *5*, e109. [CrossRef] [PubMed]

35. Staser, K.; Yang, F.-C.; Clapp, D.W. Mast cells and the neurofibroma microenvironment. *Blood* **2010**, *116*, 157–164. [CrossRef] [PubMed]

36. Rigoni, A.; Colombo, M.P.; Pucillo, C. The Role of Mast Cells in Molding the Tumor Microenvironment. *Cancer Microenviron. Off. J. Int. Cancer Microenviron. Soc.* **2015**, *8*, 167–176. [CrossRef]

37. de Vasconcelos, R.A.T.; Guimarães Coscarelli, P.; Vieira, T.M.; Noguera, W.S.; Rapozo, D.C.M.; Acioly, M.A. Prognostic significance of mast cell and microvascular densities in malignant peripheral nerve sheath tumor with and without neurofibromatosis type 1. *Cancer Med.* **2019**, *8*, 972–981. [CrossRef]

38. Nagashima, M.; Koyanagi, M.; Arimura, Y. Comparative Analysis of Bone Marrow-derived Mast Cell Differentiation in C57BL/6 and BALB/c Mice. *Immunol. Investig.* **2019**, *48*, 303–320. [CrossRef]

39. Noguchi, J.; Kuroda, E.; Yamashita, U. Strain difference of murine bone marrow-derived mast cell functions. *J. Leukoc. Biol.* **2005**, *78*, 605–611. [CrossRef]

40. Pae, S.; Cho, J.Y.; Dayan, S.; Miller, M.; Pemberton, A.D.; Broide, D.H. Chronic allergen challenge induces bronchial mast cell accumulation in BALB/c but not C57BL/6 mice and is independent of IL-9. *Immunogenetics* **2010**, *62*, 499–506. [CrossRef]

41. Laviron, M.; Boissonnas, A. Ontogeny of Tumor-Associated Macrophages. *Front. Immunol.* **2019**, *10*, 1799. [CrossRef]

42. Bennett, B.J.; Farber, C.R.; Orozco, L.; Kang, H.M.; Ghazalpour, A.; Siemers, N.; Neubauer, M.; Neuhaus, I.; Yordanova, R.; Guan, B.; et al. A high-resolution association mapping panel for the dissection of complex traits in mice. *Genome Res.* **2010**, *20*, 281–290. [CrossRef] [PubMed]

43. Buscher, K.; Ehinger, E.; Gupta, P.; Pramod, A.B.; Wolf, D.; Tweet, G.; Pan, C.; Mills, C.D.; Lusis, A.J.; Ley, K. Natural variation of macrophage activation as disease-relevant phenotype predictive of inflammation and cancer survival. *Nat. Commun.* **2017**, *8*, 16041. [CrossRef] [PubMed]

44. Martinez, F.O.; Gordon, S. The M1 and M2 paradigm of macrophage activation: Time for reassessment. *F1000prime Rep.* **2014**, *6*, 13. [CrossRef] [PubMed]

45. Murray, P.J.; Allen, J.E.; Biswas, S.K.; Fisher, E.A.; Gilroy, D.W.; Goerdt, S.; Gordon, S.; Hamilton, J.A.; Ivashkiv, L.B.; Lawrence, T.; et al. Macrophage activation and polarization: Nomenclature and experimental guidelines. *Immunity* **2014**, *41*, 14–20. [CrossRef] [PubMed]

46. Sherr, C.J. The INK4a/ARF network in tumour suppression. *Nat. Rev. Mol. Cell Biol.* **2001**, *2*, 731–737. [CrossRef]

47. Mock, B.A.; Krall, M.M.; Dosik, J.K. Genetic mapping of tumor susceptibility genes involved in mouse plasmacytomagenesis. *Proc. Natl. Acad. Sci. USA* **1993**, *90*, 9499–9503. [CrossRef]

48. Zhang, S.; Ramsay, E.S.; Mock, B.A. Cdkn2a, the cyclin-dependent kinase inhibitor encoding p16INK4a and p19ARF, is a candidate for the plasmacytoma susceptibility locus, Pctr1. *Proc. Natl. Acad. Sci. USA* **1998**, *95*, 2429–2434. [CrossRef]

49. Sittig, L.J.; Carbonetto, P.; Engel, K.A.; Krauss, K.S.; Barrios-Camacho, C.M.; Palmer, A.A. Genetic Background Limits Generalizability of Genotype-Phenotype Relationships. *Neuron* **2016**, *91*, 1253–1259. [CrossRef]

50. Tsang, J.S.; Schwartzberg, P.L.; Kotliarov, Y.; Biancotto, A.; Xie, Z.; Germain, R.N.; Wang, E.; Olnes, M.J.; Narayanan, M.; Golding, H.; et al. Global analyses of human immune variation reveal baseline predictors of postvaccination responses. *Cell* **2014**, *157*, 499–513. [CrossRef]
51. Casanova, J.-L.; Abel, L. The human model: A genetic dissection of immunity to infection in natural conditions. *Nat. Rev. Immunol.* **2004**, *4*, 55–66. [CrossRef]

Article

Unmasking Intra-Tumoral Heterogeneity and Clonal Evolution in NF1-MPNST

Chang-In Moon [1,†], William Tompkins [2,†], Yuxi Wang [1], Abigail Godec [3], Xiaochun Zhang [1], Patrik Pipkorn [4,5], Christopher A. Miller [5,6], Carina Dehner [7], Sonika Dahiya [5,7] and Angela C. Hirbe [1,5,*]

[1] Division of Medical Oncology, Department of Medicine, Washington University School of Medicine, St. Louis, MO 63110, USA; moonchangin@wustl.edu (C.-I.M.); yuxi.w@wustl.edu (Y.W.); zhang.x@wustl.edu (X.Z.)
[2] Washington University School of Medicine, St. Louis, MO 63110, USA; wtompkins@wustl.edu
[3] College of Human Medicine, Michigan State University, East Lansing, MI 48824, USA; godecabi@msu.edu
[4] Department of Otolaryngology, Division of Head and Neck Surgery, Washington University School of Medicine, St. Louis, MO 63110, USA; ppipkorn@wustl.edu
[5] Siteman Cancer Center, St. Louis, MO 63110, USA; c.a.miller@wustl.edu (C.A.M.); sdahiya@wustl.edu (S.D.)
[6] McDonnell Genome Institute, Division of Oncology—Stem Cell Biology, Department of Medicine, Washington University School of Medicine, St. Louis, MO 63110, USA
[7] Department of Pathology and Immunology, Washington University School of Medicine, St. Louis, MO 63110, USA; cdehner@wustl.edu
* Correspondence: hirbea@wustl.edu; Tel.: +1-314-747-3096
† These authors contributed equally.

Received: 6 March 2020; Accepted: 30 April 2020; Published: 1 May 2020

Abstract: Sarcomas are highly aggressive cancers that have a high propensity for metastasis, fail to respond to conventional therapies, and carry a poor 5-year survival rate. This is particularly true for patients with neurofibromatosis type 1 (NF1), in which 8%–13% of affected individuals will develop a malignant peripheral nerve sheath tumor (MPNST). Despite continued research, no effective therapies have emerged from recent clinical trials based on preclinical work. One explanation for these failures could be the lack of attention to intra-tumoral heterogeneity. Prior studies have relied on a single sample from these tumors, which may not be representative of all subclones present within the tumor. In the current study, samples were taken from three distinct areas within a single tumor from a patient with an NF1-MPNST. Whole exome sequencing, RNA sequencing, and copy number analysis were performed on each sample. A blood sample was obtained as a germline DNA control. Distinct mutational signatures were identified in different areas of the tumor as well as significant differences in gene expression among the spatially distinct areas, leading to an understanding of the clonal evolution within this patient. These data suggest that multi-regional sampling may be important for driver gene identification and biomarker development in the future.

Keywords: NF1; MPNST; genomics; heterogeneity

---

1. Introduction

Malignant peripheral nerve sheath tumor (MPNSTs) is the sixth most common soft tissue sarcoma [1] and has an incidence rate of 0.1–0.2 per 100,000 persons per year [2]. MPNSTs are often associated with neurofibromatosis type 1 (NF1). The incidence rate of MPNSTs in patients with NF1 is much higher than that of the general population, estimated to be 1.6 per 1000 per year, or a lifetime risk of 8–13% [3]. Approximately 50% of MPNSTs occur in patients with neurofibromatosis [4–7], and the other 50% of MPNSTs occur sporadically or in the setting of previous radiation therapy [4,6]. In the setting of NF1, MPNSTs often arise within a pre-existing benign nerve sheath tumor (plexiform neurofibroma) [4,7].

Prognosis remains poor for patients with MPNST despite multi-modality therapy [2,5–10]. In the setting of metastatic disease, treatment is limited to cytotoxic chemotherapy, typically consisting of single agent doxorubicin or a combination of doxorubicin and ifosfamide [11–13].

A number of different genes have been implicated in the development of MPNSTs. One of the most commonly used models for preclinical testing was developed by Cichowski et al. and Vogel et al; they demonstrated that mice with germline variants in *Nf1* and *Tp53* develop MPNSTs, supporting a cooperative and causal role for these tumor suppressors in the context of MPNST formation [14,15]. Other groups have found a reduction in expression of *PTEN*, a tumor suppressor in the *PI3K/AKT/mTOR* pathway, in MPNSTs compared to benign nerve sheath tumors in a manner that is not regulated by *NF1* [16]. Keng et al. went on to demonstrate the cooperative roles of *Pten* and *Nf1* in the tumorigenesis of MPNSTs in vivo with transgenic mouse models [17]. Gregorian et al. further elucidated the cooperative relationship between *k-ras* activation and *Pten* deletion, showing that both variants in combination led to 100% penetrable development of MPNSTs [18]. Another gene implicated in MPNST pathogenesis is *INK4A*, a tumor suppressor encoding both *p16* and *p19*. Deletions in this gene have been identified in MPNSTs but not in benign neurofibromas [19]. Lu et al. demonstrated a difference in aberrant expression of *ATRX*, a DNA helicase that plays a role in chromatin regulation and maintenance of telomeres, between MPNSTs and benign neurofibromas [20]. Additionally, variants in *EED* and *SUZ12* have been observed in MPNST. These genes code for components of the PRC2 complex which is involved in transcriptional repression. Lee et al. showed loss-of-function somatic alterations of PRC2 components in 92% of sporadic, 70% of NF1-associated and 90% of radiotherapy-associated MPNSTs. Further, introduction of the lost PRC2 component in a PRC2-deficient MPNST cell line decreased cell growth [21]. Others have found alterations such as structural alterations of *PDGFRA* (platelet-derived growth factor-α) in 26% of MPNST samples [22]; increased expression of *EGF-R* (epidermal growth factor receptor) by immunohistochemistry in MPNSTs [23]; and *IGFR1* gene amplification in 24% of MPNSTs [24].

Despite all of this research, no effective therapies have emerged from recent clinical studies based on this genomic data and subsequent preclinical studies. Intra-tumoral heterogeneity is a possible reason for these shortcomings. Prior studies have relied on a single sample from these tumors. All the subclones within a tumor may not be captured by this approach. Our aim in this study is to investigate intra-tumoral heterogeneity more thoroughly through analysis of samples taken from multiple sites of the same MPNST.

2. Materials and Methods

2.1. Study Approvals

Blood and tumor were obtained from an individual diagnosed with NF1 according to established criteria [25] and treated for a MPNST at Washington University/St. Louis Children's Hospital NF Clinical Program (St. Louis, MO, USA). The human tumor samples were collected under an approved IRB protocol (#201203042) at Washington University, and the patient was appropriately consented.

2.2. Sample Collection

Samples were taken from three distinct areas within a single tumor from a patient with an NF1-MPNST immediately after surgical resection with guidance from a pathologist (SD). While area "1" represented solid, tan homogeneous tumor lacking hemorrhage and/or necrosis, areas "2" and "3" of the tumor grossly appeared necrotic and hemorrhagic respectively. 20 g of tissue was taken from each area. Each area was then divided to be used for RNA extraction, DNA extraction, and slide preparation to analyze the histology. A gross image of the tumor was taken at this time and is shown as Figure 1.

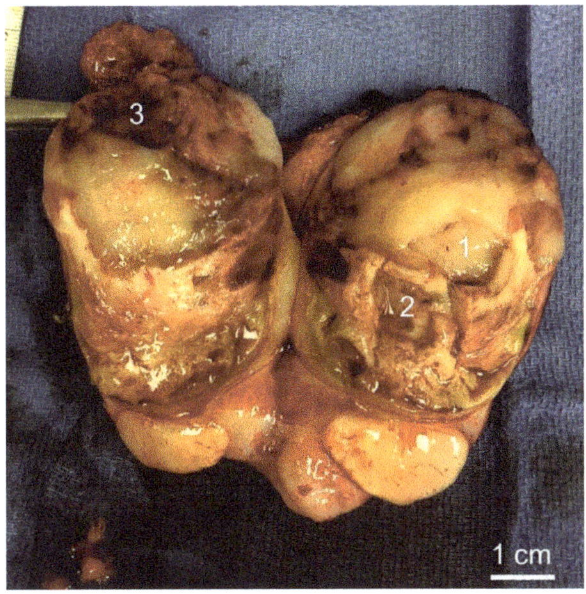

Figure 1. Malignant peripheral nerve sheath tumor (MPNST) sampled areas. Area 1 shows an area centrally located in MPNST, Area 2 an area of hemorrhage, and Area 3 an area of necrosis.

2.3. Histology

Images of the hematoxylin-eosin sections were taken (20X magnification) using an Olympus BX-51 microscope using an Olympus DP71 digital camera, and DP Controller software. Tumor purity was estimated based on morphologic review of the entire hematoxylin-eosin stained section estimating the number of tumor cells, stromal cells, lymphocytes, and extravasated red blood cells. Two pathologists reviewed these slides independently providing an estimated percentage of total tumors cells per slide.

2.4. Sequencing and Bioinformatics Analysis

Whole exome sequencing (WES), RNA sequencing (RNA-Seq), and copy number analysis (CNVkit) [26] were performed on each sample and compared to a blood sample as a germline DNA control. Both Illumina Whole Genome Sequencing (eWGS) of 3 tumor samples and 1 PBMC normal sample, and Illumina RNA Sequencing of the 3 tumor samples were generated from the sampled areas.

2.4.1. Library Construction and Sequencing

Each tumor had 2 enriched libraries constructed (n = 6), and the PBMCs had a single enriched library constructed (n = 1). Exome libraries were captured with an IDT exome reagent, then pooled with a WGS library for sequencing on an Illumina HiSeq4000 with at least 1000x coverage. RNA was prepared with a TrueSeq stranded total RNA library kit, then sequenced on an Illumina HISeq4000 with 72M reads per sample.

2.4.2. IDT Exome Sequencing Variant Detection

Genomic data were aligned against reference sequence hg38 via BWA-MEM [27] with Base Quality Score Recalibration (BQSR). Structural variants (SVs) and large indels were detected using manta [28]. SNVs and small indels were detected using VarScan2 [29], Strelka2 [30], MuTect2 [31], and Pindel [32] via the somatic pipelines available at https://github.com/genome/analysis-workflows, which includes best-practice variant filtering and annotation with VEP (Variant Effect Predictor, version 95) [33].

Manual review was used to remove additional sequencing artifacts. Germline variants and somatic variants reported on variant detecting pipeline were compared to see any intersection of variants. Any intersecting variants were removed from the somatic variant gene list, thus filtering out the germline variants. Common variants with 1000 genome MAF (minor allele frequency) > 0.05 were filtered out. Waterfall somatic variant plots were created with GenVisR [34] by including somatic variants that occurred in each area. Variants reported on the waterfall plot are most likely to be pathogenic, which is reported via VEP. These variants were not reported as a somatic variant in COSMIC (Catalogue Of Somatic Mutations In Cancer) [35] and ClinVar [36] archive, thus these variants are best classified as variants with unknown significance. In order to predict clinical significance and predictions of the functional effects of these variants, each variant was reviewed on SIFT [37] and Polyphen [38]. IMPACT rating was determined by VEP for each non-coding variant.

2.4.3. Copy Number Analysis

CNVkit was used to infer and visualize copy number from high-throughput DNA sequencing data. Coverage for each bait position in the exome reagent was calculated, then segments of constant copy number were identified using circular binary segmentation. Data were plotted to provide visualization of CNVs.

2.4.4. Inference of Clonal Phylogeny

SciClone [39] and ClonEvol [40] were utilized to attempt to perform a phylogeny inference. However, the analysis was complicated by the abundance of copy number-altered regions in these tumors, and these standard algorithms were unable to automatically perform that inference. Manual review of the shared and private single nucleotide variants and large copy number altered areas, though, revealed only one possible phylogeny for this tumor.

2.4.5. RNA Sequence Preprocessing

RNA-Sequence (RNA-seq) was trimmed from 3′-end with a minimum quality Phred score of 20 and aligned against hg38—Ensembl Transcripts release 99 via BWA-MEM. Pre/post quality control and full expectation-maximization (EM) quantification were run via Partek® Flow® [41]. Gene counts and transcript counts were normalized by CPM (counts per million) by using edgeR [42] package. Heatmap visualizations were created using gplots [43] R package (Warnes, G.R. Seattle, WA, USA).

2.4.6. Gene Differential Expression Analysis

The gene-specific analysis (GSA) method was used to test for differential expression of genes or transcript between sample regions in Partek® Flow® [44]. Differential expressed genes were defined as the following statistic parameters: p-value <= 0.05; FDR step up <= 0.05; Fold Change < −2 or >2. From differentially expressed genes, a GO enrichment test was used to functionally profile this set of genes, to determine which GO terms appear more frequently than would be expected by chance when examining the set of terms annotated to the input genes, each associated with a p-value.

2.4.7. Pathway Analysis

A list of genes in copy number aberrant (CNA) regions was extracted. CNA regions were defined as copy number regions greater than 3 or copy number regions less than 1. For each area, we intersected the list of genes that are located in the CNA regions with the differentially expressed gene list reported in the RNA differential expression analysis (p-value <= 0.05). PantherDB [45] was utilized to discover GO terms and pathways that may be affected by these genes.

3. Results

3.1. Patient Information

Patient characteristics can be seen in Table 1. The patient was a male with a history significant for a clinical diagnosis of neurofibromatosis type 1—patient had a plexiform neurofibroma, spinal neurofibromas, café au lait macules, and multiple first-degree relatives with neurofibromatosis type 1—and was 40 years old at the time of diagnosis of MPNST. He presented with a large tumor located in the left neck. Resection showed a high-grade malignant peripheral nerve sheath tumor, 10.2 cm in the largest dimension, with negative margins. The patient did not receive any adjuvant therapy for his MPNST following initial resection due to poor performance status. He recurred 21 months after the initial diagnosis and ultimately died secondary to complications from metastatic disease (33 months after initial diagnosis). Samples were taken in three different locations within the primary tumor immediately following the initital resection for the purpose of this study.

Table 1. Patient Characteristics.

Age at Diagnosis, Years	Sex	Tumor Location	Tumor Size/Grade	Surgical Margin Status	Disease Status	Metastasis	Adjuvant Treatment	OS *, Months
40	Male	Left neck	10.2 cm, Grade 3 [1]	Negative	Recurred	Lung	None	33

[1] By French Federation of Cancer Centers Sarcoma Group Grading System (FNCLCC) [46]; * OS = Overall Survival-time from diagnosis of MPNST to death.

3.2. Histology of Biopsy Sites

We first reviewed the H&E images of the tumor to correlate histology to the gross images of the tumor. H&E stained sections in Figure 2 show representative images of the three sampled areas. Area #1 demonstrates tissue of a spindle cell neoplasm of neural differentiation arranged in fascicles with elongated hyperchromatic nuclei and a mild to moderate amount of cytoplasm. The tumor purity of this sample was >95%. Area #2 shows spindled cells in a background of hemorrhage, a finding commonly seen in these high-grade tumors with a tumor purity of >95%. Area #3 represents an area of necrosis, another characteristic finding for MPNST. This sample showed >95% tumor purity.

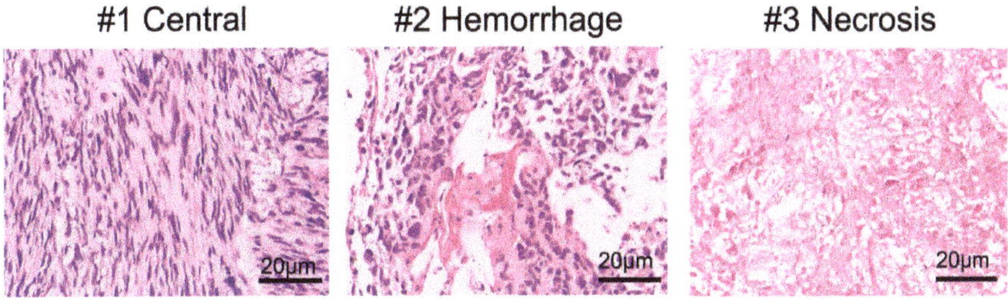

Figure 2. H&E stained sections of the biopsy sites. H&E stained sections (20X) show areas (#1) of relatively uniform, spindled cells with fascicular growth pattern, characteristic for MPNST. Sampled area #2 shows evidence of hemorrhage within the tumor, a feature commonly seen in MPNST. Area #3 shows abundant tumor necrosis.

3.3. Whole Exome Sequencing (WES), RNA Sequencing (RNA-Seq), and Copy Number Analysis

We first interrogated the sequencing data to identify the germline NF1 variant within this tumor. Figure 3 shows a lollipop plot identifying the patient's likely NF1 germline variant based on exclusion of any variants with minor allele frequency >0.05 in the 1000 genomes database. Next, to investigate intra-tumoral heterogeneity within the sample, RNA sequencing of the three sample sites was performed and is shown in Figure 4.

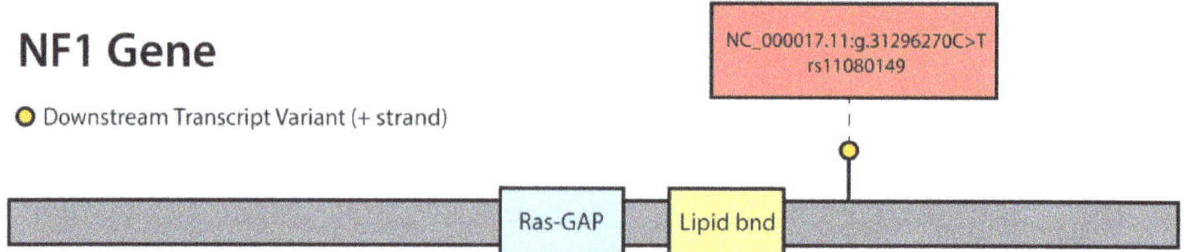

Figure 3. Location of NF1 germline variant. One intronic germline variant, NC_000017.11:g.31296270C>T (rs11080149). was identified and is depicted in this figure.

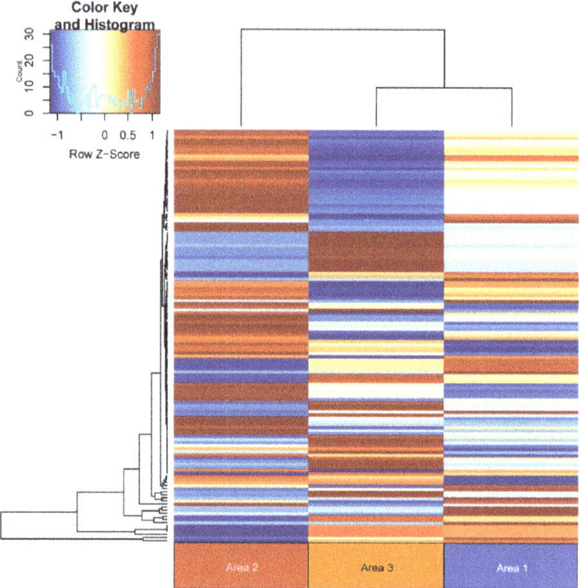

Figure 4. RNA-Seq Heatmap. Normalized read counts by counts per million (CPM) in differentially expressed genes are depicted here. Distinct gene expression profiles can be appreciated in each biopsied area. Each column is depicted as list of genes.

Distinct gene expression profiles were observed in each of the areas sampled. The top 16 differentially expressed genes are listed in Table 2 and include a number of genes involved in transcription and translation. We next performed a copy number analysis of the three biopsy sites to determine whether or not different copy number alterations were observed in each area (Figure 5). Distinct copy number signatures can be appreciated in each of the three samples further illustrating intra-tumoral heterogeneity. Additionally, we evaluated the single nucleotide variants found in each of the samples. This broad overview of all somatic variants is depicted in the waterfall plot in Figure 6. Again, distinct somatic variants can be appreciated across different areas. We next explored the potential significance of these variants through further bioinformatics analysis. While the biological significance of each of these variants is uncertain, there is evidence that some of these variants may play a role in the pathogenesis. For each variant in a coding region, CBioPortal [47] was queried for each gene to determine if the somatic variant was in a functional domain. Additionally, the RNAseq data was queried to determine if the variant in a specific area of the tumor influenced the gene expression of that gene in a specific area. Finally, SIFT and Polyphen were used to predict pathogenicity. Table 3a,b list the somatic variants in the coding region that may play a role in the pathogenesis of this tumor based on the above criteria. For those mutaions in non-coding regions, the Ensembl Variant Effect Predictor [33] was used to determine whether or not the variant would be predicted to affect gene expression. All of the identified variants were classified as modifiers, indicating that pathogenicity prediction is difficult, thus the effects of these variants are unclear. (Table 3c). Further details of the somatic variants can be found in Supplemental Table S1. Next, a gene ontology analysis was performed. To do this, a list of genes in copy number aberrant (CNA) regions was extracted. For each area, the list of genes located in the CNA regions intersected with the differentially expressed gene list reported in the RNA differential expression analysis, and PantherDB [45] was utilized to identify pathways that may be affected by these genes. Table 4 displays the unique genes in each area with copy number aberrations and alterations in gene expression. Genes depicted in Area 1 have been reported in the literature to serve a myriad of functions in tumorigenesis, including base excision repair, nucleotide excision repair, and alternative splicing [48–55]. Those in Area 2 are involved in several different pathways, including transcriptional regulation in addition to ribosomal and proteasomal function [56–60]. Finally, the genes in Area 3 consist of several ribosomal subunits and small nucleolar RNAs, suggesting that both translation and transcription are uniquely affected compared to other areas [61–63]. This analysis suggests that there may be different functional programs at play across the three areas. Next, we manually reviewed the data to look for changes in other known drivers of MPNST including TP53, ATRX, EED, SUZ12, and CDKN2A. There were no copy number changes or somatic mutions in any of these genes. Finally, we performed a careful manual review of all of the shared and unique somatic variants and copy number alterations in each area in order to develop a predicted clonal evolution. Figure 7 depicts the predicted phylogenetic tree of the subclones from each area, representing the likely clonal evolution of the tumor.

Table 2. Top Differentially Expressed Genes. The gene-specific analysis was used to test for differential expression of genes or transcript between sample regions in Partek® Flow®. Statistical cutoff are made by these following parameters: p-value <= 0.05; FDR step up <= 0.05; Fold Change <−2 or >2.

Gene Symbol	p-Value (1 vs. 2)	Fold Change (1 vs. 2)	p-Value (1 vs. 3)	Fold Change (1 vs. 3)	p-Value (2 vs. 3)	Fold Change (2 vs. 3)
EEF1A1	2.04×10^{-84}	−3.32	3.33×10^{-16}	2.20	1.35×10^{-119}	7.31
RPS27	4.32×10^{-24}	−2.51	7.64×10^{-13}	3.01	4.27×10^{-46}	7.55
RPS27A	1.69×10^{-12}	−2.62	9.42×10^{-05}	2.27	4.16×10^{-21}	5.95
H3C3	7.46×10^{-12}	−4.51	5.05×10^{-04}	11.2	5.54×10^{-09}	50.6
RPLP1	2.36×10^{-10}	−2.57	7.25×10^{-04}	2.13	2.43×10^{-17}	5.48
SNORD13	3.24×10^{-10}	3.00	8.25×10^{-62}	−4.91	3.52×10^{-66}	−14.8
RPLP0	1.05×10^{-09}	−2.26	1.60×10^{-04}	2.09	1.73×10^{-18}	4.72
TPI1	1.65×10^{-08}	−2.27	5.61×10^{-04}	2.08	6.52×10^{-16}	4.72
RPL23AP42	3.77×10^{-07}	−2.21	8.40×10^{-04}	2.16	8.65×10^{-14}	4.78
RPS23	5.34×10^{-06}	−2.46	1.17×10^{-03}	2.92	9.16×10^{-11}	7.19
MT-TI	4.64×10^{-05}	3.44	1.19×10^{-15}	−3.67	6.36×10^{-20}	−12.6
SNORA81	2.28×10^{-04}	33.3	4.00×10^{-11}	−3.39	5.12×10^{-07}	−11.3
RNY1	2.45×10^{-04}	2.65	4.67×10^{-24}	−4.71	8.10×10^{-27}	−12.5
RNVU1-31	5.00×10^{-04}	−4.18	3.83×10^{-14}	−17.7	7.07×10^{-13}	−4.23
MT-TM	6.37×10^{-04}	3.70	2.29×10^{-07}	−2.89	2.69×10^{-11}	−10.7
TMSB4XP6	1.16×10^{-03}	3.19	2.89×10^{-04}	−2.15	7.91×10^{-09}	−6.87

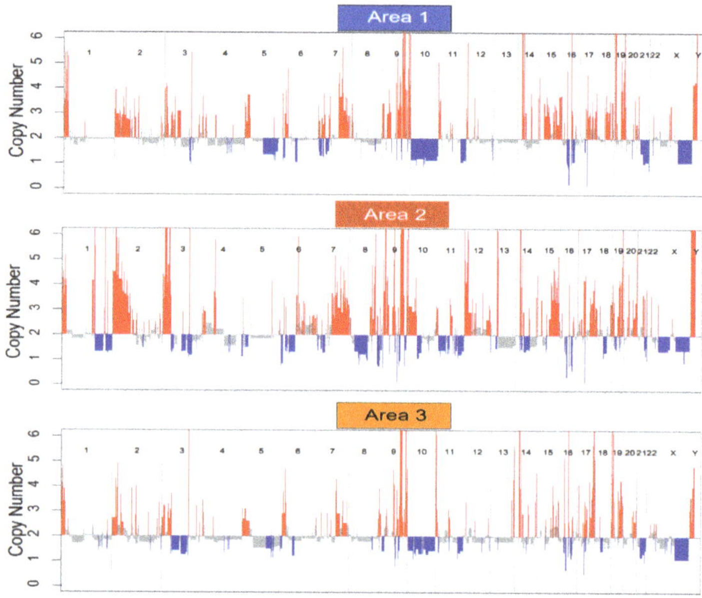

Figure 5. Copy Number Variation Plot. Copy number variation plots for each biopsied site demonstrate distinct copy number signatures.

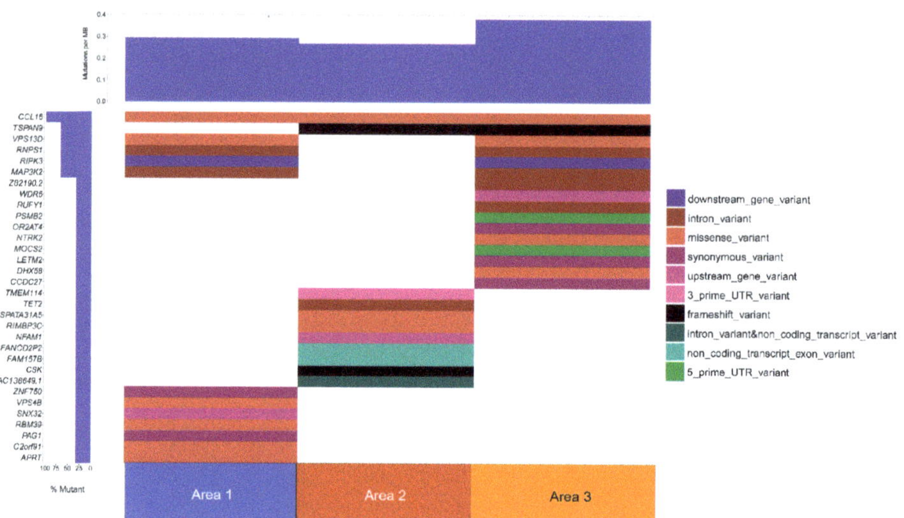

Figure 6. Somatic Variant Waterfall Plot. All somatic variants displayed on a waterfall plot. Each row represents a gene. Distinct somatic variant signatures are appreciated.

Table 3. (a) Details of the Tumor Related Somatic Variants in Coding Regions. Each gene with a somatic point variant is listed along with the area in which the variant occurred, the type of variant, the amino acid change, and whether or not the variant occurs in a putative functional domain. The final column lists whether or not the gene expression is altered in the area in which the variant occurred. The magnitude of gene expression is expressed as any of the following: NA indicates no change; "-" indicates 1-2 fold decrease in gene expression; "- -" indicates greater than 2 fold decrease in gene expression; "+" indicates 1-2 fold increase in gene expression; "+ +" indicates greater than 2 fold increase in gene expression compared to two other areas. Pathogenicity predictions are made based on SIFT and PolyPhen scores. (b) Details of the Tumor Related Somatic Frameshift Variants in Coding Regions. Each gene with a somatic point variant resulting in a frameshift is listed along with the area in which the frameshift variant occurred and whether or not the frameshift variant occurs in a putative functional domain. (c) Details of the Tumor Related Somatic Variants in Non-coding Transcript Exons, Untranslated Regions, Introns, and Upstream and Downstream Genes. Each gene with a somatic point variant is listed along with the area in which the variant occurred, the genomic location, the type of variant, and whether or not the gene expression is altered in the area in which the variant occurred. The magnitude of gene expression is expressed as: NA indicates no change; "-" indicates 1-2 fold decrease in gene expression; "- -" indicates greater than 2 fold decrease in gene expression; "+" indicates 1-2 fold increase in gene expression; "+ +" indicates greater than 2 fold increase in gene expression compared to two other areas. The final column lists the potential impact rating as evaluated by VEP. All of these variants are listed as "modifier" indicating that predictions are difficult or there is no evidence of impact.

(a)

Gene	Area	Genomic Location	Variant	Amino Acid Change	Functional Domain Affected	Gene Expression Altered	Pathogenicity Prediction
C2orf91	1	Chr2:41953024	missense	p.(Arg91Ile)	N	NA	Possibly damaging
CCL16	1	Chr17:35978161	missense	p.(Cys60Ser)	Y	NA	Probably damaging
PAG1	1	Chr8:80984896	synonymous	p.(Pro252=)	Y	NA	Unknown
VPS13D	1	Chr1:12283596	missense	p.(Phe1832Val)	N	-	Probably damaging
VPS4B	1	Chr18:63400074	missense	p.(Lys255Thr)	Y	NA	Probably damaging
ZNF750	1	Chr17:82830337	synonymous	p.(Pro659=)	N	NA	Unknown
RIMBP3C	2	Chr22:21546513	missense	p.(Arg1488Ser)	N	NA	Possibly Damaging
SPATA31A5	2	Chr9:60919364	missense	p.(Leu970Phe)	N	-	Possibly Damaging
CCDC27	3	Chr1:3752496	synonymous	p.(Ile5=)	N	+	Unknown
LETM2	3	Chr8:38400906	synonymous	p.(Leu279=)	Y	+	Unknown
NTRK2	3	Chr9:84670796	missense	p.(Trp16Cys)	N	NA	Possibly Damaging

(b)

Gene	Area	Genomic Location	Variant	Amino Acid Change	Functional Domain Affected
CSK	2	Chr15:74798671	frameshift	p.(Glu25fs)	Y
TSPAN9	2	Chr12:3283047	frameshift	p.(Leu218fs)	Y

Table 3. *Cont.*

(c)

Gene	Area	Genomic Location	Variant	Gene Expression Altered	IMPACT
MAP3K2	1	Chr2:127387525	intron	-	Modifier
RIPK3	1	Chr14:24332669 or Chr14:24332869	downstream gene	+	Modifier
RNPS1	1	Chr16:2266329	intron	-	Modifier
SNX32	1	Chr11:65832561	upstream gene	-	Modifier
AC138649.1	2	Chr15:22768761	intron	NA	Modifier
FAM157B	2	Chr9:138231054	non-coding transcript exon	+	Modifier
FANCD2P2	2	Chr3:11871392	non-coding transcript exon	+	Modifier
LAIR1	2	Chr19:54358582	intron	NA	Modifier
NFAM1	2	Chr22:42432412	upstream gene	+	Modifier
TET2	2	Chr4:105241954	intron	NA	Modifier
TMEM114	2	Chr16:8569715	3 prime UTR	NA	Modifier
MOCS2	3	Chr5:53109455	5 prime UTR	-	Unknown
PSMB2	3	Chr1:35641574	5 prime UTR	NA	Modifier
RUFY1	3	Chr5:179608552	intron	-	Modifier
WDR6	3	Chr3:49005134	upstream gene	NA	Modifier
Z82190.2	3	Chr22:31821630	intron	NA	Modifier

Table 4. Differentially Expressed Gene Pathway Analysis. These genes were located in copy number aberrant regions defined as copy number more than 3 or lower 1 and also demonstrated differential expression by RNA seq. Different pathways are implicated in the distinct sections.

Location	Chromosome	Start Position	End Position	Raw Copy Number	Genes	Role in Tumorigenesis
Area1	chr17	81509970	81523847	3.151914	*ACTG1*	Anti-apoptosis, motility [64,65]
Area1	chr17	81887843	81891586	3.151914	*ALYREF*	Genomic stability [66]
Area1	chr14	20455190	20457772	4.883921	*APEX1*	Base-excision repair [49]
Area1	chr17	81867720	81871406	3.151914	*ARHGDIA*	Invasiveness, metastasis [67]
Area1	chr12	7080208	7092607	5.842557	*C1R*	Inflammation [68]
Area1	chr17	79778131	79787983	3.109085	*CBX2*	Transcription [69]
Area1	chr17	50183288	50201632	3.060268	*COL1A1*	Metastasis [70]
Area1	chr17	82078332	82098332	3.562293	*FASN*	Metabolism [71]
Area1	chr7	128830376	128859274	3.66148	*FLNC*	Invasiveness [72]
Area1	chr17	82050690	82057470	3.562293	*GPS1*	COP9 signalosome subunit/ubiquitin-proteasome pathway
Area1	chr19	11164266	11197791	7.563794	*KANK2*	Cytoskeleton formation [73]
Area1	chrX	54807598	54816012	3.320925	*MAGED2*	Cell-cycle regulator [74]
Area1	chrX	55452104	55453566	3.320925	*MAGEH1*	Proliferation [75]
Area1	chr7	100092727	100101940	4.16605	*MCM7*	Proliferation [76]
Area1	chr14	22836556	22849027	4.136412	*MMP14*	Invasiveness, metastasis [77]
Area1	chr14	39175182	39183218	3.038443	*PNN*	Splicing [51]
Area1	chr9	107283136	107332194	14.61502	*RAD23B*	Nucleotide-excision repair [53]
Area1	chr18	49488452	49492523	3.095593	*RPL17*	Ribosome biogenesis, protein translation [61]
Area1	chrX	54814369	54814497	3.320925	*SNORA11*	Maturation of ribosomal RNA [62]
Area1	chr7	102194075	102194164	4.159154	*SNORA48*	Maturation of ribosomal RNA
Area1	chr2	5692666	5701385	3.929294	*SOX11*	Transcription
Area1	chr17	76734114	76737374	3.109085	*SRSF2*	Splicing [54]
Area1	chr9	35099775	35103195	3.374564	*STOML2*	Anti-apoptosis [78]
Area1	chr17	61399895	61409466	3.52571	*TBX2*	Transcription [79]
Area1	chr19	58544090	58550722	3.012426	*TRIM28*	Proliferation [80]
Area1	chr9	35056063	35073249	3.374564	*VCP*	Protein degradation [81]
Area1	chr7	101162508	101165593	4.159154	*VGF*	Transcription [82]
Area2	chr2	47335314	47335514	4.114423	*BCYRN1*	Transcription [56]
Area2	chr6	73515749	73523797	3.582945	*EEF1A1*	Translation [57]
Area2	chr19	3976055	3985469	3.359182	*EEF2*	Translation [58]
Area2	chr1	150574550	150579738	4.140715	*MCL1*	Anti-apoptosis [83]
Area2	chr1	151399533	151401944	4.140715	*PSMB4*	Proteasomal function [59]
Area2	chr11	67583594	67586660	3.211531	*GSTP1*	Metabolism [84]
Area2	chr15	65296050	65296166	3.976034	*RNU5A-1*	RNA processing
Area2	chr15	65304676	65304792	3.976034	*RNU5B-1*	RNA processing
Area2	chr7	148983754	148983856	3.383375	*RNY3*	RNA processing
Area2	chr13	27251308	27256691	6.141368	*RPL21*	Ribosome biogenesis, protein translation
Area2	chr9	19375714	19380254	3.739665	*RPS6*	Ribosome biogenesis, protein translation
Area2	chr2	24273613	24273741	4.326829	*SCARNA21*	RNA processing

Table 4. *Cont.*

Location	Chromosome	Start Position	End Position	Raw Copy Number	Genes	Role in Tumorigenesis
Area2	chr15	78091171	78091297	3.898802	SNORA63	Maturation of ribosomal RNA
Area2	chr1	12221147	12221271	3.552826	SNORA70	Maturation of ribosomal RNA
Area2	chr2	10446713	10446849	4.496897	SNORA80B	Maturation of ribosomal RNA
Area2	chr12	124911603	124917368	3.034233	UBC	Ubiquitin homeostasis [85]
Area3	chr16	28823034	28837237	5.159031	ATXN2L	Stress granule regulator [86]
Area3	chr9	136862118	136866286	3.830137	EDF1	Transcription
Area3	chr11	2129111	2141238	7.774932	IGF2	Proliferation [87]
Area3	chr11	2608327	2699994	7.774932	KCNQ1OT1	Transcription [88]
Area3	chr11	2134133	2134209	7.774932	MIR483	Transcription [89]
Area3	chr9	127447673	127451405	3.212283	RPL12	Ribosome biogenesis, protein translation
Area3	chr19	49487553	49492308	3.051258	RPL13A	Ribosome biogenesis, protein translation
Area3	chr19	48615327	48619536	3.174325	RPL18	Ribosome biogenesis, protein translation
Area3	chr1	6181268	6209389	3.792397	RPL22	Ribosome biogenesis, protein translation
Area3	chr17	74203581	74210655	3.363835	RPL38	Ribosome biogenesis, protein translation
Area3	chr11	809646	812880	3.117378	RPLP2	Ribosome biogenesis, protein translation
Area3	chr19	49496364	49499689	3.051258	RPS11	Ribosome biogenesis, protein translation
Area3	chr19	39433206	39435948	3.408557	RPS16	Ribosome biogenesis, protein translation
Area3	chr16	1962051	1964860	3.301972	RPS2	Ribosome biogenesis, protein translation
Area3	chr19	8321157	8323340	3.044231	RPS28	Ribosome biogenesis, protein translation
Area3	chr17	76557765	76565348	3.374444	SNHG16	Transcription [90]
Area3	chr16	1962333	1962466	3.301972	SNORA10	Maturation of ribosomal RNA
Area3	chr2	30187433	30187566	3.83836	SNORA10B	Maturation of ribosomal RNA
Area3	chr9	136726104	136726234	3.830137	SNORA17B	Maturation of ribosomal RNA
Area3	chrY	16138247	16138379	3.968437	SNORA20	Maturation of ribosomal RNA
Area3	chr16	1965183	1965310	3.301972	SNORA78	Maturation of ribosomal RNA
Area3	chr19	10109756	10109835	5.45924	SNORD105B	Ribosomal RNA modification [63]
Area3	chr19	49490614	49490699	3.051258	SNORD33	Ribosomal RNA modification
Area3	chr14	21397291	21397401	3.835309	SNORD8	Ribosomal RNA modification
Area3	chr14	21392149	21392253	3.835309	SNORD9	Ribosomal RNA modification

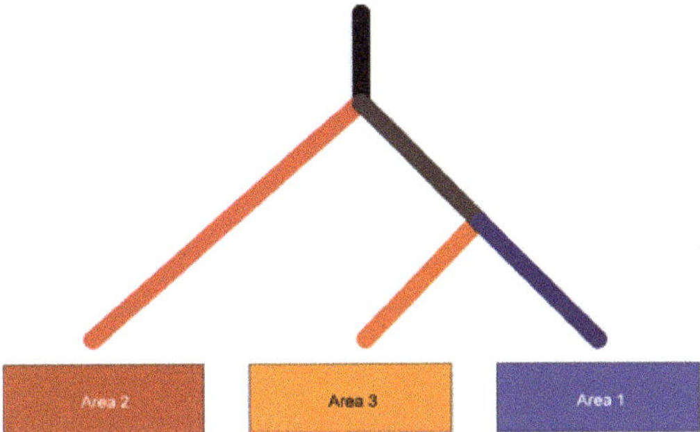

Figure 7. Phylogenetic Tree. A predicted phylogenetic tree of the tumor subclones.

4. Discussion

Despite advances in our understanding of the pathobiology of MPNST and the identification of seemingly promising therapeutic targets using a single model system in preclinical studies, no investigational agents have demonstrated efficacy following translation to human clinical trials. One element that has largely been ignored in the study of MPNST has been the possible existence of intra-tumoral heterogeneity. No single study in MPNST has focused on intra-tumoral heterogeneity. However, spatial intra-tumoral heterogeneity has become an area of interest in the study of other solid malignancies to begin to understand clonal evolution [91–95]. Within the NF1 field, researchers are beginning to appreciate the importance of understanding spatial and temporal heterogeneity. For example, Peacock et al. performed a genomic analysis of serial samples from one patient who developed an MPNST. Samples were taken at four timepoints (benign plexiform neurofibroma, MPNST pre-treatment, MPNST post-treatment, and MPNST at time of metastasis) [96]. They observed early hemizygous microdeletions in *NF1* and *TP53* with progressive amplifications of *MET*, *HGF*, and *EGFR*, highlighting the potential role of these pathways in progression. Additionally, Carrió et al. have started to examine intra-tumoral heterogeneity in PNF (plexiform neurofibromas), ANF (atypical neurofibroma) and ANNUBP (atypical neurofibromatous neoplasms with uncertain biological potential), the precursors to MPNST. They performed SNP-array analysis and exome sequencing on multiple biopsies of eight PNF, of which some had areas consistent with ANF or ANNUBP. Their data suggested that loss of a single copy of *CDKN2A/B* in *NF1* null cells is sufficient to start ANF development and that total inactivation of both copies is necessary to form ANNUBP [97]. Our study represents the first look at spatial intra-tumoral heterogeneity within an MPNST. We have demonstrated differing mutational profiles, copy number alteration signatures, and gene expression profiles within the three areas sampled. The differing mutation profile includes a variety of single nucleotide variants, including missense, frameshift, and synonymous variants. The role of synonymous variants in the tumorigenesis of MPNST is uncertain. However, there is increasing evidence that synonymous variants can alter gene expression and protein function and thus cannot be simply disregarded [98–101]. Additionally, several of the genes in Table 3a,b have previously been implicated in cancer [102–115]. For example, in Area 2, *CSK* was found to have a frameshift variant in its functional domain. *CSK* encodes a C-terminal Src kinase that has previously been found to act as a tumor suppressor in both breast cancer and prostate cancer [112–114]. Interestingly, in the context of breast cancer, Smith et al. showed that C-terminal Src kinase loss facilitated tumorigenesis by altering expression of the *PRC2* complex subunits, *EZH2* and *SUZ12* [113]. Based on these data, it is possible that alterations in *CSK* could be another way in which the *PRC2* complex is affected in MPNST. Another gene, *CCL16*, is involved in chemotaxis of human monocytes and lymphocytes. This chemokine was shown to delay mammary tumor growth and reduce rates of metastasis in mouse models [115], raising the possibility of decreased immune surveillance of our patient's MPNST secondary to a non-functional *CCL16*. In addition to the differences in single nucleotide variants, there were differences in copy number alterations across the three areas with Area

2 showing the most distinct signature in terms of copy number gains and losses. The degree to which each somatic variant, differentially expressed gene, and copy number aberration contributes to the biologic heterogeneity of the tumor remains uncertain. However, future work in our lab will be geared at elucidating this information. Finally, there was a distinct difference in gene expression among the three areas with gene ontology studies pointing toward differences in translation and protein targeting.

Taken together, these data point toward the existence of intra-tumoral heterogeneity and suggest that further investigation into this phenomenon is warranted. Additionally, these data suggest that there should be some caution taken in interpreting sequencing that comes from a single biopsy site. The advent of single cell sequencing has allowed for more rigorous evaluation of intra-tumoral heterogeneity in other cancers including acute leukemias [116,117], as well as in some solid malignancies [118,119]. Future work will be geared at using this data as the foundation to better understand clonal heterogeneity along with single cell sequencing to comprehensively evaluate intra-tumoral heterogeneity and clonal evolution of MPNST.

5. Conclusions

Significant intra-tumoral heterogeneity exists and may be a barrier to our ability to improve outcomes in patients with NF1-MPNST. These data suggest that multi-regional sampling may be necessary to understand clonal evolution, and for driver gene identification and biomarker development in the future.

Supplementary Materials: The following are available online at http://www.mdpi.com/2073-4425/11/5/499/s1, Supplemental Table S1: Comprehensive Genomic Information for Single Nucleotide Variants.

Author Contributions: Conceptualization, A.C.H.; Formal analysis, C.-I.M., Y.W., C.D., and A.G.; Funding acquisition, A.C.H.; Investigation, C.-I.M., W.T., C.D., Y.W. and X.Z.; Resources, A.G., X.Z., P.P. and S.D.; Software, C.-I.M.; C.A.M. Supervision, A.C.H.; Writing—original draft, C.-I.M. and W.T.; Writing—review & editing, Y.W., C.D., A.G., X.Z., P.P., S.D., C.A.M. and A.C.H. All authors have read and agreed to the published version of the manuscript.

Funding: This work was funded by the St. Louis Men's Group Against Cancer. Hirbe is funded by a Francis Collins Scholar Award through NTAP.

Conflicts of Interest: The authors declare no potential conflicts of interest.

References

1. Eilber, F.C.; Brennan, M.F.; Eilber, F.R.; Dry, S.M.; Singer, S.; Kattan, M.W. Validation of the Postoperative Nomogram for 12-Year Sarcoma-Specific Mortality. *Cancer* **2004**, *101*, 2270–2275. [CrossRef] [PubMed]

2. Ng, V.Y.; Scharschmidt, T.J.; Mayerson, J.L.; Fisher, J.L. Incidence and Survival in Sarcoma in the United States: A Focus on Musculoskeletal Lesions. *Anticancer Res.* **2013**, *33*, 2597–2604. [PubMed]

3. Evans, D.G.R.; Baser, M.E.; McGaughran, J.; Sharif, S.; Howard, E.; Moran, A. Malignant peripheral nerve sheath tumours in neurofibromatosis 1. *J. Med. Genet.* **2002**, *39*, 311–314. [CrossRef] [PubMed]

4. Ducatman, B.S.; Scheithauer, B.W.; Piepgras, D.G.; Reiman, H.M.; Ilstrup, D.M. Malignant Peripheral Nerve Sheath Tumors. A Clinicopathologic Study of 120 Cases. *Cancer* **1986**, *57*, 2006–2021. [CrossRef]

5. Porter, D.E.; Prasad, V.; Foster, L.; Dall, G.F.; Birch, R.; Grimer, R.J. Survival in malignant peripheral nerve sheath tumours: A comparison between sporadic and neurofibromatosis type 1-associated tumours. *Sarcoma* **2009**, *2009*, 1–5. [CrossRef]

6. Zou, C.; Smith, K.D.; Liu, J.; Lahat, G.; Myers, S.; Wang, W.L.; Zhang, W.; McCutcheon, I.E.; Slopis, J.M.; Lazar, A.J.; et al. Clinical, pathological, and molecular variables predictive of malignant peripheral nerve sheath tumor outcome. *Ann. Surg.* **2009**, *249*, 1014–1022. [CrossRef]

7. LaFemina, J.; Qin, L.X.; Moraco, N.H.; Antonescu, C.R.; Fields, R.C.; Crago, A.M.; Brennan, M.F.; Singer, S. Oncologic outcomes of sporadic, neurofibromatosis-associated, and radiation-induced malignant peripheral nerve sheath tumors. *Ann. Surg. Oncol.* **2013**, *20*, 66–72. [CrossRef]

8. Farid, M.; Demicco, E.G.; Garcia, R.; Ahn, L.; Merola, P.R.; Cioffi, A.; Maki, R.G. Malignant Peripheral Nerve Sheath Tumors. *Oncologist* **2014**, *19*, 193–201. [CrossRef]

9. Anghileri, M.; Miceli, R.; Fiore, M.; Mariani, L.; Ferrari, A.; Mussi, C.; Lozza, L.; Collini, P.; Olmi, P.; Casali, P.G.; et al. Malignant Peripheral Nerve Sheath Tumors: Prognostic Factors and Survival in a Series of Patients Treated at a Single Institution. *Cancer* **2006**, *107*, 1065–1074. [CrossRef]

10. Stucky, C.C.; Johnson, K.N.; Gray, R.J.; Pockaj, B.A.; Ocal, I.T.; Rose, P.S.; Wasif, N. Malignant peripheral nerve sheath tumors (MPNST): The Mayo Clinic experience. *Ann. Surg. Oncol.* **2012**, *19*, 878–885. [CrossRef]

11. Ferner, R.E.; Gutmann, D.H. International Consensus Statement on Malignant Peripheral Nerve Sheath Tumors in Neurofibromatosis. *Cancer Res.* **2002**, *62*, 1573–1577.

12. Kroep, J.R.; Ouali, M.; Gelderblom, H.; Le Cesne, A.; Dekker, T.J.; Van Glabbeke, M.; Hogendoorn, P.C.; Hohenberger, P. First-Line Chemotherapy for Malignant Peripheral Nerve Sheath Tumor (MPNST) versus Other Histological Soft Tissue Sarcoma Subtypes and as a Prognostic Factor for MPNST: An EORTC Soft Tissue and Bone Sarcoma Group Study. *Ann. Oncol.* **2011**, *22*, 207–214. [CrossRef] [PubMed]

13. James, A.W.; Shurell, E.; Singh, A.; Dry, S.M.; Eilber, F.C. Malignant Peripheral Nerve Sheath Tumor. *Surg. Oncol. Clin. N. Am.* **2016**, *25*, 789–802. [CrossRef] [PubMed]

14. Cichowski, K.; Shih, T.S.; Schmitt, E.; Santiago, S.; Reilly, K.; McLaughlin, M.E.; Bronson, R.T.; Jacks, T. Mouse Models of Tumor Development in Neurofibromatosis Type 1. *Science* **1999**, *286*, 2172–2176. [CrossRef] [PubMed]

15. Vogel, K.S.; Klesse, L.J.; Velasco-Miguel, S.; Meyers, K.; Rushing, E.J.; Parada, L.F. Mouse Tumor Model for Neurofibromatosis Type 1. *Science* **1999**, *286*, 2176–2179. [CrossRef]

16. Bradtmoller, M.; Hartmann, C.; Zietsch, J.; Jäschke, S.; Mautner, V.F.; Kurtz, A.; Park, S.J.; Baier, M.; Harder, A.; Reuss, D.; et al. Impaired Pten Expression in Human Malignant Peripheral Nerve Sheath Tumours. *PLoS ONE* **2012**. [CrossRef]

17. Keng, V.W.; Rahrmann, E.P.; Watson, A.L.; Tschida, B.R.; Moertel, C.L.; Jessen, W.J.; Rizvi, T.A.; Collins, M.H.; Ratner, N.; Largaespada, D.A. PTEN and NF1 inactivation in Schwann cells produces a severe phenotype in the peripheral nervous system that promotes the development and malignant progression of peripheral nerve sheath tumors. *Cancer Res.* **2012**, *72*, 3405–3413. [CrossRef]

18. Gregorian, C.; Nakashima, J.; Dry, S.M.; Nghiemphu, P.L.; Smith, K.B.; Ao, Y.; Dang, J.; Lawson, G.; Mellinghoff, I.K.; Mischel, P.S.; et al. PTEN dosage is essential for neurofibroma development and malignant transformation. *Proc. Natl. Acad. Sci. USA* **2009**, *106*, 19479–19484. [CrossRef]

19. Kourea, H.P.; Orlow, I.; Scheithauer, B.W.; Cordon-Cardo, C.; Woodruff, J.M. Deletions of the INK4A gene occur in malignant peripheral nerve sheath tumors but not in neurofibromas. *Am. J. Path.* **1999**, *155*, 1855–1860. [CrossRef]

20. Lu, H.C.; Eulo, V.; Apicelli, A.J.; Pekmezci, M.; Tao, Y.; Luo, J.; Hirbe, A.C.; Dahiya, S. Aberrant ATRX protein expression is associated with poor overall survival in NF1-MPNST. *Oncotarget* **2018**, *9*, 23018–23028. [CrossRef]

21. Lee, W.; Teckie, S.; Wiesner, T.; Ran, L.; Prieto Granada, C.N.; Lin, M.; Zhu, S.; Cao, Z.; Liang, Y.; Sboner, A.; et al. PRC2 is recurrently inactivated through EED or SUZ12 loss in malignant peripheral nerve sheath tumors. *Nat. Genet.* **2014**, *46*, 1227–1232. [CrossRef] [PubMed]

22. Holtkamp, N.; Okuducu, A.F.; Mucha, J.; Afanasieva, A.; Hartmann, C.; Atallah, I.; Estevez-Schwarz, L.; Mawrin, C.; Friedrich, R.E.; Mautner, V.F.; et al. Mutation and expression of PDGFRA and KIT in malignant peripheral nerve sheath tumors, and its implications for imatinib sensitivity. *Carcinogenesis* **2006**, *27*, 664–671. [CrossRef] [PubMed]

23. DeClue, J.E.; Heffelfinger, S.; Benvenuto, G.; Ling, B.; Li, S.; Rui, W.; Vass, W.C.; Viskochil, D.; Ratner, N. Epidermal growth factor receptor expression in neurofibromatosis type 1-related tumors and NF1 animal models. *J. Clin. Investig.* **2000**, *105*, 1233–1241. [CrossRef] [PubMed]

24. Yang, J.; Ylipää, A.; Sun, Y.; Zheng, H.; Chen, K.; Nykter, M.; Trent, J.; Ratner, N.; Lev, D.C.; Zhang, W. Genomic and molecular characterization of malignant peripheral nerve sheath tumor identifies the IGF1R pathway as a primary target for treatment. *Clin. Cancer Res.* **2011**, *17*, 7563–7573. [CrossRef]

25. Symposium on Linkage of von Recklinghausen Neurofibromatosis (NF1). Closing in on the gene for von Recklinghausen neurofibromatosis. *Genomics* **1987**, *1*, 335–383.

26. Talevich, E.; Shain, A.H.; Botton, T.; Bastian, B.C. CNVkit: Genome-Wide Copy Number Detection and Visualization from Targeted DNA Sequencing. *PLoS Comput. Biol.* **2016**, *12*, e1004873. [CrossRef]

27. Li, H. Aligning sequence reads, clone sequences and assembly contigs with BWA-MEM. *arXiv Preprint* **2013**, arXiv:1303.3997.

28. Chen, X.; Schulz-Trieglaff, O.; Shaw, R.; Barnes, B.; Schlesinger, F.; Källberg, M.; Cox, A.J.; Kruglyak, S.; Saunders, C.T. Manta: Rapid detection of structural variants and indels for germline and cancer sequencing applications. *Bioinformatics* **2015**, *32*, 1220–1222. [CrossRef]

29. Koboldt, D.C.; Zhang, Q.; Larson, D.E.; Shen, D.; McLellan, M.D.; Lin, L.; Miller, C.A.; Mardis, E.R.; Ding, L.; Wilson, R.K. VarScan 2: Somatic mutation and copy number alteration discovery in cancer by exome sequencing. *Genome Res.* **2012**, *22*, 568–576. [CrossRef]

30. Kim, S.; Scheffler, K.; Halpern, A.L.; Bekritsky, M.A.; Noh, E.; Källberg, M.; Chen, X.; Kim, Y.; Beyter, D.; Krusche, P.; et al. Strelka2: Fast and accurate calling of germline and somatic variants. *Nat. Methods* **2018**, *15*, 591–594. [CrossRef]

31. Cibulskis, K.; Lawrence, M.S.; Carter, S.L.; Sivachenko, A.; Jaffe, D.; Sougnez, C.; Gabriel, S.; Meyerson, M.; Lander, E.S.; Getz, G. Sensitive detection of somatic point mutations in impure and heterogeneous cancer samples. *Nat. Biotechnol.* **2013**, *31*, 213–219. [CrossRef] [PubMed]

32. Ye, K.; Schulz, M.H.; Long, Q.; Apweiler, R.; Ning, Z. Pindel: A pattern growth approach to detect break points of large deletions and medium sized insertions from paired-end short reads. *Bioinformatics (Oxford, England)* **2009**, *25*, 2865–2871. [CrossRef] [PubMed]

33. McLaren, W.; Gil, L.; Hunt, S.E.; Riat, H.S.; Ritchie, G.R.; Thormann, A.; Flicek, P.; Cunningham, F. The Ensembl Variant Effect Predictor. *Genome Biol.* **2016**, *17*. [CrossRef] [PubMed]

34. Bioconductor: Genomic Visualizations in R. Available online: https://bioconductor.org/packages/release/bioc/html/GenVisR.html (accessed on 24 February 2020).

35. Bamford, S.; Dawson, E.; Forbes, S.; Clements, J.; Pettett, R.; Dogan, A.; Flanagan, A.; Teague, J.; Futreal, P.A.; Stratton, M.R.; et al. The COSMIC (Catalogue of Somatic Mutations in Cancer) database and website. *Br. J. Cancer* **2004**, *91*, 355–358. [CrossRef]

36. Landrum, M.J.; Lee, J.M.; Benson, M.; Brown, G.; Chao, C.; Chitipiralla, S.; Gu, B.; Hart, J.; Hoffman, D.; Hoover, J.; et al. ClinVar: Public archive of interpretations of clinically relevant variants. *Nucleic Acids Res.* **2016**, *44*, 862–868. [CrossRef]

37. Pauline, C.; Ng, S.H. SIFT: Predicting amino acid changes that affect protein function. *Nucleic Acids Res.* **2003**, *31*, 3812–3814.

38. Adzhubei, I.A.; Schmidt, S.; Peshkin, L.; Ramensky, V.E.; Gerasimova, A.; Bork, P.; Kondrashov, A.S.; Sunyaev, S.R. A method and server for predicting damaging missense mutations. *Nat Methods* **2010**, *7*, 248–249. [CrossRef]

39. Miller, C.A.; White, B.S.; Dees, N.D.; Griffith, M.; Welch, J.S.; Griffith, O.L.; Vij, R.; Tomasson, M.H.; Graubert, T.A.; Walter, M.J.; et al. SciClone: Inferring clonal architecture and tracking the spatial and temporal patterns of tumor evolution. *PLoS Comput. Biol.* **2014**. [CrossRef]

40. Dang, H.X.; White, B.S.; Foltz, S.M.; Miller, C.A.; Luo, J.; Fields, R.C.; Maher, C.A. ClonEvol: Clonal ordering and visualization in cancer sequencing. *Ann. Oncol.* **2017**, *28*, 3076–3082. [CrossRef]

41. Xing, Y.; Yu, T.; Wu, Y.N.; Roy, M.; Kim, J.; Lee, C. An expectation-maximization algorithm for probabilistic reconstructions of full-length isoforms from splice graphs. *Nucleic Acids Res.* **2006**, *34*, 3150–3160. [CrossRef]

42. Robinson, M.D.; McCarthy, D.J.; Smyth, G.K. edgeR: A Bioconductor package for differential expression analysis of digital gene expression data. *Bioinformatics* **2010**, *26*, 139–140. [CrossRef] [PubMed]

43. Warnes, G.R.; Bolker, B.; Bonebakker, L.; Gentleman, R.; Huber, W.; Liaw, A.; Lumley, T.; Maechler, M.; Magnusson, A.; Moeller, S.; et al. gplots: Various R Programming Tools for Plotting Data. Seattle, WA, USA, 2015. Available online: https://www.scienceopen.com/document?vid=0e5d8e31-1fe4-492f-a3d8-8cd71b2b8ad9 (accessed on 29 April 2020).

44. Partek Flow Documentation: Gene-specific Analysis. Available online: https://documentation.partek.com/display/FLOWDOC/Gene-specific+Analysis (accessed on 24 February 2020).

45. Thomas, P.D.; Campbell, M.J.; Kejariwal, A.; Mi, H.; Karlak, B.; Daverman, R.; Diemer, K.; Muruganujan, A.; Narechania, A. PANTHER: A library of protein families and subfamilies indexed by function. *Genome Res.* **2003**, *13*, 2129–2141. [CrossRef] [PubMed]

46. Guillou, L.; Coindre, J.M.; Bonichon, F.; Nguyen, B.B.; Terrier, P.; Collin, F.; Vilain, M.O.; Mandard, A.M.; Le Doussal, V.; Leroux, A.; et al. Comparative study of the National Cancer Institute and French Federation of Cancer Centers Sarcoma Group grading systems in a population of 410 adult patients with soft tissue sarcoma. *J. Clin. Oncol.* **1997**, *15*, 350–362. [CrossRef] [PubMed]

47. Gao, J.; Aksoy, B.A.; Dogrusoz, U.; Dresdner, G.; Gross, B.; Sumer, S.O.; Sun, Y.; Jacobsen, A.; Sinha, R.; Larsson, E.; et al. Integrative analysis of complex cancer genomics and clinical profiles using the cBioPortal. *Sci. Signal* **2013**. [CrossRef]

48. Wang, T.; Wang, H.; Yang, S.; Guo, H.; Zhang, B.; Guo, H.; Wang, L.; Zhu, G.; Zhang, Y.; Zhou, H.; et al. Association of APEX1 and OGG1 gene polymorphisms with breast cancer risk among Han women in the Gansu Province of China. *BMC Med. Genet.* **2018**, *19*, 67. [CrossRef]

49. Kim, H.B.; Lim, H.J.; Lee, H.J.; Park, J.H.; Park, S.G. Evaluation and Clinical Significance of Jagged-1-activated Notch Signaling by APEX1 in Colorectal Cancer. *Anticancer Res.* **2019**, *39*, 6097–6105. [CrossRef]

50. Kim, H.B.; Cho, W.J.; Choi, N.G.; Kim, S.S.; Park, J.H.; Lee, H.J.; Park, S.G. Clinical implications of APEX1 and Jagged1 as chemoresistance factors in biliary tract cancer. *Ann. Surg. Treat. Res.* **2017**, *92*, 15–22. [CrossRef]

51. Blazquez, L.; Emmett, W.; Faraway, R.; Pineda, J.M.B.; Bajew, S.; Gohr, A.; Haberman, N.; Sibley, C.R.; Bradley, R.K.; Irimia, M.; et al. Exon Junction Complex Shapes the Transcriptome by Repressing Recursive Splicing. *Mol. Cell* **2018**, *72*, 496–509. [CrossRef]

52. Shen, Y.N.; Bae, I.S.; Park, G.H.; Choi, H.S.; Lee, K.H.; Kim, S.H. MicroRNA-196b enhances the radiosensitivity of SNU-638 gastric cancer cells by targeting RAD23B. *Biomed. Pharmacother.* **2018**, *105*, 362–369. [CrossRef]

53. Linge, A.; Maurya, P.; Friedrich, K.; Baretton, G.B.; Kelly, S.; Henry, M.; Clynes, M.; Larkin, A.; Meleady, P. Identification and functional validation of RAD23B as a potential protein in human breast cancer progression. *J. Proteome Res.* **2014**, *13*, 3212–3222. [CrossRef]

54. Luo, C.; Cheng, Y.; Liu, Y.; Chen, L.; Liu, L.; Wei, N.; Xie, Z.; Wu, W.; Feng, Y. SRSF2 Regulates Alternative Splicing to Drive Hepatocellular Carcinoma Development. *Cancer Res.* **2017**, *77*, 1168–1178. [CrossRef] [PubMed]

55. Liang, Y.; Tebaldi, T.; Rejeski, K.; Joshi, P.; Stefani, G.; Taylor, A.; Song, Y.; Vasic, R.; Maziarz, J.; Balasubramanian, K. SRSF2 mutations drive oncogenesis by activating a global program of aberrant alternative splicing in hematopoietic cells. *Leukemia* **2018**, *32*, 2659–2671. [CrossRef] [PubMed]

56. Gu, L.; Lu, L.; Zhou, D.; Liu, Z. Long Noncoding RNA BCYRN1 Promotes the Proliferation of Colorectal Cancer Cells via Up-Regulating NPR3 Expression. *Cell Physiol. Biochem.* **2018**, *48*, 2337–2349. [CrossRef]

57. Li, X.; Li, J.; Li, F. P21 activated kinase 4 binds translation elongation factor eEF1A1 to promote gastric cancer cell migration and invasion. *Oncol. Rep.* **2017**, *37*, 2857–2864. [CrossRef]

58. Shi, N.; Chen, X.; Liu, R.; Wang, D.; Su, M.; Wang, Q.; He, A.; Gu, H. Eukaryotic elongation factors 2 promotes tumor cell proliferation and correlates with poor prognosis in ovarian cancer. *Tissue Cell* **2018**, *53*, 53–60. [CrossRef] [PubMed]

59. Wang, H.; He, Z.; Xia, L.; Zhang, W.; Xu, L.; Yue, X.; Ru, X.; Xu, Y. PSMB4 overexpression enhances the cell growth and viability of breast cancer cells leading to a poor prognosis. *Oncol. Rep.* **2018**, *40*, 2343–2352. [CrossRef]

60. Liu, R.; Lu, S.; Deng, Y.; Yang, S.; He, S.; Cai, J.; Qiang, F.; Chen, C.; Zhang, W.; Zhao, S.; et al. PSMB4 expression associates with epithelial ovarian cancer growth and poor prognosis. *Arch. Gynecol. Obstet.* **2016**, *293*, 1297–1307. [CrossRef]

61. Xu, X.; Xiong, X.; Sun, Y. The role of ribosomal proteins in the regulation of cell proliferation, tumorigenesis, and genomic integrity. *Sci. China Life Sci.* **2016**, *59*, 656–672. [CrossRef]

62. Nallar, S.C.; Kalvakolanu, D.V. Regulation of snoRNAs in Cancer: Close Encounters with Interferon. *J. Interferon Cytokine Res.* **2013**, *33*, 189–198. [CrossRef]

63. Falaleeva, M.; Welden, J.R.; Duncan, M.J.; Stamm, S. C/D-box snoRNAs form methylating and non-methylating ribonucleoprotein complexes: Old dogs show new tricks. *Bioessays* **2017**, *39*. [CrossRef]

64. Dong, X.; Han, Y.; Sun, Z.; Xu, J. Actin γ 1, a new skin cancer pathogenic gene, identified by the biological feature-based classification. *J. Cell Biochem.* **2018**, *119*, 1406–1419. [CrossRef] [PubMed]

65. Luo, Y.; Kong, F.; Wang, Z.; Chen, D.; Liu, Q.; Wang, T.; Xu, R.; Wang, X.; Yang, J.Y. Loss of ASAP3 destabilizes cytoskeletal protein ACTG1 to suppress cancer cell migration. *Mol. Med. Rep.* **2014**, *9*, 387–394. [CrossRef] [PubMed]

66. Munschauer, M.; Nguyen, C.T.; Sirokman, K.; Hartigan, C.R.; Hogstrom, L.; Engreitz, J.M.; Ulirsch, J.C.; Fulco, C.P.; Subramanian, V.; Chen, J.; et al. The NORAD lncRNA assembles a topoisomerase complex critical for genome stability. *Nature* **2018**, *561*, 132–136. [CrossRef] [PubMed]

67. Liang, L.; Li, Q.; Huang, L.Y.; Li, D.W.; Wang, Y.W.; Li, X.X.; Cai, S.J. Loss of ARHGDIA expression is associated with poor prognosis in HCC and promotes invasion and metastasis of HCC cells. *Int. J. Oncol.* **2014**, *45*, 659–666. [CrossRef] [PubMed]

68. Riihilä, P.; Viiklepp, K.; Nissinen, L.; Farshchian, M.; Kallajoki, M.; Kivisaari, A.; Meri, S.; Peltonen, J.; Peltonen, S.; Kähäri, V.M. Tumour-cell-derived complement components C1r and C1s promote growth of cutaneous squamous cell carcinoma. *Br. J. Dermatol.* **2020**, *182*, 658–670. [CrossRef] [PubMed]

69. Wheeler, L.J.; Watson, Z.L.; Qamar, L.; Yamamoto, T.M.; Post, M.D.; Berning, A.A.; Spillman, M.A.; Behbakht, K.; Bitler, B.G. CBX2 identified as driver of anoikis escape and dissemination in high grade serous ovarian cancer. *Oncogenesis* **2018**, *7*, 92. [CrossRef]

70. Liu, J.; Shen, J.X.; Wu, H.T.; Li, X.L.; Wen, X.F.; Du, C.W.; Zhang, G.J. Collagen 1A1 (COL1A1) promotes metastasis of breast cancer and is a potential therapeutic target. *Discov. Med.* **2018**, *25*, 211–223.

71. Menendez, J.A.; Lupu, R. Fatty acid synthase and the lipogenic phenotype in cancer pathogenesis. *Nat. Rev. Cancer* **2007**, *7*, 763–777. [CrossRef]

72. Kamil, M.; Shinsato, Y.; Higa, N.; Hirano, T.; Idogawa, M.; Takajo, T.; Minami, K.; Shimokawa, M.; Yamamoto, M.; Kawahara, K.; et al. High filamin-C expression predicts enhanced invasiveness and poor outcome in glioblastoma multiforme. *Br. J. Cancer* **2019**, *120*, 819–826. [CrossRef]

73. Zhu, Y.; Kakinuma, N.; Wang, Y.; Kiyama, R. Kank proteins: A new family of ankyrin-repeat domain-containing proteins. *Biochim. Biophys. Act* **2008**, *1780*, 128–133. [CrossRef]

74. Trussart, C.; Pirlot, C.; Di Valentin, E.; Piette, J.; Habraken, Y. Melanoma antigen-D2 controls cell cycle progression and modulates the DNA damage response. *Biochem. Pharmacol.* **2018**, *153*, 217–229. [CrossRef] [PubMed]

75. Wang, P.C.; Hu, Z.Q.; Zhou, S.L.; Zhan, H.; Zhou, Z.J.; Luo, C.B.; Huang, X.W. Downregulation of MAGE family member H1 enhances hepatocellular carcinoma progression and serves as a biomarker for patient prognosis. *Future Oncol.* **2018**, *14*, 1177–1186. [CrossRef] [PubMed]

76. Qu, K.; Wang, Z.; Fan, H.; Li, J.; Liu, J.; Li, P.; Liang, Z.; An, H.; Jiang, Y.; Lin, Q.; et al. MCM7 promotes cancer progression through cyclin D1-dependent signaling and serves as a prognostic marker for patients with hepatocellular carcinoma. *Cell Death Dis* **2017**, *8*, e2603. [CrossRef] [PubMed]

77. Nguyen, A.T.; Chia, J.; Ros, M.; Hui, K.M.; Saltel, F.; Bard, F. Organelle Specific O-Glycosylation Drives MMP14 Activation, Tumor Growth, and Metastasis. *Cancer Cell* **2017**, *32*, 639–653. [CrossRef]

78. Hu, G.; Zhang, J.; Xu, F.; Deng, H.; Zhang, W.; Kang, S.; Liang, W. Stomatin-like protein 2 inhibits cisplatin-induced apoptosis through MEK/ERK signaling and the mitochondrial apoptosis pathway in cervical cancer cells. *Cancer Sci.* **2018**, *109*, 1357–1368. [CrossRef]

79. Du, W.L.; Fang, Q.; Chen, Y.; Teng, J.W.; Xiao, Y.S.; Xie, P.; Jin, B.; Wang, J.Q. Effect of silencing the T-Box transcription factor TBX2 in prostate cancer PC3 and LNCaP cells. *Mol. Med. Rep.* **2017**, *16*, 6050–6058. [CrossRef]

80. Czerwińska, P.; Mazurek, S.; Wiznerowicz, M. The complexity of TRIM28 contribution to cancer. *J. Biomed. Sci.* **2017**, *29*, 63. [CrossRef]

81. Lan, B.; Chai, S.; Wang, P.; Wang, K. VCP/p97/Cdc48, A Linking of Protein Homeostasis and Cancer Therapy. *Curr. Mol. Med.* **2017**, *17*, 608–618. [CrossRef]

82. Hwang, W.; Chiu, Y.F.; Kuo, M.H.; Lee, K.L.; Lee, A.C.; Yu, C.C.; Chang, J.L.; Huang, W.C.; Hsiao, S.H.; Lin, S.E.; et al. Expression of Neuroendocrine Factor VGF in Lung Cancer Cells Confers Resistance to EGFR Kinase Inhibitors and Triggers Epithelial-to-Mesenchymal Transition. *Cancer Res.* **2017**, *77*, 3013–3026. [CrossRef]

83. De Blasio, A.; Vento, R.; Di Fiore, R. Mcl-1 targeting could be an intriguing perspective to cure cancer. *J. Cell Physiol.* **2018**, *233*, 8482–8498. [CrossRef]

84. Louie, S.M.; Grossman, E.A.; Crawford, L.A.; Ding, L.; Camarda, R.; Huffman, T.R.; Miyamoto, D.K.; Goga, A.; Weerapana, E.; Nomura, D.K. GSTP1 Is a Driver of Triple-Negative Breast Cancer Cell Metabolism and Pathogenicity. *Cell Chem. Biol.* **2016**, *23*, 567–578. [CrossRef] [PubMed]

85. Bianchi, M.; Crinelli, R.; Giacomini, E.; Carloni, E.; Radici, L.; Scarpa, E.S.; Tasini, F.; Magnani, M. A negative feedback mechanism links UBC gene expression to ubiquitin levels by affecting RNA splicing rather than transcription. *Sci. Rep.* **2019**, *9*, 18556. [CrossRef] [PubMed]

86. Lin, L.; Li, X.; Pan, C.; Lin, W.; Shao, R.; Liu, Y.; Zhang, J.; Luo, Y.; Qian, K.; Shi, M.; et al. ATXN2L upregulated by epidermal growth factor promotes gastric cancer cell invasiveness and oxaliplatin resistance. *Cell Death Dis.* **2019**, *10*, 173. [CrossRef] [PubMed]

87. Livingstone, C. IGF2 and cancer. *Endocr. Relat. Cancer* **2013**, *20*, 321–339. [CrossRef] [PubMed]

88. Feng, W.; Wang, C.; Liang, C.; Yang, H.; Chen, D.; Yu, X.; Zhao, W.; Geng, D.; Li, S.; Chen, Z.; et al. The Dysregulated Expression of KCNQ1OT1 and Its Interaction with Downstream Factors miR-145/CCNE2 in Breast Cancer Cells. *Cell Physiol. Biochem.* **2018**, *49*, 432–446. [CrossRef]

89. Zhang, Y.; Hu, J.F.; Wang, H.; Cui, J.; Gao, S.; Hoffman, A.R.; Li, W. CRISPR Cas9-guided chromatin immunoprecipitation identifies miR483 as an epigenetic modulator of IGF2 imprinting in tumors. *Oncotarget* **2017**, *8*, 34177–34190. [CrossRef]

90. Gong, C.Y.; Tang, R.; Nan, W.; Zhou, K.S.; Zhang, H.H. Role of SNHG16 in human cancer. *Clin. Chim. Acta* **2020**, *503*, 175–180. [CrossRef]

91. Gerlinger, M.; Horswell, S.; Larkin, J.; Rowan, A.J.; Salm, M.P.; Varela, I.; Fisher, R.; McGranahan, N.; Matthews, N.; Santos, C.R.; et al. Genomic architecture and evolution of clear cell renal cell carcinomas defined by multiregion sequencing. *Nat. Genet.* **2014**, *46*, 225–233. [CrossRef]

92. Yates, L.R.; Gerstung, M.; Knappskog, S.; Desmedt, C.; Gundem, G.; Van Loo, P.; Aas, T.; Alexandrov, L.B.; Larsimont, D.; Davies, H.; et al. Subclonal diversification of primary breast cancer revealed by multiregion sequencing. *Nat. Med.* **2015**, *21*, 751–759. [CrossRef]

93. Hao, J.J.; Lin, D.C.; Dinh, H.Q.; Mayakonda, A.; Jiang, Y.Y.; Chang, C.; Jiang, Y.; Lu, C.C.; Shi, Z.Z.; Xu, X.; et al. Spatial intratumoral heterogeneity and temporal clonal evolution in esophageal squamous cell carcinoma. *Nat. Genet.* **2016**, *48*, 1500–1507. [CrossRef]

94. Jamal-Hanjani, M.; Wilson, G.A.; McGranahan, N.; Birkbak, N.J.; Watkins, T.B.K.; Veeriah, S.; Shafi, S.; Johnson, D.H.; Mitter, R.; Rosenthal, R.; et al. Tracking the evolution of non-small-cell lung cancer. *N. Engl. J. Med.* **2017**, *376*, 2109–2121. [CrossRef] [PubMed]

95. Harbst, K.; Lauss, M.; Cirenajwis, H.; Isaksson, K.; Rosengren, F.; Törngren, T.; Kvist, A.; Johansson, M.C.; Vallon-Christersson, J.; Baldetorp, B.; et al. Multiregion whole-exome sequencing uncovers the genetic evolution and mutational heterogeneity of early-stage metastatic melanoma. *Cancer Res.* **2016**, *76*, 4765–4774. [CrossRef] [PubMed]

96. Peacock, J.D.; Pridgeon, M.G.; Tovar, E.A.; Essenburg, C.J.; Bowman, M.; Madaj, Z.; Koeman, J.; Boguslawski, E.A.; Grit, J.; Dodd, R.D.; et al. Genomic Status of MET Potentiates Sensitivity to MET and MEK Inhibition in NF1-Related Malignant Peripheral Nerve Sheath Tumors. *Cancer Res.* **2018**, *78*, 3672–3678. [CrossRef] [PubMed]

97. Carrió, M.; Gel, B.; Terribas, E.; Zucchiatti, A.C.; Moliné, T.; Rosas, I.; Teulé, Á.; Ramón, Y.; Cajal, S.; López-Gutiérrez, J.C.; et al. Analysis of intratumor heterogeneity in Neurofibromatosis type 1 plexiform neurofibromas and neurofibromas with atypical features: Correlating histological and genomic findings. *Hum. Mutat.* **2018**, *39*, 1112–1125. [CrossRef] [PubMed]

98. Nackley, A.G.; Shabalina, S.A.; Tchivileva, I.E.; Satterfield, K.; Korchynskyi, O.; Makarov, S.S.; Maixner, W.; Diatchenko, L. Human catechol-O-methyltransferase haplotypes modulate protein expression by altering mRNA secondary structure. *Science* **2006**, *314*, 1930–1933. [CrossRef] [PubMed]

99. Kimchi-Sarfaty, C.; Oh, J.M.; Kim, I.W.; Sauna, Z.E.; Calcagno, A.M.; Ambudkar, S.V.; Gottesman, M.M. A "silent" polymorphism in the MDR1 gene changes substrate specificity. *Science* **2007**, *315*, 525–528. [CrossRef]

100. Kudla, G.; Murray, A.W.; Tollervey, D.; Plotkin, J.B. Coding-sequence determinants of gene expression in Escherichia coli. *Science* **2009**, *324*, 255–258. [CrossRef]

101. Sauna, Z.E.; Kimchi-Sarfaty, C. Understanding the contribution of synonymous mutations to human disease. *Nat. Rev. Genet.* **2011**, *12*, 683–691. [CrossRef]

102. Pey, J.; San José-Eneriz, E.; Ochoa, M.C.; Apaolaza, I.; de Atauri, P.; Rubio, A.; Cendoya, X.; Miranda, E.; Garate, L.; Cascante, M.; et al. In-silico gene essentiality analysis of polyamine biosynthesis reveals APRT as a potential target in cancer. *Sci. Rep.* **2017**, *7*, 14358. [CrossRef]

103. Shen, L.; Ke, Q.; Chai, J.; Zhang, C.; Qiu, L.; Peng, F.; Deng, X.; Luo, Z. PAG1 promotes the inherent radioresistance of laryngeal cancer cells via activation of STAT3. *Exp. Cell Res.* **2018**, *370*, 127–136. [CrossRef]

104. Agarwal, S.; Ghosh, R.; Chen, Z.; Lakoma, A.; Gunaratne, P.H.; Kim, E.S.; Shohet, J.M. Transmembrane adaptor protein PAG1 is a novel tumor suppressor in neuroblastoma. *Oncotarget* **2016**, *7*, 24018–24026. [CrossRef] [PubMed]

105. Jiang, D.; Hu, B.; Wei, L.; Xiong, Y.; Wang, G.; Ni, T.; Zong, C.; Ni, R.; Lu, C. High expression of vacuolar protein sorting 4B (VPS4B) is associated with accelerated cell proliferation and poor prognosis in human hepatocellular carcinoma. *Pathol. Res. Pract.* **2015**, *211*, 240–247. [CrossRef] [PubMed]

106. Liu, Y.; Lv, L.; Xue, Q.; Wan, C.; Ni, T.; Chen, B.; Liu, Y.; Zhou, Y.; Ni, R.; Mao, G. Vacuolar protein sorting 4B, an ATPase protein positively regulates the progression of NSCLC via promoting cell division. *Mol. Cell Biochem.* **2013**, *381*, 163–171. [CrossRef] [PubMed]

107. Lin, H.H.; Li, X.; Chen, J.L.; Sun, X.; Cooper, F.N.; Chen, Y.R.; Zhang, W.; Chung, Y.; Li, A.; Cheng, C.T.; et al. Identification of an AAA ATPase VPS4B-dependent pathway that modulates epidermal growth factor receptor abundance and signaling during hypoxia. *Mol. Cell Biol.* **2012**, *32*, 1124–1138. [CrossRef] [PubMed]

108. Hazawa, M.; Lin, D.C.; Handral, H.; Xu, L.; Chen, Y.; Jiang, Y.Y.; Mayakonda, A.; Ding, L.W.; Meng, X.; Sharma, A.; et al. ZNF750 is a lineage-specific tumour suppressor in squamous cell carcinoma. *Oncogene* **2017**, *36*, 2243–2254. [CrossRef] [PubMed]

109. Zhang, P.; He, Q.; Lei, Y.; Li, Y.; Wen, X.; Hong, M.; Zhang, J.; Ren, X.; Wang, Y.; Yang, X.; et al. m^{6}A-mediated ZNF750 repression facilitates nasopharyngeal carcinoma progression. *Cell Death Dis.* **2018**, *5*, 1169. [CrossRef]

110. Feng, T.; Sun, L.; Qi, W.; Pan, F.; Lv, J.; Guo, J.; Zhao, S.; Ding, A.; Qiu, W. Prognostic significance of Tspan9 in gastric cancer. *Mol. Clin. Oncol.* **2016**, *5*, 231–236. [CrossRef]

111. Qi, Y.; Lv, J.; Liu, S.; Sun, L.; Wang, Y.; Li, H.; Qi, W.; Qiu, W. TSPAN9 and EMILIN1 synergistically inhibit the migration and invasion of gastric cancer cells by increasing TSPAN9 expression. *BMC Cancer* **2019**, *19*, 630. [CrossRef]

112. Xiao, T.; Li, W.; Wang, X.; Xu, H.; Yang, J.; Wu, Q.; Huang, Y.; Geradts, J.; Jiang, P.; Fei, T.; et al. Estrogen-regulated feedback loop limits the efficacy of estrogen receptor–targeted breast cancer therapy. *Proc. Natl. Acad. Sci. USA* **2018**, *115*, 7869–7878. [CrossRef]

113. Smith, H.W.; Hirukawa, A.; Sanguin-Gendreau, V.; Nandi, I.; Dufour, C.R.; Zuo, D.; Tandoc, K.; Leibovitch, M.; Singh, S.; Rennhack, J.P.; et al. An ErbB2/c-Src axis links bioenergetics with PRC2 translation to drive epigenetic reprogramming and mammary tumorigenesis. *Nat. Commun.* **2019**, *10*. [CrossRef]

114. Yang, C.C.; Fazli, L.; Loguercio, S.; Zharkikh, I.; Aza-Blanc, P.; Gleave, M.E.; Wolf, D.A. Downregulation of c-SRC kinase CSK promotes castration resistant prostate cancer and pinpoints a novel disease subclass. *Oncotarget* **2015**, *6*, 22060–22071. [CrossRef] [PubMed]

115. Guiducci, C.; Di Carlo, E.; Parenza, M.; Hitt, M.; Giovarelli, M.; Musiani, P.; Colombo, M.P. Intralesional injection of adenovirus encoding CC chemokine ligand 16 inhibits mammary tumor growth and prevents metastatic-induced death after surgical removal of the treated primary tumor. *J. Immunol.* **2004**, *172*, 4026–4036. [CrossRef] [PubMed]

116. Paulsson, K. Genomic heterogeneity in acute leukemia. *Cytogenet. Genome Res.* **2013**, *139*, 174–180. [CrossRef] [PubMed]

117. Saadatpour, A.; Guo, G.; Orkin, S.H.; Yuan, G.C. Characterizing heterogeneity in leukemic cells using single-cell gene expression analysis. *Genome Biol.* **2014**, *15*, 525. [CrossRef]

118. Navin, N.; Kendall, J.; Troge, J.; Andrews, P.; Rodgers, L.; McIndoo, J.; Cook, K.; Stepansky, A.; Levy, D.; Esposito, D.; et al. Tumor evolution inferred by single cell sequencing. *Nature* **2011**, *472*, 90–94. [CrossRef]

119. Xu, X.; Hou, Y.; Yin, X.; Bao, L.; Tang, A.; Song, L.; Li, F.; Tsang, S.; Wu, K.; Wu, H.; et al. Single-cell exome sequencing reveals single-nucleotide mutation characteristics of a kidney tumor. *Cell* **2012**, *148*, 886–895. [CrossRef]

Communication

Genomics of MPNST (GeM) Consortium: Rationale and Study Design for Multi-Omic Characterization of NF1-Associated and Sporadic MPNSTs

David T. Miller [1,*], Isidro Cortés-Ciriano [2], Nischalan Pillay [3,4], Angela C. Hirbe [5], Matija Snuderl [6], Marilyn M. Bui [7], Katherine Piculell [1], Alyaa Al-Ibraheemi [8], Brendan C. Dickson [9], Jesse Hart [10], Kevin Jones [11], Justin T. Jordan [12], Raymond H. Kim [13], Daniel Lindsay [4], Yoshihiro Nishida [14], Nicole J. Ullrich [15], Xia Wang [16], Peter J. Park [17], Adrienne M. Flanagan [3,4] and on behalf of the Genomics of MPNST (GeM) Consortium

[1] Division of Genetics and Genomics, Boston Children's Hospital, Boston, MA 02115, USA; katherine.piculell@childrens.harvard.edu

[2] European Molecular Biology Laboratory, European Bioinformatics Institute, Hinxton, Cambridge CB10 1SD, UK; icortes@ebi.ac.uk

[3] Department of Pathology, University College London Cancer Institute, Bloomsbury, London WC1E 6BT, UK; n.pillay@ucl.ac.uk (N.P.); a.flanagan@ucl.ac.uk (A.M.F.)

[4] Royal National Orthopaedic Hospital, Brockley Hill, Stanmore, Middlesex HA7 4LP, UK; daniel.lindsay1@nhs.net

[5] Oncology Division, Department of Medicine, Washington University School of Medicine in St. Louis, St. Louis, MO 63110, USA; hirbea@wustl.edu

[6] Department of Pathology, New York University Langone Health, New York City, NY 10016, USA; matija.snuderl@nyulangone.org

[7] Department of Pathology, Moffitt Cancer Center, Tampa, FL 33612, USA; marilyn.bui@moffitt.org

[8] Department of Pathology, Boston Children's Hospital, Boston, MA 02115, USA; Alyaa.Al-Ibraheemi@childrens.harvard.edu

[9] Department of Pathology and Laboratory Medicine, Mt. Sinai Hospital, Toronto, ON M5G 1XF, Canada; Brendan.Dickson@sinaihealth.ca

[10] Department of Pathology, Lifespan Laboratories, Rhode Island Hospital, Providence, RI 02903, USA; jhart5@lifespan.org

[11] Departments of Orthopaedics and Oncological Sciences; Huntsman Cancer Institute, University of Utah, Salt Lake City, UT 84112, USA; kevin.jones@hci.utah.edu

[12] Pappas Center for Neuro-Oncology, Massachusetts General Hospital, Boston, MA 02114, USA; jtjordan@mgh.harvard.edu

[13] Department of Medical Oncology, Princess Margaret Cancer Center, University Health Network, Toronto, ON M5G 2C1, Canada; raymond.kim@uhn.ca

[14] Department of Rehabilitation, Nagoya University Hospital, Nagoya 466-8550, Aichi, Japan; ynishida@med.nagoya-u.ac.jp

[15] Department of Neurology, Boston Children's Hospital, Boston, MA 02115, USA; nicole.ullrich@childrens.harvard.edu

[16] GeneHome, Moffitt Cancer Center, Tampa, FL 33612, USA; xia.wang@moffitt.org

[17] Department of Biomedical Informatics, Harvard Medical School, Boston, MA 02115, USA; peter_park@hms.harvard.edu

* Correspondence: david.miller2@childrens.harvard.edu; Tel.: +1-617-355-8221

Received: 10 March 2020; Accepted: 31 March 2020; Published: 2 April 2020

Abstract: The Genomics of Malignant Peripheral Nerve Sheath Tumor (GeM) Consortium is an international collaboration focusing on multi-omic analysis of malignant peripheral nerve sheath tumors (MPNSTs), the most aggressive tumor associated with neurofibromatosis type 1 (NF1). Here we present a summary of current knowledge gaps, a description of our consortium and the cohort we have assembled, and an overview of our plans for multi-omic analysis of these tumors. We propose that our analysis will lead to a better understanding of the order and timing of genetic events related to MPNST initiation and progression. Our ten institutions have assembled 96 fresh frozen NF1-related (63%) and sporadic MPNST specimens from 86 subjects with corresponding clinical and pathological data. Clinical data have been collected as part of the International MPNST Registry. We will characterize these tumors with bulk whole genome sequencing, RNAseq, and DNA methylation profiling. In addition, we will perform multiregional analysis and temporal sampling, with the same methodologies, on a subset of nine subjects with NF1-related MPNSTs to assess tumor heterogeneity and cancer evolution. Subsequent multi-omic analyses of additional archival specimens will include deep exome sequencing (500×) and high density copy number arrays for both validation of results based on fresh frozen tumors, and to assess further tumor heterogeneity and evolution. Digital pathology images are being collected in a cloud-based platform for consensus review. The result of these efforts will be the largest MPNST multi-omic dataset with correlated clinical and pathological information ever assembled.

Keywords: genomics; MPNST; tumor evolution; neurofibromatosis; pathology; next generation sequencing; clinical genetics

---

1. The Complex Genomic Landscape of MPNST (Malignant Peripheral Nerve Sheath Tumors)

Malignant Peripheral Nerve Sheath Tumors (MPNST) confer high morbidity and currently has limited treatment options. Patients with neurofibromatosis type 1 (NF1) have an 8–13% lifetime risk of developing malignant peripheral nerve sheath tumor (MPNST) which is the most frequent cause of early death [1]. Surgical resection with negative margins is the principal curative therapeutic modality, but is not always feasible [2,3]. Radiation and/or chemotherapy are often used in the adjuvant or neoadjuvant setting. As there are no randomized trials for MPNST to justify this treatment, recommendations are based on data from the high grade soft tissue sarcoma group as a whole, including both sporadic and NF1-associated MPNST [4–6]. The 5 year overall survival rate is modest, with one meta-analysis estimating survival at only 26–39%, and with high rates of metastasis, morbidity, and mortality [2]. In the setting of metastatic disease, treatments are limited to palliative chemotherapy and clinical trials [7,8]. Despite a recent increase in the knowledge of molecular aberrations underpinning MPNST there have not been any new effective therapeutic options developed; this may be explained by the rarity of these tumors.

Prior studies assessing the MPNST genomic landscape have been relatively small. This consortium therefore sees an opportunity to expand upon this foundation of knowledge. Somatic loss of either *TP53* or *CDKN2A* has been demonstrated in essentially all MPNST, either via somatic copy number alterations (SCNAs) or single nucleotide variants (SNVs) [9]. The genetic complexity of MPNST was better understood after two independent studies highlighted the prominent role of Polycomb repressor complex 2 (PRC2) inactivation in the development of MPNST through somatic inactivating mutations or deletions in *SUZ12, EED or EZH2* [10,11]. A subset of MPNSTs with PRC2 loss shows loss of trimethylation at lysine 27 of histone H3 (H3K27me3) [10]. This consortium will leverage a larger sample size to determine what proportion of MPNST show PRC2 loss, and how that may correlate with other aspects of the data.

This consortium also represents an opportunity to expand upon prior studies demonstrating altered methylation and gene expression patterns in MPNST development. For example, a project on soft-tissue sarcomas conducted by The Cancer Genome Atlas (TCGA) performed multi-omic analysis of over 200 sarcoma specimens (n = 5 MPNST) found differential patterns of methylation and gene expression in certain sarcoma types [9]. Additional studies of altered gene expression in MPNST have been recently reviewed [12]. Similarly, a methylation classifier analysis of 171 peripheral nerve sheath tumors that included 28 conventional high-grade MPNST, six atypical neurofibromas and other related tumors, such as neurofibromas and schwannomas, demonstrated patterns that helped differentiate the high grade tumors [13]. More specifically, by unsupervised hierarchical clustering, atypical neurofibromas and low-grade MPNST were indistinguishable, and also harbored frequent *CDKN2A* deletions. High-grade MPNST formed two distinct methylation groups which shared a frequent loss of the *NF1* locus, and also showed some differences based on anatomical location. This highlights the potential emerging role of DNA methylation profiles for diagnosis and categorization of nerve sheath tumors. Our consortium is optimistic that further exploration of the role of epigenetics, and related alterations in gene expression, in the pathogenesis of MPNST will prove fruitful. In support of this view, a recent report showed that methylated *RASSF1A* in MPNSTs identified patients with *NF1* silencing and an inferior prognosis, suggesting that methylation at a specific locus may correlate with clinical behavior [14].

2. Knowledge Gaps in the Understanding of Tumor Drivers and Evolution of MPNST

Prior studies have identified a number of recurring molecular events that appear in a majority of MPNSTs, but there is lack of uniformity of any of these molecular markers across all tumors in this histological group. Our consortium thinks that this knowledge gap is due primarily to the low overall incidence of MPNST, and therefore relatively low numbers of viable samples included in prior studies. For example, NF1-associated MPNSTs typically arise by malignant transformation of an existing plexiform, or nodular or atypical/ANNUBP neurofibroma [9]. The development of plexiform neurofibromas (PN) follows Knudsen's two-hit hypothesis with loss of heterozygosity of the *NF1* tumor suppressor gene, the likely initiating rate-limiting event for tumorigenesis [15]. However, loss of function of the second *NF1* allele has not been observed in all specimens studied, suggesting that other mechanisms are likely important in PN development [16]. Since the *NF1* gene is large, a second genetic event affecting *NF1* may not always be easy to detect [16].

The genomic landscape is more complex for MPNSTs, as compared to PN and ANNUBP; this is reflected in the acquisition of additional mutations, genomic rearrangements and copy number alterations (CNA) as the histological appearance of the tumor progresses [17]. Whole genome sequencing is expected to provide the best opportunity to detect these types of mechanisms, but only a small number of MPNST whole genomes from patients with NF1 have been published and no pathognomonic chromosomal translocations have been identified [10,11,18]. Collectively, these studies identified frequent somatic loss of *NF1, CDNKN2A, TP53,* and genes from the PRC2 complex, specifically *SUZ12* and *EED*. A variety of other genes have been implicated in the progression from benign to malignant peripheral nerve sheath tumors, including candidate driver genes such as *EPC1, CHD4, AEBP2* and *ATRX*. These genes have been implicated in MPNST primarily because of their critical interaction with molecules in the PRC2 complex [11]. Other research has implicated additional pathways (e.g., Hippo/LATS), but the relative contributions of these in the pathogenesis of the disease is not well understood, perhaps due to the relatively low number of samples studied overall [19]. Furthermore, copy number alterations (CNA) on several chromosomes have been identified through array comparative genomic hybridization (aCGH) studies, and these have been reviewed in detail elsewhere [12,17].

Intra-tumor heterogeneity is also a challenge to understanding the molecular drivers of tumorigenesis and disease progression in MPNST [17,20]. There is also substantial interpatient tumor heterogeneity. These findings highlight that such genetic variability within and between tumors plays a critical role in clinical management and treatment resistance. Hence, there is a need to catalogue the molecular events in the primary tumor and understand how these change over time and with treatment (e.g., mutations acquired during chemotherapy that lead to drug resistance).

3. Establishment of the Genomics of MPNST Consortium to Address Knowledge Gaps

Our overarching goal for the GeM(Genomics of MPNST) Consortium is to accelerate the identification of diagnostic and prognostic markers, and potential therapeutic targets for MPNST, through comprehensive molecular profiling of these rare tumors and international sharing of clinical and genomic datasets across multiple institutions worldwide. A consortium-based approach was deemed necessary in order to facilitate collection of a sufficiently large number of samples. Consequently, the Genomics of MPNST (GeM) Consortium was initiated in 2017 by the NF Research Initiative (NFRI), a philanthropically-funded translational research program at Boston Children's Hospital. The overarching focus of the GeM Consortium is to facilitate the collection and sharing of molecular and clinical data on rare NF1-related malignant and pre-malignant tumors related to MPNST, such as atypical neurofibromatous neoplasms of uncertain biological potential (ANNUBP), and, as a lower priority, sporadic MPNST, with the goal of facilitating more rapid progress in translational research to improve clinical outcomes.

Several factors influenced our decision to pursue a genomics project related to MPNST. We were motivated by the energy of a small but dedicated interdisciplinary community of experts with enthusiasm for pursuing genomics as a mechanism to identify potential therapies. This was manifested by our effort to pursue the aim set out at a 2001 MPNST consensus conference "to establish an international, multidisciplinary consortium of experts on MPNST and NF1, to collate the known clinical and genetic information about these tumors and to establish a database to record information in a uniform manner" [7]. The desire to accomplish that goal was reignited at the 2016 MPNST "State of the Science" meeting at the National Cancer Institute, in which some current GeM Consortium members participated, and led to the formation of the current GeM Consortium [15].

The GeM Coordinating Center began recruiting collaborators internationally in the Summer of 2017 through promotion on the NFRI website. In addition, a request for applications distributed at the Children's Tumor Foundation's annual meeting in June and by email to members of the NF research community. By the Fall of 2017, 13 founding member institutions, representing five countries, had established the GeM Consortium. Ultimately, four sites had to withdraw due to inability to provide the required specimens. Each site nominated one representative from their institution to serve on the GeM Steering Committee (SC), a multidisciplinary group that provides oversight to all aspects of this collaborative effort. The GeM Steering Committee established three Working Groups, composed of SC members and experts from their respective sites, to address logistical issues related to the following Consortium functions: Genomics and Informatics, Oncology and Pathology, and Data Use and Publications. This organizational structure allows for equal representation from all GeM member institutions, and multidisciplinary input into the creation of policy and research strategy.

4. Specimen and Clinical Data Collection, and Specimen Processing

The GeM Consortium Coordinating Center at Boston Children's Hospital and Dana-Farber/ Harvard Cancer Center established a non-human subjects research protocol to allow for the aggregation and analysis of de-identified clinical and genetic data, and specimens from GeM collaborators collected under pre-existing IRB-approved protocols at each participating institution. These pre-existing IRB protocols already permitted specimen and data collection, sharing with outside investigators, and comprehensive molecular testing. We also explored the possibility of establishing a central IRB for prospective collection, but collecting specimens via existing IRB protocols was preferable because MPNST is a rare tumor, and prospectively collecting enough samples would not have been possible in the 1-2 years allotted for establishing this collection. For example, the samples aggregated for our consortium were collected over a span of almost 20 years, indicating that it would have taken approximately that long to establish a similar sample size through prospective collection.

Specimens sent to the Coordinating Center include MPNST and related neurofibroma as fresh frozen, paraffin-embedded tissue, tissue microarrays, or isolated DNA/RNA along with paired normal samples such as peripheral nerve or blood. Collection and processing of specimens, followed by nucleic acid extraction, was coordinated among four Pathology Departments selected from the participating sites (Boston Children's Hospital, Moffitt Cancer Center, Mt. Sinai Hospital Toronto, and University College London). For comprehensive molecular analyses, the first step included pathological analysis of sectioned, fresh frozen tumor specimens to select the most viable areas from high quality tumor samples judged by cellularity, lack of necrosis and areas with little contaminating non-neoplastic tissue. The GeM Consortium's Standard Operating Procedure (SOP) for tissue processing, pathology review and molecular analysis is modelled on the Royal National Orthopaedic Hospital's SOP for the 100,000 Genomes Project, founded by England's National Health Service in 2012. Additional tissue sections from the regions selected for multi-omic analysis were collected in order to perform immuno-histochemical classification. In addition, whole slide digital pathology images have been collected to facilitate a cloud-based histology review and correlation with molecular markers in tissue sections.

The GeM Consortium partnered with the international MPNST Registry at Washington University School of Medicine (WUSM) for the collection of comprehensive clinical data and diagnostic imaging reports. All data are collected and housed in REDCap (https://www.project-redcap.org/). This worldwide database collects the clinical data in a comprehensive and standardized manner for each participant from diagnosis forward. Data include demographic information, disease course, tumor size/anatomical location, histological/immuno-histochemical characteristics, diagnostic imaging, surgical procedures, systemic treatment information, neoadjuvant therapy, toxicity, clinical outcomes and survival. Logistic regression models will be used to correlate clinical outcome with MPNST features. A summary of important clinical variables associated with the collected tumor specimens is presented in Table 1.

Table 1. Clinical Variables of Fresh Frozen MPNST Collected by GeM Consortium.

	NF1-Related	Sporadic or Unknown Diagnosis	Total MPNST
	n (%)	*n* (%)	*n* (%)
Fresh Frozen MPNST	60 (62.5%)	36 (39.6%)	96 (100%)
Tumor Grade			
Low Grade	5 (5.2%)	2 (2.1%)	7 (7.3%)
High Grade	49 (51.0%)	32 (33.3%)	81 (84.4%)
Unknown	6 (6.3%)	2 (2.1%)	8 (8.3%)
Neo-Adjuvant Treatment			
Chemotherapy	8 (8.3%)	5 (5.2%)	13 (13.5%)
Radiation	3 (3.1%)	9 (9.4%)	12 (12.5%)
Chemotherapy and Radiation	2 (2.1%)	2 (2.1%)	4 (4.2%)
No neo-adjuvant treatment	44 (45.8%)	20 (20.8%)	64 (66.7%)
Unknown	3 (3.1%)	0 (0%)	3 (3.1%)
Tumor Anatomic Location			
Head/Neck/Face	4 (4.2%)	0 (0%)	4 (4.2%)
Lower Limb	18 (18.8%)	18 (18.8%)	36 (37.5%)
Upper Limb	14 (14.6%)	12 (12.5%)	26 (27.1%)
Brachial Plexus	3 (3.1%)	1 (1.0%)	4 (4.2%)
Lumbosacral Plexus	5 (5.2%)	3 (3.1%)	8 (8.3%)
Trunk	8 (8.3%)	1 (1.0%)	9 (9.4%)
Retroperitoneum	1 (1.0%)	0 (0%)	1 (1.0%)
Other	7 (7.3%)	1 (1.0%)	8 (8.3%)
Total MPNST	60 (62.5%)	36 (37.5%)	96 (100%)

We ultimately collected 96 fresh frozen MPNST (60 NF1-related MPNST and 36 non-NF1 or unknown) with paired normal specimens (i.e., peripheral blood) from 86 subjects (51 with confirmed NF1 diagnosis; 35 non-NF1 or unknown) (Figure 1). The size of our final cohort was more influenced by the availability of viable biological specimens as compared to the availability of detailed clinical data. For example, when we initially established the consortium, there were 14 participating sites, estimating that approximately 215 unique MPNST specimens would be available for study. Subsequently, three sites had to withdraw due to inability to obtain permission from their home institution to share samples. Further, there was one site that had to withdraw due to nonviable samples. Among the ten active sites, we collectively estimated that there would be 165 MPNST specimens. Unfortunately, several samples were nonviable, either due to low DNA quantity or quality. Ultimately, our original estimate of 215 tumors decreased to 165 tumors after four sites withdrew, and then decreased further to 96 tumors after accounting for samples that were nonviable due to either low DNA quantity or quality.

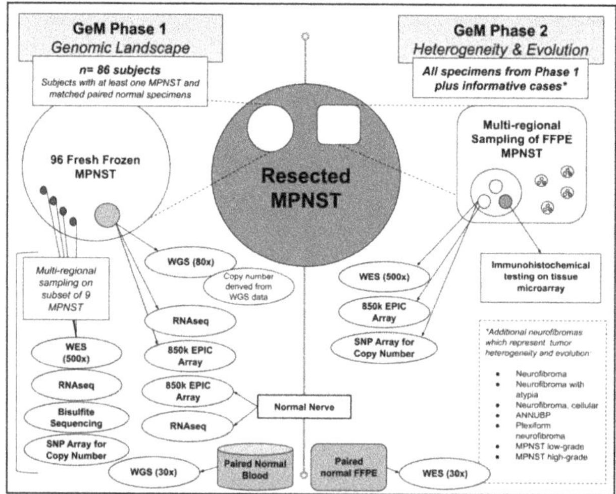

Figure 1. The GeM(Genomics of MPNST) Consortium will conduct multi-omic analyses on both fresh frozen and paraffin-embedded tissue samples of resected MPNSTs(Malignant Peripheral Nerve Sheath Tumors) and related neurofibroma, and normal nerve.

5. Plan for Multi-Omic Characterization of MPNST

5.1. Phase 1: Multi-Omic Profiling on Frozen Tumor Material to Study MPNST Genomic Complexity, Tumor Drivers, and Tumor Heterogeneity

Based on the aforementioned expert Pathology review, these tumors had a high, but variable, purity of approximately 40–70%. Whole genome sequencing (WGS) is being performed on all fresh frozen tumor samples at 80x coverage using libraries created with the TruSeq DNA-PCR free kit (Illumina Inc., San Diego, CA, USA). Given the known tumor heterogeneity and variable purity, we elected to perform deeper sequencing than previously reported to identify drivers of tumorigenesis. WGS will also be performed on paired normal germline DNA samples at 30x coverage. Although it requires more DNA, a PCR-free library preparation was selected to minimize artifacts of both single nucleotide variants and copy number variants that may arise from PCR amplification, and this is particularly important for the tumor-derived DNA. In addition, each of these bulk frozen tumor specimens is being analyzed by RNAseq with the TruSeq Transcriptome kit (Illumina Inc., San Diego, CA, USA) and epigenetic profiling with DNA methylation analyses using the Illumina EPIC array platform (Methylation EPIC 850k BeadChip). Control data for RNAseq analysis has been collected on the same platform from nine samples of healthy peripheral nerve frozen tissue collected from a subset of individuals in this cohort. A smaller number of high quality samples derived from snap frozen DNA will undergo whole genome bisulfite sequencing to achieve a genome-wide view of DNA methylation with a higher resolution compared to the EPIC array. Finally, multi-regional sampling was also performed on a subset of nine fresh frozen NF1-related MPNST specimens to assess intra-tumor variation (i.e., tumor heterogeneity) by performing 500× exome sequencing, RNA-seq, and epigenetic profiling.

5.2. Phase 2: Extensive Characterization of Tumor Heterogeneity and Evolution Using FFPE MPNST Samples from Phase 1 and Additional Informative Cases

To study intra-tumor (i.e., heterogeneity) and inter-tumor (i.e., evolution) variability, the GeM Consortium will conduct additional molecular analyses on formalin-fixed paraffin embedded (FFPE) MPNST specimens from pathology archives. The first priority will be to analyze FFPE specimens from tumors and subjects represented in the first phase of the project (WGS on fresh frozen tumor and normal DNA). The consortium will also identify potentially more informative cases where there are multiple tumors from the same person (e.g., an MPNST and precursor lesion such as ANNUBP or metastases), and tumors within which there is more than one line of differentiation (for example, nerve sheath and osteosarcomatous differentiation) with the hope that multi-regional sampling from multiple FFPE tumors representing differences in space and changes over time will uncover how and why precursor neurofibromas evolve into MPNST.

Having an accurate characterization of tumor type is critically important for downstream data analysis. GeM pathologists use the World Health Organization's (WHO's) histopathological and morphological classification system to establish tumor diagnosis and determine grade of MPNST, and also to confirm the diagnosis of any related precursor tumors that are available from subjects with a MPNST [21]. Immunohistochemistry studies based on tissue microarray (TMA) will add further detail to the classification of NF1-related MPNST, sporadic MPNST, and related tumors such as ANNUBP and PN. Multi-regional tumor cores ($n = 5$) from each FFPE tumor will be selected, on the basis of histological features, such as rhabdomyosarcomatous, angiosarcomatous, osteosarcomatous areas, etc. DNA will be extracted from these regional samples and will be subjected to deep (500×) bulk whole exome sequencing (WES), copy number analysis using Illumina's Omni Array and DNA methylation analysis using Illumina's EPIC array. Two adjacent cores of tumor will be taken for building a TMA. Annotation and analysis of digital images of FFPE slides will be used to collate pathological features and clinical outcome data with genomic data.

5.3. Bioinformatics Analysis for Multi-Omic Profiling of MPNSTs

Although an exhaustive list of all intended analyses would be beyond the scope of this white paper, we would like to highlight a few key areas of interest. GeM Consortium data analysis pipelines will utilize the alignment and mutation detection pipelines used by recent national and international cancer genomics consortia, such as methods for WES, WGS, and RNAseq used in the Pan-Cancer Atlas Consortium and in the Pan-Cancer Analysis of Whole Genomes (PCAWG) project [22]. This will include harmonization of third-party contributed data by lifting over to common reference, re-alignment, re-calling variants, and/or re-quantitating RNA-seq data. Our GeM analysis team has extensive experienced with these methods [23]. Uniform quality-control, alignment and processing of the WGS, WES, and RNA-seq will be performed, including the detection of somatic single-nucleotide variants, small insertions and deletions, microsatellite instability, structural variants, copy number variants, and gene fusions using WGS/WES data. Neoadjuvant treatment status will be considered during the bioinformatic analysis.

Analysis pipelines will also include gene expression quantification and gene fusion detection using RNAseq data. Epigenomic alterations will be correlated with other data types. Our analysis will also include integration of data generated outside the GeM Consortium, re-processing the data as needed. Mutational signature analysis will be performed on this dataset. Mutational signatures have emerged as a useful computational approach for identification of the biological processes that generate somatic mutations. Mutational signatures refer to patterns of nucleotide changes and their contexts, occurring due to various environmental carcinogens or endogenous DNA damage processes. A good example of this is the work of Alexandrov et al. that demonstrated tobacco induces a specific pattern of C>A transversions during the lifetime of the lung cancer cell [24]. Although these patterns can indicate specific etiological processes, they can also serve as markers of immunological response as seen in breast cancer where DNA damage response signatures were associated with lymphocytic infiltration.

There is also emerging evidence that extraction of copy number signatures from a common cancer, such as high grade serous ovarian carcinoma, provides more robust prognostic information than pathological grading or the analysis of single gene mutations [25]. In sarcomas, copy number signatures have proved useful in understanding the evolutionary trajectories of the genomically complex cancers [26]. In a meta-analysis study of >5000 samples of 12 cancer types, some patterns of somatic copy number alterations were associated with reduced expression of cytotoxic immune signatures [27]. Moreover, the copy number scores were predictive of response to immune checkpoint blockade. There is also some preliminary evidence from sequencing of osteosarcoma genomes that a homologous recombination deficiency signature may be a consistent feature of that tumor type. A similar signature of "BRCAness" in breast cancer is predictive of sensitivity to poly (ADP-ribose) polymerase (PARP) inhibitors [28]. These are just some examples of the type of analysis that will be possible with this large dataset.

5.4. GeM Data Display and Availability

The goal of the GeM Consortium, like that of the Genomics Evidence Neoplasia Information Exchange (GENIE) effort by the American Association of Cancer Research (AACR), is to enable identification of novel therapeutic targets and genomic markers of response to therapy [29]. To do so, we are developing a secure Django-based web platform for the visualization and reporting of the data generated by the GeM Consortium. De-identified molecular data derived from submitted specimens and clinical data submitted by GeM sites will be hosted in this database and may be accessed by sites and other qualified researchers from the broader community under the terms set by the Data Use and Publications Working Group and with approval of the Steering Committee. We will create a publicly accessible instance of cBioPortal to display molecular alterations that resulted from our multi-omic analysis of MPNSTs [30,31].

6. Concluding Remarks and Future Directions

The GeM Consortium's ultimate goal is to improve clinical care for patients with MPNST through a better understanding of genetic and epigenetic drivers of MPNST initiation and progression. We are motivated by our desire to provide the best care for our patients. Collectively, GeM clinicians and their multi-disciplinary teams care for approximately 2400 patients with NF1 per year. Due to the appreciable incidence of MPNST in this patient population, a better approach to treatment is needed, and we think that this will only be possible through a better understanding of the molecular drivers of MPNST development and progression. Although variation in genes such as *NF1*, *CDKN2A/B*, *TP53*, and *SUZ12* and/or *EED* are found in most MPNST, the precise order and timing of these changes remain poorly described, and this may be important for understanding the early stages of tumor development and/or the development of treatment resistance.

A comprehensive mutational, rearrangement and copy number signature analysis has not been performed in MPNST. The patients recruited through the consortium and their samples will serve as a valuable resource to facilitate the identification of such signatures through mRNA expression analysis, epigenetic profiling, and the correlation with clinical endpoints. Our hope is that this dataset will provide the most comprehensive knowledge to date and reveal previously unrecognized important pathways for MPNST initiation and progression. Beyond the currently described approach, future efforts are likely to include single-cell DNA and RNA analyses to describe better clonal drivers within MPNSTs. We hope that the knowledge gained through the GeM Consortium's efforts will inform both pre-clinical studies of MPNST and selection of candidate drugs for future clinical trials. For more information or to contact us, please visit www.NFResearch-Childrens.org.

Author Contributions: Conceptualization, A.A.-I., A.C.H., A.M.F., D.L., D.T.M., I.C.-C., M.M.B., M.S., N.P. and P.J.P.; methodology, A.A.-I., A.C.H., A.M.F., D.L., D.T.M., I.C.-C., M.M.B., M.S., N.P. and P.J.P.; software, I.C.-C. and P.J.P.; validation, I.C.-C.; investigation, A.A.-I., A.C.H., A.M.F., B.C.D., D.L., D.T.M., I.C.-C., M.M.B., N.P. and P.J.P.; resources, A.M.F., D.T.M., M.S. and P.J.P.; data curation, A.A.-I., A.C.H., I.C.-C., K.P. and P.J.P.; writing—original draft preparation, A.C.H., A.M.F., D.T.M., I.C.-C., K.P., N.P. and P.J.P.; writing—review and editing, A.A.-I., A.C.H., A.M.F., B.C.D., D.L., D.T.M., I.C.-C., J.H., J.T.J., K.J., K.P., M.M.B., M.S., N.J.U., N.P., P.J.P., R.H.K., X.W. and Y.N.; supervision, A.A.-I., A.C.H., A.M.F., B.C.D., D.L., D.T.M., M.M.B., M.S. and P.J.P.; project administration, A.A.-I., A.C.H., A.M.F., B.C.D., D.L., D.T.M., K.P., M.M.B. and P.J.P.; funding acquisition, D.T.M. All authors have read and agreed to the published version of the manuscript.

Funding: This research was primarily funded by the Neurofibromatosis Research Initiative (NFRI) at Boston Children's Hospital, made possible by an anonymous donation. N.P. is a Cancer Research UK clinician scientist, grant number 18387. A.M.F is a NIHR senior investigator; A.M.F., and N.P. are supported by the UK Department of Health's National Institute for Health Research, UCLH Biomedical Research Centre and by UCL Experimental Cancer Centre. Biobanking of U.K. patient samples was funded by the Research and Development Office of the RNOH. The methylation profiling program at NYU Langone Health has been supported by the Friedberg Charitable Foundation, the Sohn Foundation and the Making Headway Foundation (to M.S.). R.H.K. was supported by the Bhalwani Family Charitable Foundation

Acknowledgments: Sample identification and processing was supported by several individuals and core facilities, including: Sam Cano and Tamara Restrepo in the Laboratory for Advanced Molecular Pediatric Pathology (LaMPP) and Umesh Paneru in Histology at Boston Children's Hospital; Maia Roche in the Department of Histopathology at Royal National Orthopaedic Hospital; Anthony Griffin and Nalan Gokgoz at Mt. Sinai Hospital; the Collaborative Data Services Core, Tissue Core and Analytic Microscopy Core Facilities at the H. Lee Moffitt Cancer Center & Research Institute, an NCI designated Comprehensive Cancer Center (P30-CA076292); Diane Lucente, Alexandra Silva, and James Gusella at Massachusetts General Hospital; Jacqueline Hart and David Viskochil at Huntsman Cancer Institute; Sonika Dahiya and Vanessa Eulo at Washington University School of Medicine. We also thank Sarah Berns and Diane Shao at Boston Children's Hospital for help preparing an early outline of the manuscript. Administrative and regulatory support for the Coordinating Center was provided by Jacqueline Breen, Tyson Heilhecker, Noah McClanan, Sara Soto, and Tim Linkkila at Boston Children's Hospital.

Conflicts of Interest: J.T.J. is a paid consultant for Recursion Pharmaceuticals and Neurofibromatosis Network, Royalties from Elsevier, and Honoraria from American Academy of Neurology. No other author has any conflict of interest to declare. The funders had no role in the design of the study; in the collection, analyses, or interpretation of data; in the writing of the manuscript, or in the decision to publish the results

References

1. Evans, D.G.; Huson, S.M.; Birch, J.M. Malignant peripheral nerve sheath tumours in inherited disease. *Clin. Sarcoma Res.* **2012**, *2*, 17. [CrossRef] [PubMed]

2. Dunn, G.P.; Spiliopoulos, K.; Plotkin, S.R.; Hornicek, F.J.; Harmon, D.C.; Delaney, T.F.; Williams, Z. Role of resection of malignant peripheral nerve sheath tumors in patients with neurofibromatosis type 1. *J. Neurosurg.* **2013**, *118*, 142–148. [CrossRef] [PubMed]

3. Hruban, R.H.; Shiu, M.H.; Senie, R.T.; Woodruff, J.M. Malignant peripheral nerve sheath tumors of the buttock and lower extremity. A study of 43 cases. *Cancer* **1990**, *66*, 1253–1265. [CrossRef]

4. Frustaci, S.; Gherlinzoni, F.; De Paoli, A.; Bonetti, M.; Azzarelli, A.; Comandone, A.; Olmi, P.; Buonadonna, A.; Pignatti, G.; Barbieri, E.; et al. Adjuvant chemotherapy for adult soft tissue sarcomas of the extremities and girdles: Results of the Italian randomized cooperative trial. *J. Clin. Oncol.* **2001**, *19*, 1238–1247. [CrossRef] [PubMed]

5. Kaushal, A.; Citrin, D. The role of radiation therapy in the management of sarcomas. *Surg. Clin. N. Am.* **2008**, *88*, 629–646. [CrossRef] [PubMed]

6. Yang, J.C.; Chang, A.E.; Baker, A.R.; Sindelar, W.F.; Danforth, D.N.; Topalian, S.L.; DeLaney, T.; Glatstein, E.; Steinberg, S.M.; Merino, M.J.; et al. Randomized prospective study of the benefit of adjuvant radiation therapy in the treatment of soft tissue sarcomas of the extremity. *J. Clin. Oncol.* **1998**, *16*, 197–203. [CrossRef]

7. Ferner, R.E.; Gutmann, D.H. International consensus statement on malignant peripheral nerve sheath tumors in neurofibromatosis. *Cancer Res.* **2002**, *62*, 1573–1577.

8. Prudner, B.C.; Ball, T.; Rathore, R.; Hirbe, A.C. Diagnosis and management of malignant peripheral nerve sheath tumors: Current practice and future perspectives. *Neuro Oncol. Adv.* **2019**. [CrossRef]

9. Kim, A.; Stewart, D.R.; Reilly, K.M.; Viskochil, D.; Miettinen, M.M.; Widemann, B.C. Malignant Peripheral Nerve Sheath Tumors State of the Science: Leveraging Clinical and Biological Insights into Effective Therapies. *Sarcoma* **2017**, *2017*, 7429697. [CrossRef]

10. Lee, W.; Teckie, S.; Wiesner, T.; Ran, L.; Prieto Granada, C.N.; Lin, M.; Zhu, S.; Cao, Z.; Liang, Y.; Sboner, A.; et al. PRC2 is recurrently inactivated through EED or SUZ12 loss in malignant peripheral nerve sheath tumors. *Nat. Genet.* **2014**, *46*, 1227–1232. [CrossRef]

11. Zhang, M.; Wang, Y.; Jones, S.; Sausen, M.; McMahon, K.; Sharma, R.; Wang, Q.; Belzberg, A.J.; Chaichana, K.; Gallia, G.L.; et al. Somatic mutations of SUZ12 in malignant peripheral nerve sheath tumors. *Nat. Genet.* **2014**, *46*, 1170–1172. [CrossRef] [PubMed]

12. Pemov, A.; Li, H.; Presley, W.; Wallace, M.R.; Miller, D.T. Genetics of human malignant peripheral nerve sheath tumors. *Neuro Oncol. Adv.* **2019**. [CrossRef]

13. Rohrich, M.; Koelsche, C.; Schrimpf, D.; Capper, D.; Sahm, F.; Kratz, A.; Reuss, J.; Hovestadt, V.; Jones, D.T.; Bewerunge-Hudler, M.; et al. Methylation-based classification of benign and malignant peripheral nerve sheath tumors. *Acta Neuropathol.* **2016**, *131*, 877–887. [CrossRef] [PubMed]

14. Danielsen, S.A.; Lind, G.E.; Kolberg, M.; Holand, M.; Bjerkehagen, B.; Sundby Hall, K.; van den Berg, E.; Mertens, F.; Smeland, S.; Picci, P.; et al. Methylated RASSF1A in malignant peripheral nerve sheath tumors identifies neurofibromatosis type 1 patients with inferior prognosis. *Neuro Oncol.* **2015**, *17*, 63–69. [CrossRef]

15. Pemov, A.; Li, H.; Patidar, R.; Hansen, N.F.; Sindiri, S.; Hartley, S.W.; Wei, J.S.; Elkahloun, A.; Chandrasekharappa, S.C.; Program, N.C.S.; et al. The primacy of NF1 loss as the driver of tumorigenesis in neurofibromatosis type 1-associated plexiform neurofibromas. *Oncogene* **2017**, *36*, 3168–3177. [CrossRef]

16. Eisenbarth, I.; Beyer, K.; Krone, W.; Assum, G. Toward a survey of somatic mutation of the NF1 gene in benign neurofibromas of patients with neurofibromatosis type 1. *Am. J. Hum. Genet.* **2000**, *66*, 393–401. [CrossRef]

17. Carroll, S.L. The Challenge of Cancer Genomics in Rare Nervous System Neoplasms: Malignant Peripheral Nerve Sheath Tumors as a Paradigm for Cross-Species Comparative Oncogenomics. *Am. J. Pathol.* **2016**, *186*, 464–477. [CrossRef]

18. De Raedt, T.; Beert, E.; Pasmant, E.; Luscan, A.; Brems, H.; Ortonne, N.; Helin, K.; Hornick, J.L.; Mautner, V.; Kehrer-Sawatzki, H.; et al. PRC2 loss amplifies Ras-driven transcription and confers sensitivity to BRD4-based therapies. *Nature* **2014**, *514*, 247–251. [CrossRef]

19. Wu, L.M.N.; Deng, Y.; Wang, J.; Zhao, C.; Wang, J.; Rao, R.; Xu, L.; Zhou, W.; Choi, K.; Rizvi, T.A.; et al. Programming of Schwann Cells by Lats1/2-TAZ/YAP Signaling Drives Malignant Peripheral Nerve Sheath Tumorigenesis. *Cancer Cell* **2018**, *33*, 292–308. [CrossRef]

20. Turajlic, S.; Sottoriva, A.; Graham, T.; Swanton, C. Resolving genetic heterogeneity in cancer. *Nat. Rev. Genet.* **2019**, *20*, 404–416. [CrossRef]

21. Zambo, I.; Vesely, K. WHO classification of tumours of soft tissue and bone 2013: The main changes compared to the 3rd edition. *Cesk. Patol.* **2014**, *50*, 64–70. [PubMed]

22. Cancer Genome Atlas Research Network. Electronic address, E.D.S.C.; Cancer Genome Atlas Research, N. Comprehensive and Integrated Genomic Characterization of Adult Soft Tissue Sarcomas. *Cell* **2017**, *171*, 950–965. [CrossRef] [PubMed]

23. Cortes-Ciriano, I.; Lee, S.; Park, W.Y.; Kim, T.M.; Park, P.J. A molecular portrait of microsatellite instability across multiple cancers. *Nat. Commun.* **2017**, *8*, 15180. [CrossRef] [PubMed]

24. Alexandrov, L.B.; Ju, Y.S.; Haase, K.; Van Loo, P.; Martincorena, I.; Nik-Zainal, S.; Totoki, Y.; Fujimoto, A.; Nakagawa, H.; Shibata, T.; et al. Mutational signatures associated with tobacco smoking in human cancer. *Science* **2016**, *354*, 618–622. [CrossRef] [PubMed]

25. Macintyre, G.; Goranova, T.E.; De Silva, D.; Ennis, D.; Piskorz, A.M.; Eldridge, M.; Sie, D.; Lewsley, L.A.; Hanif, A.; Wilson, C.; et al. Copy number signatures and mutational processes in ovarian carcinoma. *Nat. Genet.* **2018**, *50*, 1262–1270. [CrossRef]

26. Steele, C.D.; Tarabichi, M.; Oukrif, D.; Webster, A.P.; Ye, H.; Fittall, M.; Lombard, P.; Martincorena, I.; Tarpey, P.S.; Collord, G.; et al. Undifferentiated Sarcomas Develop through Distinct Evolutionary Pathways. *Cancer Cell* **2019**, *35*, 441–456. [CrossRef]

27. Davoli, T.; Uno, H.; Wooten, E.C.; Elledge, S.J. Tumor aneuploidy correlates with markers of immune evasion and with reduced response to immunotherapy. *Science* **2017**, *355*, eaaf8399. [CrossRef]

28. Kovac, M.; Blattmann, C.; Ribi, S.; Smida, J.; Mueller, N.S.; Engert, F.; Castro-Giner, F.; Weischenfeldt, J.; Kovacova, M.; Krieg, A.; et al. Exome sequencing of osteosarcoma reveals mutation signatures reminiscent of BRCA deficiency. *Nat. Commun.* **2015**, *6*, 8940. [CrossRef]

29. AACR Project GENIE Consortium. AACR Project GENIE: Powering Precision Medicine through an International Consortium. *Cancer Discov.* **2017**, *7*, 818–831. [CrossRef]

30. Cerami, E.; Gao, J.; Dogrusoz, U.; Gross, B.E.; Sumer, S.O.; Aksoy, B.A.; Jacobsen, A.; Byrne, C.J.; Heuer, M.L.; Larsson, E.; et al. The cBio cancer genomics portal: An open platform for exploring multidimensional cancer genomics data. *Cancer Discov.* **2012**, *2*, 401–404. [CrossRef]

31. Gao, J.; Aksoy, B.A.; Dogrusoz, U.; Dresdner, G.; Gross, B.; Sumer, S.O.; Sun, Y.; Jacobsen, A.; Sinha, R.; Larsson, E.; et al. Integrative analysis of complex cancer genomics and clinical profiles using the cBioPortal. *Sci. Signal.* **2013**, *6*, pl1. [CrossRef] [PubMed]

 *genes*

Article

Kinome Profiling of NF1-Related MPNSTs in Response to Kinase Inhibition and Doxorubicin Reveals Therapeutic Vulnerabilities

Jamie L. Grit [1], Matt G. Pridgeon [1,2], Curt J. Essenburg [1], Emily Wolfrum [3], Zachary B. Madaj [3], Lisa Turner [4], Julia Wulfkuhle [5], Emanuel F. Petricoin III [5], Carrie R. Graveel [1,†] and Matthew R. Steensma [1,2,6,*,†]

[1] Center for Cancer and Cell Biology, Van Andel Research Institute, Grand Rapids, MI 49503, USA; jamie.grit@vai.edu (J.L.G.); matthew.pridgeon@helendevoschildrens.org (M.G.P.); curt.essenburg@vai.org (C.J.E.); carrie.graveel@vai.org (C.R.G.)

[2] Helen DeVos Children's Hospital, Spectrum Health System, Grand Rapids, MI 49503, USA

[3] Bioinformatics & Biostatistics Core, Van Andel Research Institute, Grand Rapids, MI 49503, USA; emily.wolfrum@vai.org (E.W.); zachary.madaj@vai.org (Z.B.M.)

[4] Pathology and Biorepository Core, Van Andel Research Institute, Grand Rapids, MI 49503, USA; lisa.turner@vai.org

[5] Center for Applied Proteomics and Molecular Medicine, George Mason University, Manassas, VA 22030, USA; jwulfkuh@gmu.edu (J.W.); epetrico@gmu.edu (E.F.P.)

[6] Michigan State University College of Human Medicine, Grand Rapids, MI 49503, USA

* Correspondence: matt.steensma@vai.org

† These authors contributed equally.

Received: 1 February 2020; Accepted: 9 March 2020; Published: 20 March 2020

Abstract: Neurofibromatosis Type 1 (NF1)-related Malignant Peripheral Nerve Sheath Tumors (MPNST) are highly resistant sarcomas that account for significant mortality. The mechanisms of therapy resistance are not well-understood in MPNSTs, particularly with respect to kinase inhibition strategies. In this study, we aimed to quantify the impact of both the genomic context and targeted therapy on MPNST resistance using reverse phase phosphoproteome array (RPPA) analysis. We treated tumorgrafts from three genetically engineered mouse models using MET (capmatinib) and MEK (trametinib) inhibitors and doxorubicin, and assessed phosphosignaling at 4 h, 2 days, and 21 days. Baseline kinase signaling in our mouse models recapitulated an MET-addicted state (NF1-MET), P53 mutation (NF1-P53), and HGF overexpression (NF1). Following perturbation with the drug, we observed broad and redundant kinome adaptations that extended well beyond canonical RAS/ERK or PI3K/AKT/mTOR signaling. MET and MEK inhibition were both associated with an initial inflammatory response mediated by kinases in the JAK/STAT pathway and NFkB. Growth signaling predominated at the 2-day and 21-day time points as a result of broad RTK and intracellular kinase activation. Interestingly, AXL and NFkB were strongly activated at the 2-day and 21-day time points, and tightly correlated, regardless of the treatment type or genomic context. The degree of kinome adaptation observed in innately resistant tumors was significantly less than the surviving fractions of responsive tumors that exhibited a latency period before reinitiating growth. Lastly, doxorubicin resistance was associated with kinome adaptations that strongly favored growth and survival signaling. These observations confirm that MPNSTs are capable of profound signaling plasticity in the face of kinase inhibition or DNA damaging agent administration. It is possible that by targeting AXL or NFkB, therapy resistance can be mitigated.

Keywords: MPNST; NF1; kinase; kinome adaptation; kinome reprogramming; MET; MEK; doxorubicin; capmatinib; tram

---

1. Introduction

Targeting the RAS/ERK signaling pathway is an effective treatment for numerous cancers with hyper-activation of the RAS pathway. The most striking clinical responses to inhibitors of BRAF, MEK, and EGFR have been observed in melanoma and lung cancers where RAS pathway activation is intrinsic. Even though these targeted therapies have resulted in an extension of the overall survival in these inherently aggressive cancers, the clinical response is often transient and complete remission is rare. Resistance to kinase inhibition is a significant clinical challenge and numerous studies have identified multifactorial and heterogeneous mechanisms of resistance to kinase inhibition [1].

Malignant Peripheral Nerve Sheath Tumors (MPNSTs) are aggressive, highly chemoresistant sarcomas arising from Schwann cells that are a leading cause of death in patients with Neurofibromatosis Type 1 (NF1) [2,3]. Neurofibromatosis Type 1 (NF1) is caused by germline mutations in the *NF1* gene and is the most common single-gene disorder, affecting 1 in 3000 live births. The *NF1* gene encodes neurofibromin, a GTPase-activating protein that negatively regulates RAS (including HRAS, NRAS, and KRAS), where the loss of NF1 leads to deregulated RAS signaling. Deregulated RAS signaling caused by the loss of neurofibromin is both permissive and instructive for MPNST progression (3–5). Recent clinical trials have focused on targeting members of the RAS signaling pathway or the PI3K/mTOR pathway. To date, these trials have failed to identify consistent therapeutic vulnerabilities in MPNSTs; however, few studies have examined why these therapies failed. These clinical results highlight our limited knowledge of the mechanisms that drive resistance to kinase inhibition in MPNSTs.

In addition to loss of the *NF1* gene, NF1-related MPNSTs exhibit highly complex genomic alterations that result in substantial tumor suppressor gene loss and oncogene copy number variations [4,5]. How MPNST genomic alterations affect therapy resistance is currently unclear. Recently, we performed a genomic analysis of longitudinally collected MPNST samples. This study revealed the early concomitant presence of *MET*, *HGF*, and *EGFR* amplifications, as well as the site-specific expansion of these loci over time and treatment. These data point to an adaptive mechanism involving RTK signaling for both malignant transformation and clonal selection in MPNSTs [6]. To advance our understanding of the MPNST therapeutic response and resistance to RAS pathway inhibition, we developed diverse preclinical NF1-related MPNST models, including an "MET-addicted" model of NF1-related MPNSTs (NF1-MET), an *Nf1/Trp53*-deficient model (NF1-P53), and an NF1 model (P53^{WT}, *Hgf*-amplified) [7–9]. Using these MPNST models, we determined that P53 deficiency significantly exacerbates resistance to MEK inhibition; however, combined MEK and MET inhibition overcame therapy resistance [6]. Importantly, these results demonstrated that NF1-related MPNSTs maintain multiple signaling dependencies beyond RAS, and that genomic determinants, such as P53 and RTK genomic alterations, profoundly influence the therapy response.

Kinome reprogramming is a powerful barrier to a durable treatment response to kinase inhibition [10–12]. These signaling adaptations occur as a result of the compensatory activation of evolutionarily conserved signaling pathways that drive growth and proliferation, especially when central pathways such as RAS/MEK and PI3K/mTOR are blocked by drugs [13]. Kinome adaptation leads to diverse mechanisms of therapy resistance that can be classified as three categories, all of which can occur simultaneously during treatment [1]. The most common resistance mechanism is defined as "pathway reactivation", which can occur through multiple mechanisms that reinforce oncogenic signaling in the face of strong target inhibition. The second common mechanism of resistance is "pathway bypass", which occurs when oncogenic pathways are activated at a downstream convergence point by a parallel pathway, despite effective upstream inhibition (e.g., PI3K/AKT/ERK activation during MEK inhibition). The third resistance mechanism is "pathway indifference", where the cancer cells transition to an alternative survival state that is independent of the targeted oncogenic pathway. Each of these resistance mechanisms have been observed in response to kinase inhibition in cancers with RAS-activation dependencies. Efforts to delineate patterns of therapy resistance have been valuable in understanding the treatment response and the identification of targets for salvage therapy. Even though genetic mechanisms of resistance (e.g., somatic mutations or gene amplifications) have been a major focus of resistance research, the impact of phosphoproteomic changes in therapeutic resistance has been increasingly acknowledged [1].

The mechanisms that regulate adaptive kinome reprogramming in NF1-deficient cancers are not well-elucidated. In this study, we aimed to define both mechanisms and categories of resistance to standard chemotherapy and targeted kinase inhibition in NF1-related MPNSTs in order to identify 1) novel therapeutic strategies that correlate with the genomic status or 2) salvage therapies that are focused on the emerging 'resistant' tumor populations. Kinome reprogramming can be heavily influenced by genomic alterations (i.e., amplification and mutation) of kinases, the overexpression of other kinases, or ligand activation [13–16]. Our genomic analysis of MPNST progression identified genomic events during human MPNST progression that heighten RTK signaling, AKT/mTOR activation, and cell survival [6]. To understand how these genomic contexts influence the therapy response to kinase inhibition, we conducted a phosphoproteomic analysis using a reverse phase protein array (RPPA) in our established MPNST mouse models. RPPA is a powerful research tool that simultaneously interrogates a broad range of phosphosites across the entire kinome, including activating, inactivating, and alternative sites. Using this comprehensive analytical technique, we confirmed substantial signaling heterogeneity following kinase inhibition in NF1-related MPNSTs leading to therapy resistance, including examples of pathway reactivation, bypass, and indifference. Our data verifies that a broad range of signaling pathways are activated as a result of effective MET, MEK, and MET/MEK blockade, most notably, AXL, NFκB, RAS/RAF/MEK, and AKT/mTOR pathways. Interestingly, the patterns of kinome adaptation were distinct based on the kinase target and the genomic context of the tumor. We also demonstrated that administration of the DNA damaging reagent doxorubicin resulted in a distinct pattern of kinome adaptation that was partially, but not fully, mitigated by MET and MEK inhibitors. Categorizing these patterns of therapy resistance is valuable for inferring patient stratification and the identification of targets for salvage therapy.

2. Materials and Methods

2.1. Murine MPNST Tumorgrafts and Treatment

Immediately following the euthanasia of tumor-bearing mice, 15–25 mg portions of each tumor were transplanted into the flank of NSG-SCID mice using a 10 gauge trochar. Tumors were measured twice weekly and euthanized when the tumor size exceeded 2500 mm^3. When the tumor volume reached approximately 150 mm^3, mice were randomized into treatment groups, treated, and euthanized as independent groups at 4-h, 2-day, or 21-day time points (or until mice reached the euthanasia criteria). Respective doses across all treatment combinations were capmatinib (30 mg/kg twice daily vial oral gavage), trametinib (1 mg/kg daily via oral gavage), and doxorubicin (1 mg/kg once via subcutaneous injection). For the 4-h time point, all animals were euthanized 4 h after a single dose. For the 2-day time point, capmatinib-treated animals received three total doses and were euthanized 4 h after the final treatment, whereas trametinib-treated animals received two total doses before euthanasia. Tumors were immediately harvested and either snap-frozen or formalin-fixed for further analysis. Three representative tumors were assessed by RPPA for each time point, treatment, and genotype group. Capmatinib and trametinib were obtained from Novartis. Doxorubicin was obtained from LC Laboratories. All animal experimentation in this study was approved by the Van Andel Institute's Internal Animal Care and Use Committee (XPA-19-04-001).

2.2. Sample Collection and Preparation for RPPA Downstream Analysis

Samples were frozen in liquid nitrogen within 20 min upon surgical resection to preserve the integrity of the phosphoproteome. Specimens were then embedded in an optimal cutting temperature compound (Sakura Finetek, Torrance, CA, USA), cut into 8 μm cryo-sections, mounted on uncharged glass slides, and stored at −80 °C until microdissected. Each slide was fixed in 70% ethanol (Sigma Aldrich, Darmstadt, Germany), washed in deionized water, stained with hematoxylin (Sigma Aldrich,

Darmstadt, Germany) and blued in Scott's Tap Water substitute (Electron Microscopy Sciences), and dehydrated through an ethanol gradient (70%, 95%, and 100%) and xylene (Sigma Aldrich, Darmstadt, Germany). In order to prevent protein degradation, complete protease inhibitor cocktail tablets (Roche Applied Science, Basel, Switzerland) were added to the ethanol, water, hematoxylin, and Scott's Tap Water substitute [17]. For each sample, an average of 15,000 tumor cells was isolated from the surrounding microenvironment using a Pixcell II LCM system (Arcturus, Mountain View, CA, USA). Microdissected cells were lysed in a 1:1 solution of 2× Tris-Glycine SDS Sample buffer (Invitrogen Life Technologies, Carlsbad, CA, USA) and Tissue Protein Extraction Reagent (Pierce, Waltham, MA, USA) supplemented with 2.5% of 2-mercaptoethanol (Sigma Aldrich, Darmstadt, Germany). Cell lysates were boiled for 8 min and stored at −80 °C.

2.3. Reverse Phase Protein Microarray Construction and Immunostaining

Using an Aushon 2470 arrayer (Aushon BioSystems, Billerica, MA, USA) equipped with 185 μm pins, samples and standard curves for internal quality assurance were printed in triplicate onto Oncyte Avid nitrocellulose-coated slides (Grace Bio-labs, Bend, OR, USA), as previously described [17]. A Sypro Ruby Protein Blot Stain (Molecular Probes, Eugene, OR, USA) protocol was used to stain selected arrays to quantify the total amount of protein within each sample.

Before immunostaining, each array was first incubated with Reblot Antibody stripping solution (Chemicon) for 15 min at room temperature, followed by two washes in PBS. To minimize potential nonspecific bindings, arrays were then incubated in I-block solution (Invitrogen Life Technologies, Carlsbad, CA, USA) for 1 h. Each array was tested with a single primary antibody using an automated system (Dako, Santa Clara, CA, USA). The antibody specificity was tested by immunoblotting using a wide panel of cell lysates, as previously described [17,18]. Negative control arrays were incubated with the anti-rabbit secondary antibody only to account for unspecific binding and background noise. A commercially available catalyzed signal amplification system (Dako, Santa Clara, CA, USA) coupled with a biotinylated anti-rabbit secondary antibody (Vector Laboratories) and a streptavidin-conjugated IRDye680 (LI-COR Biosciences, Lincoln, NE, USA) were used for the amplification and detection of the fluorescent signal. Arrays were probed with a total of 99 antibodies targeting protein kinases involved in major cellular functions, and the results of the broad screening were previously published.

Antibody and Sypro Ruby-stained arrays were scanned using a laser-based PowerScanner (TECAN, Mönnedorf, Switzerland). Acquired images were analysed using the MicroVigene software version 5.1 (Vigene Tech, Carlisle, MA, USA). This commercially available software performs spot finding, averages the triplicates, subtracts the background from the negative control slide(s), and normalizes each sample to the corresponding amount of total protein measured by Sypro Ruby staining. Intra- and inter-assay reproducibility of the RPPA platform has been previously reported [19,20].

2.4. Statistical Methods

Tumor growth analysis: Linear mixed-effects models, with random slopes and intercepts, and false discovery rate-adjusted contrasts, were used to estimate and compare tumor growth rates for the different mono and combo therapies. For visualization of the changes in tumor growth, the tumor volume was imputed using the last observation carried forward, until the animal was euthanized. Curves terminated once >50% of mice had been euthanized in the respective treatment group. All analyses were conducted using R v3.2.2 (https://cran.r-project.org/), with an assumed level of significance of $\alpha = 0.05$.

Proteomic analysis: Fold change in expression for each phosphosite was calculated by $\log_2$ transformation of the treatment relative to the vehicle mean for that genotype group. Fold change was ranked-ordered by the median of the treatment group for each genotype and for each condition, the top and bottom 15 proteins from the ranked list were plotted in balloon plots. A total of 98 protein sites passed quality control metrics and were used for analysis. Fold change in the expression for each protein was calculated as the protein expression relative to the vehicle mean for that genotype-time group. Proteins were rank-ordered on the *y*-axis by the median transformed fold change for each treatment-genotype-time group and plotted from highest to lowest fold change. Each column represents a single animal. For the 4-h and 2-day time points, animals were plotted randomly on the x-axis. For the 21-day time point, animals were plotted on the *x*-axis based on tumor size (largest to smallest) at sacrifice, corresponding to tumors 4–6, respectively, in the associated tumor-graft plots. The balloon color indicates the $\log_2$ fold change in protein expression. Proteins with a greater than 4-fold increase or decrease in expression relative to the vehicle were plotted as $\log_2$ fold change >2 or <−2, respectively. The balloon size indicates the absolute protein expression normalized to the total protein input and background. Head-tail balloon-maps were created by plotting the 30 proteins with the highest and lowest fold change in expression for each treatment-genotype-time group. Plots were generated using R v3.6.1. For pospho-AXL and phospho-NFkB correlation analysis, normality was first assessed by a Shapiro–Wilk normality test. Correlations were analyzed by two-sided Spearman's rank correlation rho. For correlation plots, data were fit by stat_smooth using loess with span = 1 using ggplot2 (v3.2.1). Missing data points (due to a failure to meet RPPA quality control standards) were omitted from the analysis. All analyses were done using R (v3.6.2).

3. Results

3.1. Distinct Kinome Response to MET Inhibition Present in MET-Addicted MPNSTs

To understand how RTK amplification and enhanced RTK signaling impact the MPNST kinome, we assessed the influence of both the *MET* copy number and MET kinase inhibition on the drug response and resistance. Both *MET* and its ligand, hepatocyte growth factor (HGF), are implicated in NF1-related MPNST initiation and progression [21–23]. Previously, our genomic analysis of human MPNST progression revealed that *MET* and *HGF* copy number gains are present at the earliest stage of neurofibroma transformation and increase during metastasis and resistance [6]. Moreover, studies in other cancers have demonstrated that aberrant MET signaling can drive malignant progression in a variety of RAS-deregulated human tumors and augment the oncogenic effects of RAS activation [24,25]. To understand the impact of the MET genomic status on kinome adaptations, we evaluated the response and resistance to the potent and selective MET inhibitor capmatinib in three diverse models of NF1-related MPNSTs, including an "MET-addicted" model (NF1-MET), an *Nf1/Trp53*-deficient model (NF1-P53), and an NF1 model ($P53^{WT}$, MET^{WT}, *Hgf*-amplified). As we previously showed, NF1-MET MPNSTs were uniformly sensitive to MET inhibition, whereas a heterogeneous response to MET inhibition was observed in NF1-P53 and NF1 MPNSTs (Figure 1A–C) [6]. To characterize the kinome response to MET inhibition, we performed pathway activation mapping of 98 proteins and phosphoproteins. This was a targeted pathway activation analysis focused on actionable targets of RTK-mediated signaling, downstream PI3K-mTOR signaling, downstream RAS-ERK signaling, and motility/adhesion signaling. To assess the immediate, early, and late kinome responses to kinase inhibition, we profiled the tumor phosphoproteome after 4 h, 2 days, and 21 days of treatment. With these time points, we anticipated that both innate and acquired kinome adaptations would be observed in the various genomic backgrounds. Changes in the expression relative to the vehicle were plotted in rank order for each timepoint. For the 21-day RPPA analysis of each MPNST model, we analyzed tumors that had diverse treatment responses, while avoiding tumors that exhibited grossly anomalous growth patterns compared to the mean growth curve (see individual tumor annotations in Figure 1A–C). By including diverse tumors, we anticipated that we would detect the heterogeneity of mechanisms underlying drug resistance. For example, because the NF1-MET tumors are "*Met*-addicted", substantial growth inhibition was present at 21 days and minimal heterogeneity in the drug response was observed (Figure 1A) This homogeneous response was not observed in the other MPNST tumor-graft lines (Figure 1B,C). Correspondingly, we observed a more homogeneous kinome response in NF1-MET tumors in comparison to the responses observed in NF1-P53 and NF1 tumors (Figure 1D–F; Figure S1).

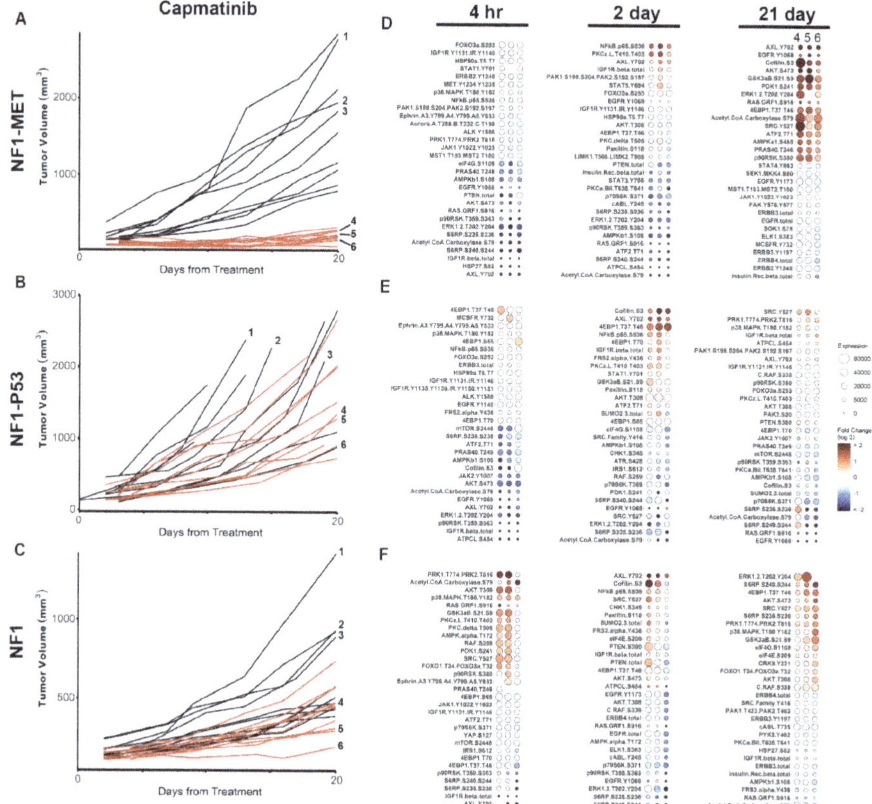

Figure 1. MET inhibition reveals differential innate and adaptive kinome reprogramming. Individual tumor growth curves for (**A**) Neurofibromatosis Type 1 (NF1)-MET, (**B**) NF1-P53, and (**C**) NF1 tumorgrafts plotted by treatment (colored lines) compared to the vehicle (black lines). The analysis of tumor growth data was previously reported [6]. The annotated tumors were analyzed by a reverse phase phosphoproteome array (RPPA) (**D–F**). The fold change relative to the mean protein expression of control tumors (i.e., #1–3) was calculated for each tumor #4–6, with the first column of Panel A at 21 days corresponding to tumor #4, the second to tumor #5, and the third to tumor #6. Ranked balloon plots of the proteins with the highest and lowest fold change in expression after 4-h, 2-day, and 21-day treatment of the NF1-MET model with capmatinib. Each column represents a single animal. Balloon color indicates the fold change in expression relative to the vehicle mean (*n* = 3) for that time point. Balloon size indicates the absolute protein expression normalized to the total protein input and background.

After 4-h capmatinib treatment, we observed a striking repression of ERK, AKT, and RTK phosphorylation that corresponded to growth reduction in the NF1-MET tumors (Figure 1D). Overall, minimal kinome activation was observed at the 4-h time point in growing NF1-MET and NF1-P53 tumors (Figure 1D,E; Figure S1B,C); however, two of three NF1 tumors had phosphorylation changes in several pathways at the 4-h time point (i.e., PRK, AKT, and p38MAPK) (Figure 1F). After 2-day capmatinib treatment, we observed increased activating phosphorylation at several sites in the NF1-P53 and NF1 tumors, including AXL (Y702), cofilin (S3), and 4EBP1 (T37/T46) (Figure 1E,F; Figure S5), which is a finding that correlated with the relatively increased capmatinib resistance at 21 days (Figure 1B,C). In the NF1-MET tumors, NFκB demonstrated the strongest increase in phosphorylation at the 2-day time point. This probe corresponds to S536 in the transactivation domain (TAD) of NFκB/p65, which leads to transactivation. Interestingly, at the 2-day time point, NFκB/p65 was in the top three most increased phoshposites in all of the tumor models after 2-day MET inhibition. Since NFκB is a master regulator of the inflammatory response, survival, and tumor proliferation [26], and a known mediator of pathway indifference [27], NFκB activation at the 48-h time point may represent a common kinome adaptation that is agnostic to the MPNST genomic context.

After 21 days of capmatinib treatment, significant tumor death was observed in the NF1-MET tumors (Figure 1A), with only a small layer of viable cells present at the edge of the tumors [6]. This is in contrast to the NF1-P53 and NF1 tumors that maintained a significant decrease in growth compared to the vehicle control. An upward growth trend was observed in the majority of tumors at the 21-day time point, despite ongoing treatment (Figure 1A–C). In the surviving, capmatinib-resistant cells present at 21 days in NF1-MET tumors, distinct changes in the kinome response were observed, comprising consistent AXL (Y702), EGFR (Y1068), cofilin (S3), and AKT (S473) activation (Figure 1D; Figure S5). These results suggest that MET-addicted MPNSTs survive MET inhibition through pathway reactivation via other RTKs (i.e., AXL and EGFR) and potentially a pathway bypass through AKT signaling. In the NF1 tumors whose ascending growth patterns indicated the beginning of drug resistance, increased phosphosite activation was observed in ERK, ribosomal protein S6 kinase (S6), 4EBP1, and AKT, yet markers of parallel RTK activation were also present. In the F1-P53 tumors, which were continuing to grow at the 21-day time point, phosphosite expression returned to levels resembling the vehicle, suggesting that broader kinome adaptation was no longer required for growth. Collectively, these data indicate that distinct mechanisms of innate and adaptive kinome reprogramming occur in genetically diverse MPNSTs.

3.2. Kinome Response to MEK Inhibition Results in Bypass Activation

The recent clinical success of MEK inhibition with selumetinib in NF1 plexiform neurofibromas and recent preclinical MPNST treatment studies highlight the therapeutic potential of targeting MEK in NF1-related MPNSTs [28–30]. To evaluate the kinome response to MEK inhibition in NF1-deficient MPNSTs with distinctive genomic backgrounds, we used the MEK inhibitor trametinib (Novartis, Cambridge, MA, USA). Trametinib is a reversible, highly selective, allosteric inhibitor of MEK1 and MEK2, which is FDA approved for melanoma, lung cancer, and anaplastic thyroid cancers with BRAF mutations. MEK inhibition significantly decreased tumor growth in all of the MPNST lines, yet substantial response heterogeneity was observed in the NF1-MET and NF1-P53 tumors (Figure 2A–C) [6]. The most uniform tumor inhibition was observed in the NF1 MPNST tumors, whereas some NF1-P53 tumors still displayed aggressive growth after 21 days. As with capmatinib, RAS and AKT pathway inactivation (i.e., ERK1/2, mTOR, S6, and p90RSK) was observed after 4 h of trametinib in the NF1-MET and NF1-P53 tumors (Figure 2D,E). Interestingly, broader kinome activation was not observed in these same genomic contexts, suggesting that NF1-MET and NF1-P53 tumors maintain a limited MEK dependency due to innate resistance (Figure 2A,B,D,E; Figure S2). Interestingly, by 2 days, trametinib treatment induced a similar response to capmatinib in the NF1-MET tumors, strongly activating EGFR (Y1068), AXL (Y702), PKCz (L410/T403), and NFκB (S536) (Figure 2D; Figure S5). In contrast, trametinib treatment resulted in the differential regulation of EGFR (Y1068) and AXL (702) in NF1-P53 tumors, as AXL (702) was upregulated, while EGFR (Y1068) was the most repressed site after 2 days (Figure 2D,E). AXL (Y07) was also highly induced after 2 days of trametinib treatment in the NF1 tumors (Figure 2F), suggesting that AXL activation may be a universal early response to MEK inhibition, regardless of the genetic context of the MPNST.

Adaptive kinome reprogramming in response to trametinib was distinct for each model. In the NF1-MET tumors, AXL (Y702) remained activated after 21 days of treatment. 4EBP1 (T37/T46), CHK1 (S345), and AKT (e.g., SGK and AKT) phosphorylation was observed in response to long-term MEK inhibition (Figure 2D; Figure S5). These results implicate a bypass mechanism of resistance to MEK inhibition, particularly in disparate signaling nodes within the AKT and mTOR pathways (i.e.,

SGK, CHK1, AKT, 4EBP1, and HSP27). Notably, after 21 days of treatment, 90% of NF1-P53 tumors had increasing growth trends and a negligible kinome response to MEK inhibition (Figure 2B,E). ERK was consistently inhibited by trametinib in these resistant tumors, confirming that ERK pathway reactivation was not required to maintain growth. In the NF1 tumors, AKT/mTOR and protein translation pathway effectors were the strongest targets activated by MEK inhibition. Collectively, the NF1-MET kinome response to both MET and MEK inhibition suggests that RTK-dependent MPNSTs may survive kinase inhibition both by the engagement of alternative RTKs (i.e., AXL and EGFR) and increasing AKT/mTOR signaling pathways.

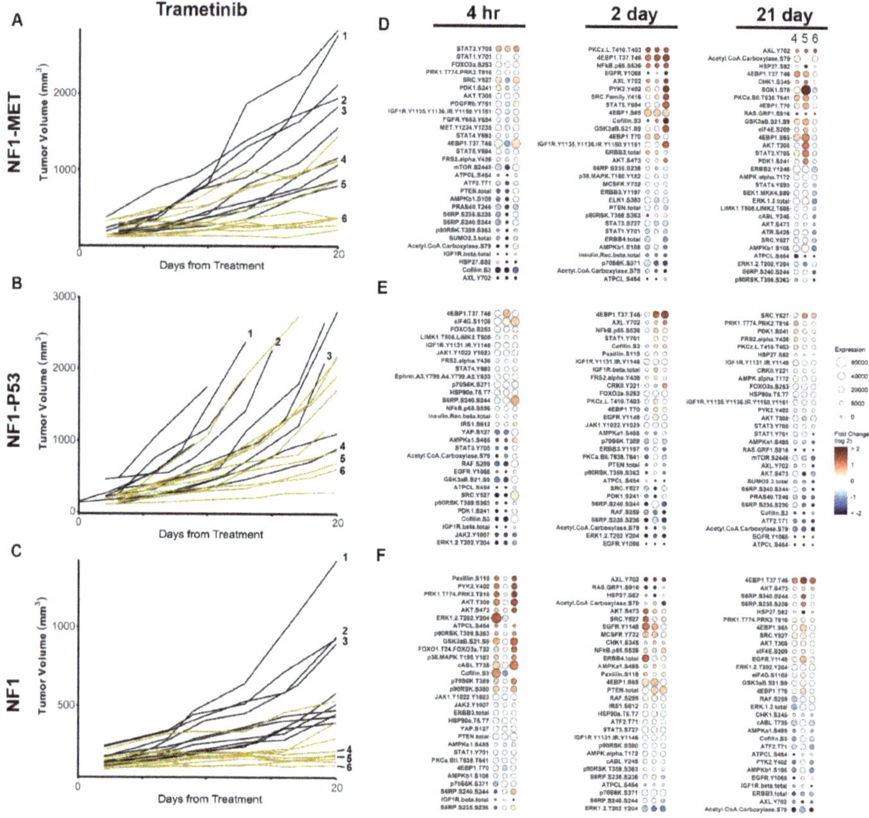

Figure 2. MEK inhibition reveals differential innate and adaptive kinome reprogramming. Individual tumor growth curves for (**A**) NF1-MET, (**B**) NF1-P53, and (**C**) NF1 tumorgrafts plotted by treatment (colored lines) compared to the vehicle (black lines). The analysis of tumor growth data was previously reported [6]. The annotated tumors were analyzed by RPPA (**D–F**). The fold change relative to the mean protein expression of control tumors (i.e., #1–3) was calculated for each tumor #4–6, with the first column of Panel A at 21 days corresponding to tumor #4, the second to tumor #5, and the third to tumor #6. Ranked balloon plots of the proteins with the highest and lowest fold change in expression after 4-h, 2-day, and 21-day treatment of the NF1-MET model with trametinib. Each column represents a single animal. Balloon color indicates fold change in expression relative to the vehicle mean (*n* = 3) for that time point. Balloon size indicates the absolute protein expression normalized to the total protein input and background.

3.3. Kinome Response to Combined MET and MEK Inhibition in NF1-Related MPNSTs

Since we observed both pathway reactivation and bypass resistance mechanisms with single-agent MET or MEK inhibition, we sought to determine whether targeting multiple signaling pathways may abrogate these kinome adaptations and achieve a more durable clinical response. Previously, we compared combined MET and MEK inhibition with monotherapy and demonstrated significant improvement in tumor inhibition and response variability with combination therapy compared to a single agent alone (Figure 3A–C) [6]. Even in NF1-P53 tumors which had the most heterogeneous responses to monotherapy with capmatinib or trametinib, we observed stable disease in all but one tumor (Figure 3B). The kinome response to combined MET-MEK inhibition exhibited striking differences in comparison to single kinase inhibition. At 4 h and 2 days of treatment, ERK1/2, S6 (S240/S244 and S235/S236), and p90RSK demonstrated the strongest decrease in phosphorylation in all of the MPNST models, suggesting that the RAS/ERK and AKT/mTOR pathways are robustly inactivated with combined MET-MEK inhibition (Figure 3D–F). As with single kinase inhibition, we observed NFκB/p65 (S536) activation at the 2-day time point. We also measured an increase in PKCζ/λ (T410/T4033) and cofilin (S3) phosphorylation at the 2-day time point with both single and combined kinase inhibition (Figure S5). Cofilin is an actin depolymerizing factor known to regulate actin dynamics and cell invasion; however, recent studies have established the role of cofilin in NFκB nuclear translocation [31,32]. The atypical protein kinase C member PKCζ is involved in several survival pathways that are deregulated in cancer and is also involved in the activation of NFκB [33,34]. Together, these findings indicate that NFκB activation is an acute response that occurs in response to monotherapy or combined kinase inhibition in MPNSTs.

At 21 days, significant tumor inhibition was observed in NF1-MET tumors and in the surviving cells, the kinome adaptations observed with single MET inhibition were intensified (Figure 3D; Figure S3A). Specifically, AXL (Y702), EGFR (Y1068), and AKT (S473) are strongly activated. Intriguingly, combined MET-MEK inhibition also resulted in the strong activation of AXL (Y702) in the NF1-P53 and NF1 tumors, which was not observed with single-agent treatment of either drug (Figure 3E,F). The NF1-P53 tumors maintained an inflammatory kinome response after 21-day treatment (PKCζ/λ, NFκB), which stands in contrast to the NF1-MET and NF1 tumors, where an inflammatory response was only observed at 4 h and 2 days (Figure 3D–F; Figure S5). Rather, after 21 days of combination therapy, the surviving cells of NF1 tumors robustly activated S6 (S240/S244 and S235/S236) and 4EBP1 (T37/T46), along with AXL (Y702) (Figure 3F). Given that AXL is activated in response to MET and MEK inhibition in all three of these genomically diverse MPNST models, AXL activation may be a common mechanism of therapeutic resistance to RAS pathway inhibitors.

3.4. Kinome Response to Doxorubicin in NF1-Related MPNSTs

Doxorubicin is a topoisomerase II inhibitor that prevents cellular replication by indirectly stabilizing double-stranded DNA breaks [35]. It has also been implicated in direct DNA damage through free radical production. Doxorubicin is currently being tested in combination with multiple kinase inhibitors for sarcoma (e.g., PDGFα inhibitor), or to treat anthracycline-resistant sarcomas [36–38]. Although the results of these trials are mixed, it is unclear whether doxorubicin resistance is mediated at least in part through kinome adaptation. How NF1-related MPNSTs confer doxorubicin resistance is likely multifactorial [39]; however, the patterns of compensatory kinase signaling have not been studied to date. Following the doxorubicin treatment of NF1-MET, NF1-P53, and NF1 tumorgrafts, significant resistance and response heterogeneity was observed (Figure 4A–C). NF1 tumor growth was significantly slower than controls; however, no tumors ultimately responded to treatment. RPPA analysis revealed a broad and diverse response to doxorubicin at early and late timepoints across all genomic contexts. Early responses at 4 h included RTK activation (EGFR, IGF1R, PDGFR, and MET), pro-inflammatory signaling mediators (p38, NFκB, and STAT3/5), and upstream kinases (SRC and RAF) (Figure 4D–F; Figure S4). Kinome responses at 2 days and 21 days of treatment were more diverse, with the emergence of increases in AXL (Y702), EGFR (Y1068), and cofilin (S3) as dominant signaling mediators (Figure S5).

Qualitatively, doxorubicin resulted in the broadest pathway responses compared to single-agent capmatinib (Figure 1), trametinib (Figure 2), and combination therapy (Figure 3). Interestingly, the PI3K/AKT/mTOR pathway response did not appear to be significantly activated in response to doxorubicin, as evidenced by the inactivation of AKT (S473), S6RP, and CHK1 (Figure 4D–F, Figure S4). These results indicate that doxorubicin treatment causes both acute and persistent kinome changes in several pathways. The diversity and perseverance of the doxorubicin-mediated kinome response may underlie innate resistance observed in sarcomas.

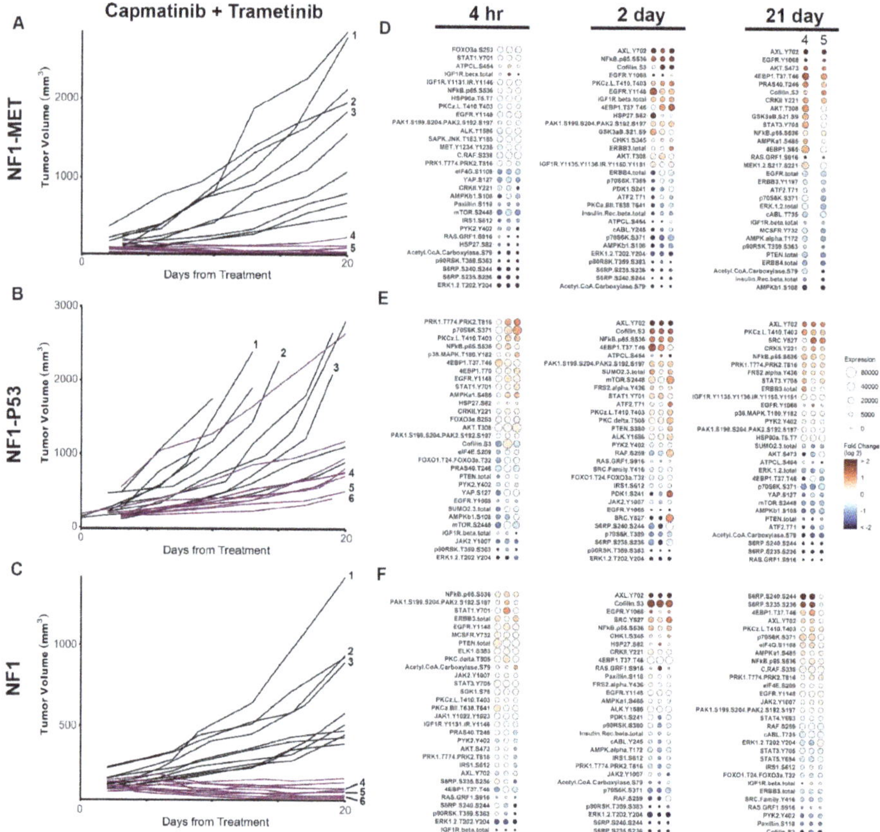

Figure 3. Combination MEK and MET inhibition reveals differential innate and adaptive kinome reprogramming. Individual tumor growth curves for (**A**) NF1-MET, (**B**) NF1-P53, and (**C**) NF1 tumorgrafts plotted by treatment (colored lines) compared to the vehicle (black lines). The analysis of tumor growth data was previously reported [6]. The annotated tumors were analyzed by RPPA (**D–F**). The fold change relative to the mean protein expression of control tumors (i.e., #1–3) was calculated for each tumor #4–6, with the first column of Panel A at 21 days corresponding to tumor #4, the second to tumor #5, and the third to tumor #6. Ranked balloon plots of the proteins with the highest and lowest fold change in expression after 4-h, 2-day, and 21-day treatment of the NF1-MET model with combination therapy. Each column represents a single animal. Balloon color indicates the fold change in expression relative to the vehicle mean (*n* = 3) for that time point. Balloon size indicates the absolute protein expression normalized to the total protein input and background.

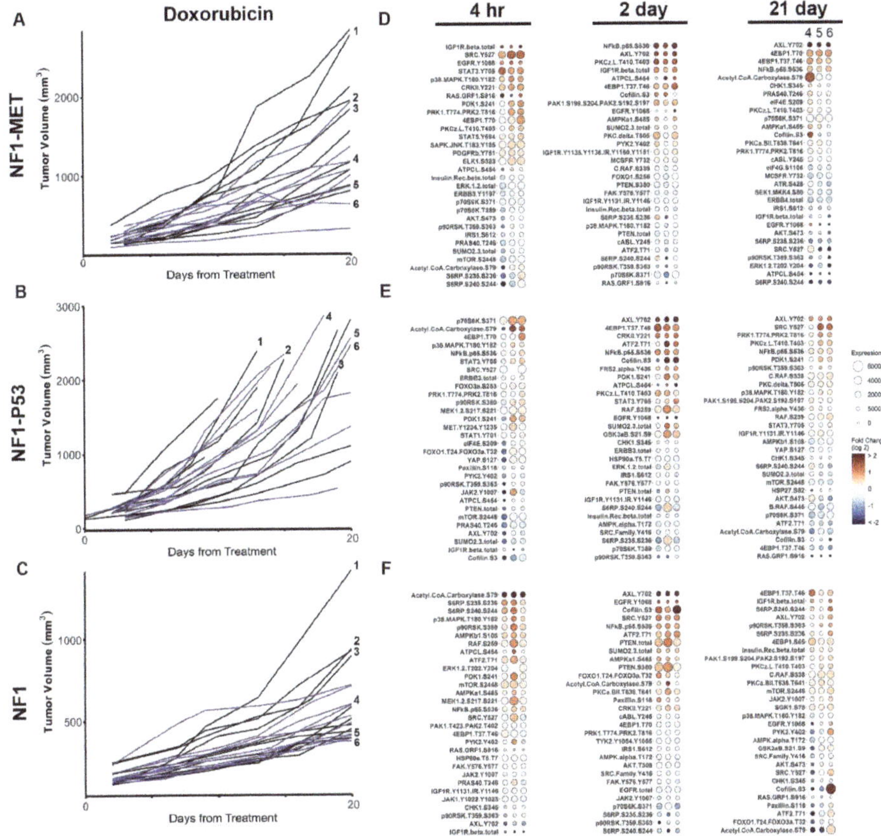

Figure 4. Doxorubicin reveals differential innate and adaptive kinome reprogramming. Individual tumor growth curves for (**A**) NF1-MET, (**B**) NF1-P53, and (**C**) NF1 tumorgrafts plotted by treatment (colored lines) compared to the vehicle (black lines). The annotated tumors were analyzed by RPPA (**D–F**). The fold change relative to the mean protein expression of control tumors (i.e., #1–3) was calculated for each tumor #4–6, with the first column of Panel A at 21 days corresponding to tumor #4, the second to tumor #5, and the third to tumor #6. Ranked balloon plots of the proteins with the highest and lowest fold change in expression after 4-h, 2-day, and 21-day treatment of the NF1-MET model with doxorubicin. Each column represents a single animal. Balloon color indicates the fold change in expression relative to the vehicle mean (*n* = 3) for that time point. Balloon size indicates the absolute protein expression normalized to the total protein input and background.

3.5. Combined MET-MEK Inhibition with Doxorubicin Decreases Response Heterogeneity

Because combined MET and MEK inhibition resulted in an improved treatment response in all MPNST lines, we investigated the efficacy of doxorubicin in combination with MET and/or MEK kinase inhibition. As discussed earlier, the kinase inhibition of MET or MEK resulted in distinct kinome adaptations compared to doxorubicin treatment. We focused our tumor growth analysis on the NF1-MET and NF1-P53 tumors since these two MPNST models had the most aggressive growth and distinctive responses to MET and MEK inhibition. In NF1-MET tumors, doxorubicin treatment caused a significant decrease in tumor growth (Figure 5A; *p* < 0.0005); however, doxorubicin treatment was inferior to capmatinib or trametinib (Figure 5B). Even though the mean growth reduction was significant, the heterogeneous response to doxorubicin was substantial (Figure 5C). Combined doxorubicin and kinase inhibition significantly improved tumor inhibition and reduced tumor heterogeneity. For example, trametinib alone resulted in moderate tumor inhibition in NF1-MET tumors (Figure 5A), yet combined trametinib + doxorubicin significantly improved the response in comparison to single-agent treatment with trametinb or doxorubicin (Figure 5B). Since these MPNST tumors are MET-addicted, capmatinib resulted in impressive tumor regression, yet combined capmatinib + trametinib was superior to capmatinib alone (Figure 5B), whereas capmatinib + doxorubicin did not significantly improve the treatment response. The treatment that resulted in the least response variability (SD = 19 mm) was the capmatinib + trametinib + doxorubicin combination, with each tumor showing consistent growth inhibition.

For the MPNST tumorgraft line, the NF1-P53 tumors had the most aggressive growth, highest response heterogeneity, and least impressive response to single-agent treatment. In NF1-P53 tumors, doxorubicin treatment did not result in tumor regression (Figure 5D–F). Combined doxorubicin and kinase inhibition reduced tumor growth in comparison to doxorubicin alone, yet this combination was not better than capmatinib or trametinib single-agent treatment (Figure 5E). The triple combination of capmatinib + trametinib + doxorubicin was not significantly better than capmatinib + trametinib; however, this treatment combination resulted in the least heterogeneity in the response (Figure 5F; SD = 343 mm^3). The heterogeneity of response and growth patterns observed correlated with the diversity and intensity of the innate and acquired kinome responses delineated in these genomically distinct MPNST tumors.

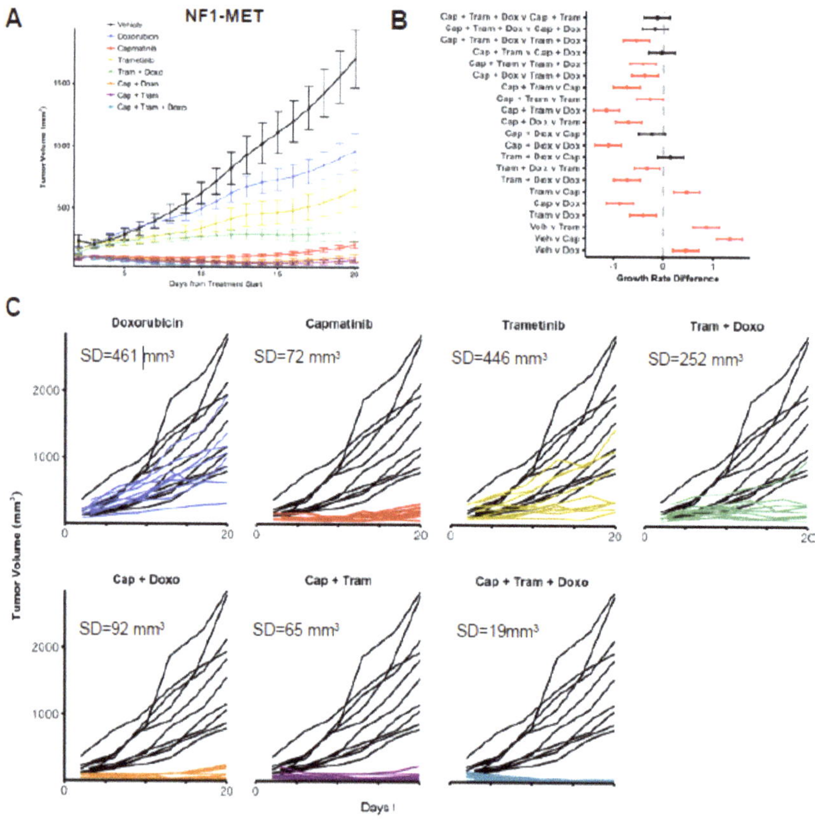

Figure 5. *Cont.*

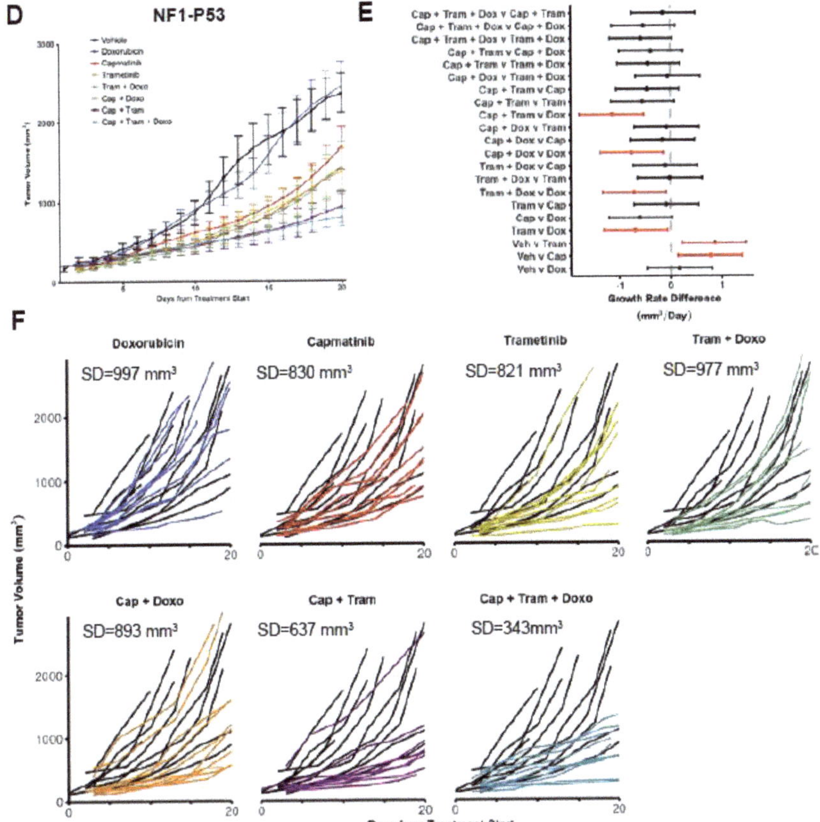

Figure 5. Combined doxorubicin, MET, and MEK inhibitor treatment reduces the response heterogeneity. Tumor growth of (**A**) NF-MET and (**D**) tumorgrafts are plotted as means with standard errors. 95% confidence intervals for the pairwise differences between the growth rates of the select treatments in the (**B**) NF1-MET and (**E**) NF1-P53 tumors, estimated and tested using linear mixed-effects models with random slopes and intercepts, and false discovery rate-adjusted contrasts. Statistically significant differences (*p*-value < 0.05) between compared therapies are highlighted in red. Individual tumor growth curves for (**C**) NF1-MET and (**F**) NF1-P53 tumorgrafts plotted by treatment (colored lines) compared to the vehicle (black lines). The analysis of tumor growth data and differences in treatment response were previously reported for single-agent treatment of capmatinib and trametinib, and combination treatment of capmatinib + trametinib [6].

3.6. ERK Reactivation is Observed in Cells Resistant to MET or MEK Inhibition

RPPA revealed the consistent de-repression or reactivation of ERK (T202/Y204) in surviving cells throughout 21 days of kinase inhibitor treatment. To determine if ERK reactivation was specific to resistant subpopulations or the entire tumor, we stained the tumors for phospho-ERK (T202/Y204) after 21 days of single or combination therapy. In vehicle-treated tumors, we observed moderate to strong pERK staining; however, distinct ERK activation patterns were observed in each tumorgraft line (Figure 6A). In the NF1-MET vehicle tumors, ERK activation was intense at the invasive edge of the tumor, whereas ERK activation was moderate to strong and uniformly expressed in the NF1-P53 and NF1 tumors (Figure 6A). After 21 days of single-agent treatment with either capmatinib or trametinib,

ERK activation was robust in the NF1-MET and NF1 tumors (Figure 6A–C). Interestingly, ERK activation decreased with single-agent treatment in the NF1-P53 tumors, except for minor cell populations at the invasive edge in some tumors (Figure 6B,C). This decrease was even more pronounced in the combined capmatinib-trametinib-treated NF1-P53 tumors, while the capmatinib-trametinib NF1-MET and NF1 tumors maintained high levels of ERK activation (Figure 6D). To directly compare RPPA and IHC and to understand how ERK phosphorylation changed over time in each model, we plotted the normalized absolute protein expression measured by RPPA for ERK (T202/Y204) for each treatment (Figure 6E). Overall, ERK maintained a similar level of activation or increased over time in the NF1-MET and NF1 models. Moreover, long-term capmatinib treatment induced strong ERK activation in these tumors, whereas NF1-P53 tumors consistently maintained lower levels of ERK activation compared to the NF1-MET and NF1 tumors (Figure 6E).

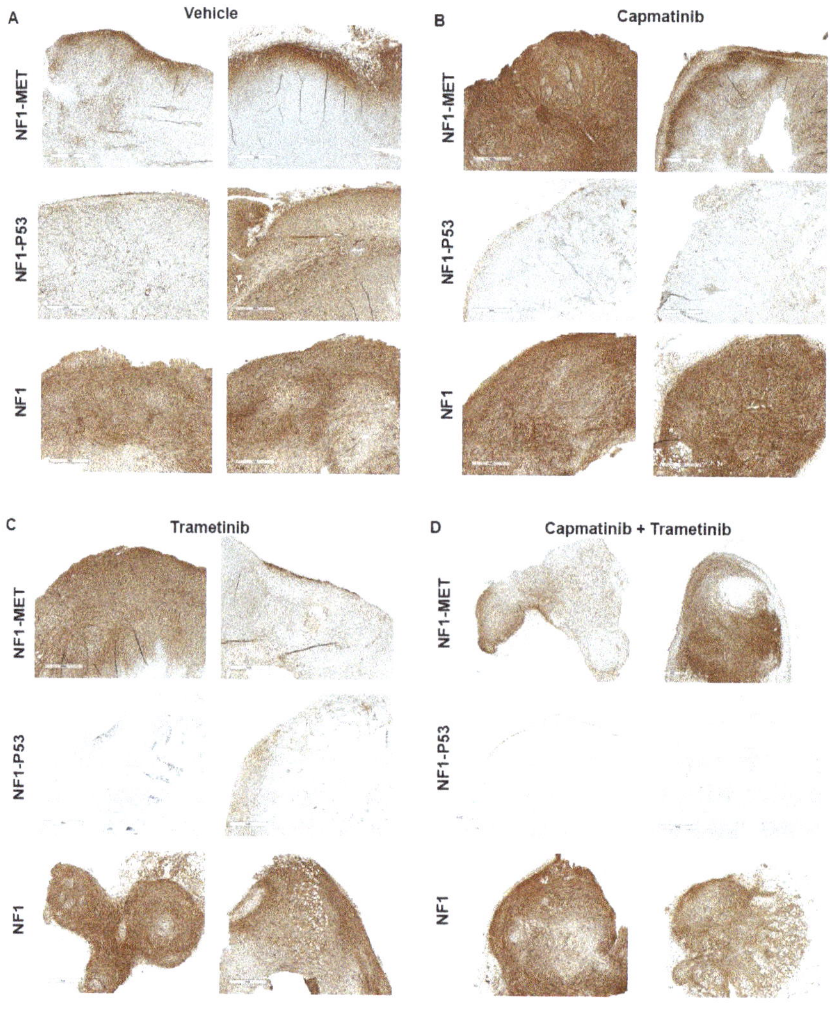

Figure 6. *Cont.*

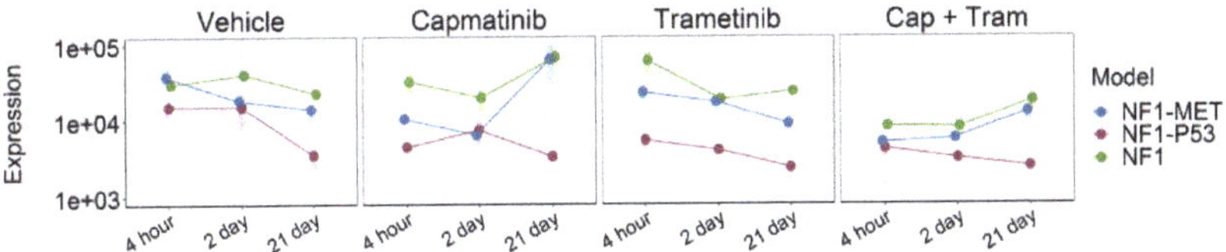

Figure 6. ERK reactivation or pathway indifference drive resistance to kinase inhibition. Phospho-ERK1/2 T202/Y204 expression in each genetic model after 21 days of (**A**) vehicle, (**B**) capmatinib, (**C**) trametinib, or (**D**) combination treatment. (**E**) phospho-ERK1/2 T202/Y204 expression values were measured by RPPA and calculated as the absolute protein expression normalized to the total protein input and background. Points represent the mean ($n = 3$) for each treatment-genotype-time group. Shaded bars represent +/− SEM.

3.7. AXL NFkB Co-Activation Associated with Therapy Resistance in MPNSTs

As both AXL and NFkB were highly activated in several treatment conditions, including therapy-resistant tumor growth, we sought to determine whether AXL and NFkB phosphorylation were correlated in our models. Recent studies in other cancer contexts suggest that AXL induces NFkB activation in response to a variety of therapies, including kinase inhibitors (22410775, 23474758, and 25568334). This novel therapy mechanism has been underreported to date. We determined that pAXL expression was tightly correlated to pNFkB (Spearmen's rank correlation rho = 0.729, p value = 3.91×10^{-21}) and grouped strongly by time point (Figure 7A,B). Expression was also grouped by treatment, as the phosphorylation of both proteins was highest in the doxorubin and combination capmatinib + trametinib treatment groups (Figure 7B), regardless of genotype (Figure 7A).

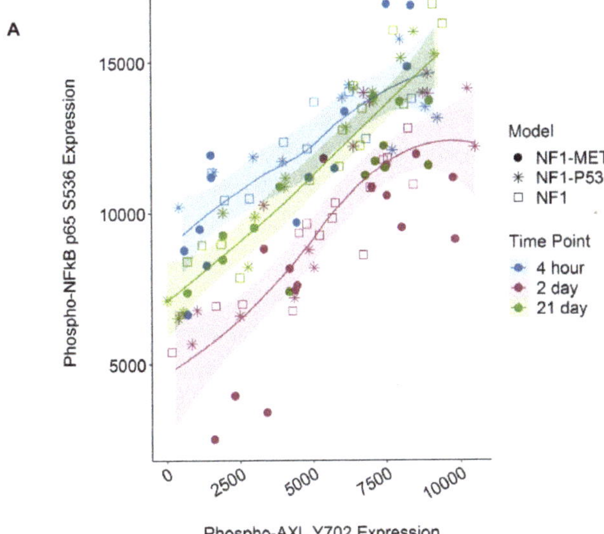

Figure 7. *Cont.*

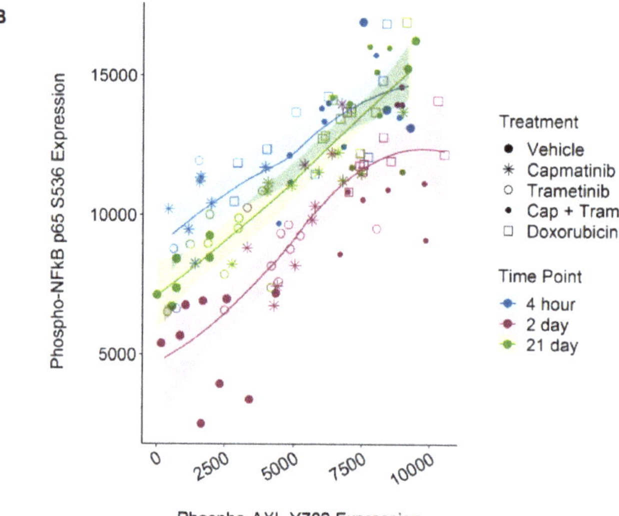

Figure 7. AXL Y702 and NFkB p65 S536 phosphorylation are highly correlated. Expression of phoshpo-AXL Y702 and phospho-NFkB p65 S536 plotted by (**A**) time and genotype group or (**B**) time and treatment group. Colors indicate treatment time and point shape indicates treatment or genotype groups. Lines indicate loess-predicted fit for each time point; shaded regions indicate 95% confidence intervals. Spearmen's rank correlation rho = 0.832, 0.835, and 0.881 with *p* value = 3.72×10^{-9}, 9.75×10^{-13}, and 6.40×10^{-15} for the 4-h, 2-day, and-21 day groups, respectively.

4. Discussion

Currently, there is no effective chemotherapy for NF1-related MPNSTs. Despite ongoing clinical trialing efforts, neither RTK nor downstream kinase inhibition has resulted in meaningful improvements in survival, despite well-founded attempts to target both the RAS/ERK and PI3K/AKT/mTOR pathways. Fundamentally, RAS deregulation as a result of *NF1* deficiency appears to be more difficult to target than constitutive RAS activation. One possible reason for this difference is that fewer discrete signaling dependencies exist with *NF1* tumor suppressor loss than cancers that are critically dependent on RAS signaling as a result of activating RAS or EGFR mutations. That is to say, NF1-related MPNSTs are less susceptible to the perturbation of oncogenic signaling, unless a genomic event such as *MET* or *HGF* amplification exhausts the negative feedback loops that drive kinome adaptation. Our data confirms substantially broad and redundant kinome adaptation in NF1-related MPNSTs in response to MET or MEK inhibition. What is even more impressive is the diversity of early and late response mediators which sit atop prominent signaling cascades that regulate growth, proliferation, inflammation, and apoptosis. These data strongly point to an evolutionary advantage in clonal selection for cell populations that maintain this degree of signaling plasticity. Based on our findings, any successful treatment strategy that relies solely on kinase inhibition will be difficult to sustain. Even in our most ideal treatment scenario where the potent MET inhibitor capmatinib suppressed growth in MET-addicted MPNSTs, kinome adaptation occurred within 21 days, leading to a slight resumption in growth by the end of the study time frame.

Proteomic profiling analysis provided significant insight into how the genomic context influenced the therapy response, particularly in the case of oncogene addiction. The "MET-addicted" NF1-MET tumor model is highly responsive to MET inhibition. We confirmed this finding with proteomic profiling by showing a sharp decrease in MET activation and downstream signaling immediately after capmatinib treatment. Even though MET inhibition was sustained at 2 days, strong AXL and AKT activation indicated the initiation of pathway reactivation and pathway bypass signaling. After 21 days of capmatinib treatment, resistant populations reprogrammed the kinome via AXL and EGFR. Besides activation of the AKT/mTOR pathway, ERK reactivation was consistently present in NF1-MET tumors, even in the minor cell populations that survived combined MET-MEK inhibition. In contrast to the MET-addicted tumors, few, if any, signaling dependencies were present in the capmatinib-treated NF1-P53 tumorgrafts. Interestingly, the pace and strength of kinome adaptations in NF1-P53 tumors were considerably reduced compared to NF1-MET and NF1 tumors. One possible reason for this observation is that MET inhibition failed to cause sufficient cellular stress to necessitate broad kinome adaptation in innately-resistant NF1-P53 tumors. MEK inhibition was confounded by a higher degree of innate resistance than MET inhibition and resulted in greater response heterogeneity and variability in kinome activation within genotype groups. As observed with MET inhibition in NF1-P53 tumors, less robust kinome activation was observed in response to trametinib in NF1-P53 tumors.

In both capmatinib- and trametinib-treated tumors, inflammatory signaling was present at the 4-h time point, whereas kinase signaling associated with proliferation and invasion dominated at 2 days and 21 days. Inflammation has not been widely studied in MPNSTs; however, it is a key determinant of neurofibroma progression and Schwann cell homeostasis [40–42]. Based on our data, all treatments were associated with an initial inflammatory response directly mediated by the kinome. Multiple targets were implicated, including key members of the JAK/STAT signaling cascade; however, NFkB was the most consistently activated target. NFkB activation results in pleiotropic effects, including broad transcriptional activation, cytokine production, and cell survival. It is an early response element to cellular stress with a known ability to activate multiple kinases. How NFkB contributes to kinome adaptation is currently unknown. These results suggest that further investigations into the inflammatory signaling and the impact of the tumor microenvironment may identify additional therapeutic targets for NF1-related MPNSTs.

AXL receptor activation was a consistent kinome adaptation observed in all of the MPNST models in response to kinase inhibition and doxorubicin treatment. Recently, AXL has been implicated in therapy resistance to multiple targeted therapies and cancer types, including MPNST. Resistance is often mediated through AXL dimerization with other RTKs, leading to the bypass of the RTK inhibitor effect [43,44]. For example, in ovarian tumors, AXL dimerizes with MET, EGFR, and HER2, leading to sustained ERK activation [45]. In response to ERK and MEK inhibition, AXL/MITF-mediated drug resistance is observed among mutant BRAF and NRAS melanoma cell lines [46]. AXL overexpression has also been observed in resistance to cytotoxic chemotherapies, such as docetaxel, in prostate cancer [47]. A principal role of AXL appears to be sustaining a mesenchymal phenotype, which is a mechanism of resistance to diverse anticancer therapies [43]. We demonstrated that AXL and NFkB activation are highly correlated, regardless of the treatment type or model genotype. These results strongly point to a unifying mechanism of therapy resistance in NF1-related MPNSTs. Therefore, further investigations into the efficacy of AXL or NFkB inhibition in conjunction with RAS pathway inhibitors in MPNSTs are warranted.

In summary, the phosphoproteomic profiling of MET and MEK inhibition revealed distinct pathways of drug resistance involving AXL activation, ERK reactivation, and inflammatory kinase signaling. As expected from our previous studies, P53-null MPNSTs were innately resistant to kinase inhibition and demonstrated the most heterogeneous kinome responses; however, combined MET-MEK inhibition exposed potential vulnerabilities in these tumors. As with other RAS-activated tumors, pathway reactivation and bypass signaling are common mechanisms of therapeutic resistance. The results in this study point toward specific vulnerable signaling nodes in MPNSTs that may be exploited through novel combination therapeutic approaches.

Supplementary Materials: The following are available online at http://www.mdpi.com/2073-4425/11/3/331/s1, Figure S1: MET inhibition reveals differential innate and adaptive kinome reprogramming. Figure S2: MEK inhibition reveals differential innate and adaptive kinome reprogramming. Figure S3: Combination MEK and MET inhibition reveals differential innate and adaptive kinome reprogramming. Figure S4: Doxorubicin reveals differential innate and adaptive kinome reprogramming. Figure S5: AXL, PKCζ/λ, p38, and NFkB phosphorylation with treatment over time.

Author Contributions: Conceptualization, J.L.G., C.R.G., and M.R.S.; methodology, J.L.G., M.G.P., C.J.E., L.T., and J.W.; formal analysis, J.L.G., E.W., and Z.B.M.; writing—original draft preparation, J.L.G., C.R.G., and M.R.S.; writing—review and editing, J.L.G., C.R.G., E.F.P., and M.R.S.; supervision, C.R.G. and M.R.S.; project administration, C.R.G. and M.R.S.; funding acquisition, J.L.G., C.R.G., and M.R.S. All authors have read and agreed to the published version of the manuscript.

Funding: Funding for this research was made possible by the Children's Tumor Foundation, NF Michigan, and the Van Andel Institute.

Acknowledgments: We would like to thank Bryn Eagleson and the VARI Vivarium for their continuous dedication. This publication was supported by an Agreement from The Johns Hopkins University School of Medicine and the Neurofibromatosis Therapeutic Acceleration Program (NTAP). Its contents are solely the responsibility of the authors and do not necessarily represent the official views of The Johns Hopkins University School of Medicine.

Conflicts of Interest: The authors declare no conflicts of interest.

References

1. Konieczkowski, D.J.; Johannessen, C.M.; Garraway, L.A. A Convergence-Based Framework for Cancer Drug Resistance. *Cancer Cell* **2018**, *33*, 801–815. [CrossRef] [PubMed]

2. Lammert, M.; Friedman, J.M.; Kluwe, L.; Mautner, V.F. Prevalence of neurofibromatosis 1 in German children at elementary school enrollment. *Arch. Dermatol.* **2005**, *141*, 71–74. [CrossRef] [PubMed]

3. Evans, D.G.; Howard, E.; Giblin, C.; Clancy, T.; Spencer, H.; Huson, S.M.; Lalloo, F. Birth incidence and prevalence of tumor-prone syndromes: Estimates from a UK family genetic register service. *Am. J. Med. Genet. A* **2010**, *152*, 327–332. [CrossRef] [PubMed]

4. Upadhyaya, M.; Spurlock, G.; Majounie, E.; Griffiths, S.; Forrester, N.; Baser, M.; Huson, S.M.; Gareth Evans, D.; Ferner, R. The heterogeneous nature of germline mutations in NF1 patients with malignant peripheral serve sheath tumours (MPNSTs). *Hum. Mutat.* **2006**, *27*, 716. [CrossRef]

5. Miller, S.J.; Rangwala, F.; Williams, J.; Ackerman, P.; Kong, S.; Jegga, A.G.; Kaiser, S.; Aronow, B.J.; Frahm, S.; Kluwe, L.; et al. Large-Scale Molecular Comparison of Human Schwann Cells to Malignant Peripheral Nerve Sheath Tumor Cell Lines and Tissues. *Cancer Res.* **2006**, *66*, 2584–2591. [CrossRef]

6. Peacock, J.D.; Pridgeon, M.G.; Tovar, E.A.; Essenburg, C.J.; Bowman, M.; Madaj, Z.; Koeman, J.; Boguslawski, E.A.; Grit, J.; Dodd, R.D.; et al. Genomic Status of MET Potentiates Sensitivity to MET and MEK Inhibition in NF1-Related Malignant Peripheral Nerve Sheath Tumors. *Cancer Res.* **2018**, *78*, 3672–3687. [CrossRef]

7. Mayes, D.A.; Rizvi, T.A.; Cancelas, J.A.; Kolasinski, N.T.; Ciraolo, G.M.; Stemmer-Rachamimov, A.O.; Ratner, N. Perinatal or adult Nf1 inactivation using tamoxifen-inducible PlpCre each cause neurofibroma formation. *Cancer Res.* **2011**, *71*, 4675–4685. [CrossRef]

8. Cichowski, K.; Shih, T.S.; Schmitt, E.; Santiago, S.; Reilly, K.; McLaughlin, M.E.; Bronson, R.T.; Jacks, T. Mouse models of tumor development in neurofibromatosis type 1. *Science* **1999**, *286*, 2172–2176. [CrossRef]

9. Brekke, H.R.; Kolberg, M.; Skotheim, R.I.; Hall, K.S.; Bjerkehagen, B.; Risberg, B.; Domanski, H.A.; Mandahl, N.; Liestol, K.; Smeland, S.; et al. Identification of p53 as a strong predictor of survival for patients with malignant peripheral nerve sheath tumors. *Neuro-Oncology* **2009**, *11*, 514–528. [CrossRef]

10. Adelaiye-Ogala, R.; Budka, J.; Damayanti, N.P.; Arrington, J.; Ferris, M.; Hsu, C.C.; Chintala, S.; Orillion, A.; Miles, K.M.; Shen, L.; et al. EZH2 Modifies Sunitinib Resistance in Renal Cell Carcinoma by Kinome Reprogramming. *Cancer Res.* **2017**, *77*, 6651–6666. [CrossRef]

11. Stuhlmiller, T.J.; Miller, S.M.; Zawistowski, J.S.; Nakamura, K.; Beltran, A.S.; Duncan, J.S.; Angus, S.P.; Collins, K.A.; Granger, D.A.; Reuther, R.A.; et al. Inhibition of Lapatinib-Induced Kinome Reprogramming in ERBB2-Positive Breast Cancer by Targeting BET Family Bromodomains. *Cell Rep.* **2015**, *11*, 390–404. [CrossRef]

12. Stuhlmiller, T.J.; Earp, H.S.; Johnson, G.L. Adaptive reprogramming of the breast cancer kinome. *Clin. Pharmacol. Ther.* **2014**, *95*, 413–415. [CrossRef] [PubMed]

13. Johnson, G.L.; Stuhlmiller, T.J.; Angus, S.P.; Zawistowski, J.S.; Graves, L.M. Molecular pathways: Adaptive kinome reprogramming in response to targeted inhibition of the BRAF-MEK-ERK pathway in cancer. *Clin. Cancer Res.* **2014**, *20*, 2516–2522. [CrossRef]

14. Wagner, J.P.; Wolf-Yadlin, A.; Sevecka, M.; Grenier, J.K.; Root, D.E.; Lauffenburger, D.A.; MacBeath, G. Receptor tyrosine kinases fall into distinct classes based on their inferred signaling networks. *Sci. Signal.* **2013**, *6*, ra58. [CrossRef]

15. Chong, C.R.; Janne, P.A. The quest to overcome resistance to EGFR-targeted therapies in cancer. *Nat. Med.* **2013**, *19*, 1389–1400. [CrossRef] [PubMed]

16. Wilson, T.R.; Fridlyand, J.; Yan, Y.; Penuel, E.; Burton, L.; Chan, E.; Peng, J.; Lin, E.; Wang, Y.; Sosman, J.; et al. Widespread potential for growth-factor-driven resistance to anticancer kinase inhibitors. *Nature* **2012**, *487*, 505–509. [CrossRef] [PubMed]

17. Pin, E.; Federici, G.; Petricoin, E.F., III. Preparation and use of reverse protein microarrays. *Curr. Protoc. Protein Sci.* **2014**, *75*, 27.7.1–27.7.29. [CrossRef]

18. Signore, M.; Reeder, K.A. Antibody validation by Western blotting. *Methods Mol. Biol.* **2012**, *823*, 139–155. [CrossRef]

19. Rapkiewicz, A.; Espina, V.; Zujewski, J.A.; Lebowitz, P.F.; Filie, A.; Wulfkuhle, J.; Camphausen, K.; Petricoin, E.F., III; Liotta, L.A.; Abati, A. The needle in the haystack: Application of breast fine-needle aspirate samples to quantitative protein microarray technology. *Cancer* **2007**, *111*, 173–184. [CrossRef]

20. Pierobon, M.; Silvestri, A.; Spira, A.; Reeder, A.; Pin, E.; Banks, S.; Parasido, E.; Edmiston, K.; Liotta, L.; Petricoin, E. Pilot phase I/II personalized therapy trial for metastatic colorectal cancer: Evaluating the feasibility of protein pathway activation mapping for stratifying patients to therapy with imatinib and panitumumab. *J. Proteome Res.* **2014**, *13*, 2846–2855. [CrossRef]

21. Rao, U.N.; Sonmez-Alpan, E.; Michalopoulos, G.K. Hepatocyte growth factor and c-MET in benign and malignant peripheral nerve sheath tumors. *Hum. Pathol.* **1997**, *28*, 1066–1070. [CrossRef]

22. Mantripragada, K.K.; Spurlock, G.; Kluwe, L.; Chuzhanova, N.; Ferner, R.E.; Frayling, I.M.; Dumanski, J.P.; Guha, A.; Mautner, V.; Upadhyaya, M. High-Resolution DNA Copy Number Profiling of Malignant Peripheral Nerve Sheath Tumors Using Targeted Microarray-Based Comparative Genomic Hybridization. *Clin. Cancer Res.* **2008**, *14*, 1015–1024. [CrossRef] [PubMed]

23. Torres, K.E.; Zhu, Q.S.; Bill, K.; Lopez, G.; Ghadimi, M.P.; Xie, X.; Young, E.D.; Liu, J.; Nguyen, T.; Bolshakov, S.; et al. Activated MET is a molecular prognosticator and potential therapeutic target for malignant peripheral nerve sheath tumors. *Clin. Cancer Res.* **2011**, *17*, 3943–3955. [CrossRef]

24. Birchmeier, C.; Birchmeier, W.; Gherardi, E.; Vande Woude, G.F. Met, metastasis, motility and more. *Nat. Rev. Mol. Cell Biol.* **2003**, *4*, 915–925. [CrossRef]

25. Orian-Rousseau, V.; Morrison, H.; Matzke, A.; Kastilan, T.; Pace, G.; Herrlich, P.; Ponta, H. Hepatocyte growth factor-induced Ras activation requires ERM proteins linked to both CD44v6 and F-actin. *Mol. Biol. Cell* **2007**, *18*, 76–83. [CrossRef]

26. Monkkonen, T.; Debnath, J. Inflammatory signaling cascades and autophagy in cancer. *Autophagy* **2018**, *14*, 190–198. [CrossRef]

27. Konieczkowski, D.J.; Johannessen, C.M.; Abudayyeh, O.; Kim, J.W.; Cooper, Z.A.; Piris, A.; Frederick, D.T.; Barzily-Rokni, M.; Straussman, R.; Haq, R.; et al. A melanoma cell state distinction influences sensitivity to MAPK pathway inhibitors. *Cancer Discov.* **2014**, *4*, 816–827. [CrossRef]

28. Dombi, E.; Baldwin, A.; Marcus, L.J.; Fisher, M.J.; Weiss, B.; Kim, A.; Whitcomb, P.; Martin, S.; Aschbacher-Smith, L.E.; Rizvi, T.A.; et al. Activity of Selumetinib in Neurofibromatosis Type 1-Related Plexiform Neurofibromas. *N. Engl. J. Med.* **2016**, *375*, 2550–2560. [CrossRef]

29. Fischer-Huchzermeyer, S.; Chikobava, L.; Stahn, V.; Zangarini, M.; Berry, P.; Veal, G.J.; Senner, V.; Mautner, V.F.; Harder, A. Testing ATRA and MEK inhibitor PD0325901 effectiveness in a nude mouse model for human MPNST xenografts. *BMC Res. Notes* **2018**, *11*, 520. [CrossRef]

30. Dodd, R.D.; Mito, J.K.; Eward, W.C.; Chitalia, R.; Sachdeva, M.; Ma, Y.; Barretina, J.; Dodd, L.; Kirsch, D.G. NF1 deletion generates multiple subtypes of soft-tissue sarcoma that respond to MEK inhibition. *Mol. Cancer Ther.* **2013**, *12*, 1906–1917. [CrossRef]

31. Coumans, J.V.F.; Davey, R.J.; Moens, P.D.J. Cofilin and profilin: Partners in cancer aggressiveness. *Biophys. Rev.* **2018**, *10*, 1323–1335. [CrossRef] [PubMed]

32. Wabnitz, G.H.; Kirchgessner, H.; Jahraus, B.; Umansky, L.; Shenolikar, S.; Samstag, Y. Protein Phosphatase 1alpha and Cofilin Regulate Nuclear Translocation of NF-kappaB and Promote Expression of the Anti-Inflammatory Cytokine Interleukin-10 by T Cells. *Mol. Cell. Biol.* **2018**, *38*, e00041-18. [CrossRef] [PubMed]

33. Smalley, T.; Islam, S.M.A.; Apostolatos, C.; Apostolatos, A.; Acevedo-Duncan, M. Analysis of PKC-zeta protein levels in normal and malignant breast tissue subtypes. *Oncol. Lett.* **2019**, *17*, 1537–1546. [CrossRef]

34. Reina-Campos, M.; Diaz-Meco, M.T.; Moscat, J. The Dual Roles of the Atypical Protein Kinase Cs in Cancer. *Cancer Cell* **2019**, *36*, 218–235. [CrossRef] [PubMed]

35. Tacar, O.; Sriamornsak, P.; Dass, C.R. Doxorubicin: An update on anticancer molecular action, toxicity and novel drug delivery systems. *J. Pharm. Pharmacol.* **2013**, *65*, 157–170. [CrossRef]

36. Tian, Z.; Wang, X.; Liu, Z.; Wang, J.; Yao, W.; Zhao, Y.; Gao, S.; Zhang, P.; Ge, H. Safety and efficacy of combination therapy with apatinib and doxorubicin in metastatic soft tissue sarcomas: An observational study from multiple institutions. *Cancer Manag. Res.* **2019**, *11*, 5293–5300. [CrossRef]

37. Mir, O.; Brodowicz, T.; Italiano, A.; Wallet, J.; Blay, J.Y.; Bertucci, F.; Chevreau, C.; Piperno-Neumann, S.; Bompas, E.; Salas, S.; et al. Safety and efficacy of regorafenib in patients with advanced soft tissue sarcoma (REGOSARC): A randomised, double-blind, placebo-controlled, phase 2 trial. *Lancet Oncol.* **2016**, *17*, 1732–1742. [CrossRef]

38. Tap, W.D.; Jones, R.L.; Van Tine, B.A.; Chmielowski, B.; Elias, A.D.; Adkins, D.; Agulnik, M.; Cooney, M.M.; Livingston, M.B.; Pennock, G.; et al. Olaratumab and doxorubicin versus doxorubicin alone for treatment of soft-tissue sarcoma: An open-label phase 1b and randomised phase 2 trial. *Lancet* **2016**, *388*, 488–497. [CrossRef]

39. Peacock, J.D.; Cherba, D.; Kampfschulte, K.; Smith, M.K.; Monks, N.R.; Webb, C.P.; Steensma, M. Molecular-guided therapy predictions reveal drug resistance phenotypes and treatment alternatives in malignant peripheral nerve sheath tumors. *J. Transl. Med.* **2013**, *11*, 213. [CrossRef]

40. Hakozaki, M.; Tajino, T.; Konno, S.; Kikuchi, S.; Yamada, H.; Yanagisawa, M.; Nishida, J.; Nagasawa, H.; Tsuchiya, T.; Ogose, A.; et al. Overexpression of cyclooxygenase-2 in malignant peripheral nerve sheath tumor and selective cyclooxygenase-2 inhibitor-induced apoptosis by activating caspases in human malignant peripheral nerve sheath tumor cells. *PLoS ONE* **2014**, *9*, e88035. [CrossRef]

41. Xu, J.; Zhang, H.; Li, C.; Du, H.; Shu, M.; Jia, J. Activation of PLCgamma/AKT/IkappaBalpha/p65 signaling increases inflammation in mast cells to promote growth of cutaneous neurofibroma. *Life Sci.* **2019**, *239*, 117079. [CrossRef] [PubMed]

42. Choi, K.; Komurov, K.; Fletcher, J.S.; Jousma, E.; Cancelas, J.A.; Wu, J.; Ratner, N. An inflammatory gene signature distinguishes neurofibroma Schwann cells and macrophages from cells in the normal peripheral nervous system. *Sci. Rep.* **2017**, *7*, 43315. [CrossRef] [PubMed]

43. Schoumacher, M.; Burbridge, M. Key Roles of AXL and MER Receptor Tyrosine Kinases in Resistance to Multiple Anticancer Therapies. *Curr. Oncol. Rep.* **2017**, *19*, 19. [CrossRef] [PubMed]

44. Hoj, J.P.; Mayro, B.; Pendergast, A.M. A TAZ-AXL-ABL2 Feed-Forward Signaling Axis Promotes Lung Adenocarcinoma Brain Metastasis. *Cell Rep.* **2019**, *29*, 3421–3434. [CrossRef]

45. Antony, J.; Tan, T.Z.; Kelly, Z.; Low, J.; Choolani, M.; Recchi, C.; Gabra, H.; Thiery, J.P.; Huang, R.Y.J. The GAS6-AXL signaling network is a mesenchymal (Mes) molecular subtype–specific therapeutic target for ovarian cancer. *Sci. Signal.* **2016**, *9*, ra97. [CrossRef]

46. Muller, J.; Krijgsman, O.; Tsoi, J.; Robert, L.; Hugo, W.; Song, C.; Kong, X.; Possik, P.A.; Cornelissen-Steijger, P.D.; Geukes Foppen, M.H.; et al. Low MITF/AXL ratio predicts early resistance to multiple targeted drugs in melanoma. *Nat. Commun.* **2014**, *5*, 5712. [CrossRef]

47. Lin, J.Z.; Wang, Z.J.; De, W.; Zheng, M.; Xu, W.Z.; Wu, H.F.; Armstrong, A.; Zhu, J.G. Targeting AXL overcomes resistance to docetaxel therapy in advanced prostate cancer. *Oncotarget* **2017**, *8*, 41064–41077. [CrossRef]

 *genes*

Article

Integrative Analysis Identifies Candidate Tumor Microenvironment and Intracellular Signaling Pathways that Define Tumor Heterogeneity in NF1

Jineta Banerjee [1,†], Robert J Allaway [1,†], Jaclyn N Taroni [2], Aaron Baker [1,3,4], Xiaochun Zhang [5], Chang In Moon [5], Christine A Pratilas [6], Jaishri O Blakeley [6,7], Justin Guinney [1], Angela Hirbe [5], Casey S Greene [2,8] and Sara JC Gosline [1,*]

1   Computational Oncology, Sage Bionetworks, Seattle, WA 98121, USA;
    jineta.banerjee@sagebionetworks.org (J.B.); robert.allaway@sagebase.org (R.J.A.);
    aabaker99@gmail.com (A.B.); justin.guinney@sagebase.org (J.G.)
2   Childhood Cancer Data Lab, Alex's Lemonade Stand Foundation, Philadelphia, PA 19102, USA;
    jaclyn.n.taroni@gmail.com (J.N.T.); greenescientist@gmail.com (C.S.G.)
3   Department of Computer Sciences, University of Wisconsin-Madison, Madison, WI 53715, USA
4   Morgridge Institute for Research, Madison, WI 53715, USA
5   Division of Oncology, Washington University Medical School, St. Louis, MO 63110, USA;
    zhang.x@wustl.edu (X.Z.); moonchangin@wustl.edu (C.I.M.); hirbea@wustl.edu (A.H.)
6   Sidney Kimmel Comprehensive Cancer Center and Department of Oncology, Johns Hopkins University
    School of Medicine, Baltimore, MD 21287, USA; cpratil1@jhmi.edu (C.A.P.); jblakel3@jhmi.edu (J.O.B.)
7   Neurology, Neurosurgery and Oncology, Johns Hopkins University, Baltimore, MD 21287, USA
8   Department of Systems Pharmacology and Translational Therapeutics, Perelman School of Medicine,
    University of Pennsylvania, Philadelphia, PA 19104, USA
*   Correspondence: sara.gosline@sagebionetworks.org; Tel.: +1-206-928-8244
†   These authors have equal contribution.

Received: 15 January 2020; Accepted: 19 February 2020; Published: 21 February 2020

Abstract: Neurofibromatosis type 1 (NF1) is a monogenic syndrome that gives rise to numerous symptoms including cognitive impairment, skeletal abnormalities, and growth of benign nerve sheath tumors. Nearly all NF1 patients develop cutaneous neurofibromas (cNFs), which occur on the skin surface, whereas 40–60% of patients develop plexiform neurofibromas (pNFs), which are deeply embedded in the peripheral nerves. Patients with pNFs have a ~10% lifetime chance of these tumors becoming malignant peripheral nerve sheath tumors (MPNSTs). These tumors have a severe prognosis and few treatment options other than surgery. Given the lack of therapeutic options available to patients with these tumors, identification of druggable pathways or other key molecular features could aid ongoing therapeutic discovery studies. In this work, we used statistical and machine learning methods to analyze 77 NF1 tumors with genomic data to characterize key signaling pathways that distinguish these tumors and identify candidates for drug development. We identified subsets of latent gene expression variables that may be important in the identification and etiology of cNFs, pNFs, other neurofibromas, and MPNSTs. Furthermore, we characterized the association between these latent variables and genetic variants, immune deconvolution predictions, and protein activity predictions.

Keywords: neurofibromatosis type 1; nerve sheath tumor; cancer; latent variables; machine learning; supervised learning; transfer learning; random forest; metaVIPER; tumor deconvolution

---

1. Introduction

Neurofibromatosis type 1 is a rare disease and a member of the family of RASopathies (diseases caused by germline mutations in genes that encode components or regulators of the

Ras/mitogen-activated protein kinase (MAPK) pathway) that occurs in approximately 1 in 3000 patients worldwide and gives rise to cognitive impairment, skeletal abnormalities, and various nerve tumors including gliomas and neurofibromas and is caused by a mutation or deletion in one NF1 allele [1–3]. Nerve sheath tumors affect more than 90% of NF1 patients, mostly in the form of cutaneous neurofibromas (cNFs). These tumors grow at the skin surface and can range in number from 10s to 100s of tumors in a given patient [4]. Neurofibromas that occur deeper in the body, including subcutaneous neurofibromas or plexiform neurofibromas (pNFs), occur in 40–60% of NF1 patients and can cause pain and disfigurement among other symptoms [1,4]. Patients with pNFs have a 10% lifetime risk of these tumors developing into malignant peripheral nerve sheath tumors (MPNSTs) which have a 5-year survival rate of 40–50% [5,6].

The rise of high-throughput genomic and transcriptomic sequencing has enabled many advances in understanding the molecular etiology of NF1 tumor types [7–11]. Genomic studies of NF1-derived tumors, particularly MPNSTs, have identified key features of tumor growth that could point to potential therapeutic avenues. For example, genomic approaches were recently used to identify the loss of function of polycomb repressor 2 complex components *EED* or *SUZ12* genes, alongside *CDKN2A* and *NF1* gene mutations as crucial co-mediators of MPNST transcriptional dysregulation, pathogenesis, and sensitivity to bromodomain (BRD4) inhibitors [9,11]. Others using genomic approaches to explore nerve sheath tumor biology identified *MET* and *HGF* gene amplifications in MPNSTs. Furthermore, models of *MET*-amplified MPNSTs were subsequently sensitive to the MET inhibitor capmatinib [7]. Transcriptomics-focused approaches have also identified molecular features such as MEK signaling, type 1 interferon signaling, and Aurora kinase A as putative therapeutic targets in NF1 tumors [12–14]. Taken together, these and other studies suggest that an integrative approach that combines multiple transcriptomic and genomic datasets might be well poised to identify new therapeutic avenues in MPNSTs and other NF1-related nerve sheath tumors.

Previous genomic profiling studies have demonstrated that many NF1 nerve sheath tumors (MPNST being the exception) are genetically quiet [3,15–17] and lack specific signatures that are predictive of drug response. An approach to compensate for the lack of genetic hotspots in NF1 tumors is to focus on combinations of transcriptomic signatures that may be unique to specific tumor types. In other tumor types, transcriptomic landscapes across cancer datasets [18,19] have shown that combining RNA-seq data from similar diseases can identify expression profiles that correlate with prognosis [20], predict drug response [21,22], or identify key tumor biology [23].

However, comprehensive analysis of genomic data in NF1 tumors is limited [3]. To enable a larger landscape analysis of NF1 nerve sheath tumors, samples from different studies were collated into a single resource as part of the NF Open Science Initiative, a collaboration between NF-related funding agencies. This resource has been made publicly available through the NF Data Portal, which houses high-throughput data for neurofibromatosis 1, neurofibromatosis 2, and schwannomatosis [24].

In this work, we reprocessed and analyzed RNA-seq data from 77 NF1 nerve sheath tumor samples to understand the biological differences that give rise to distinct tumor types in NF1 patients. Given the low sample size compared to the large feature space of RNA-seq data, we applied a transfer learning-inspired approach to meaningfully reduce the feature space with minimal decrease in information content. Transfer learning techniques can leverage large well curated datasets such as recount2 [25] to identify latent variables (LVs)—groups of genes derived from larger repositories of gene expression datasets that exhibit common transcriptomic patterns relevant to a specific subset of samples [26,27]. Although many of these LVs are composed of genes that map to known signatures (i.e., documented in Kyoto Encyclopedia of Genes and Genomes (KEGG) database, and Gene Ontology (GO) consortium database), others are uncharacterized and may allow the detection of novel and meaningful transcriptomic patterns in NF data. As a result, reduction of gene-based expression data to individual latent variables can provide multiple benefits—LVs can highlight differences in known biology in sets of samples, they can uncover previously unknown biology, and they can reduce the impact of technical and experimental differences across multiple datasets [26]. We transferred a machine learning model [27] trained on recount2 to assess LV expression in the NF1 nerve sheath tumor dataset. We then used supervised machine learning with random forests [28] to isolate combinations of such LVs to identify specific molecular signatures that may describe the underlying biology unique to each of the tumor types: cNFs, pNFs, undefined neurofibromas (NFs), and MPNSTs. Finally, we integrated this information with sample-matched variant data, immune cell signatures [29–31], and protein activity predictions [32] to provide additional biological context to the most important latent variables. This approach revealed biological patterns that underlie different NF1 nerve sheath tumor types and candidate genes and cellular signatures associated with NF1 tumor heterogeneity.

2. Materials and Methods

2.1. Materials Implementation and Data and Code Availability

All analyses were performed using the R programming language. A comprehensive list of packages used and their versions are available in an *renv* lockfile in the GitHub repository. Key packages used include "tidyverse" [33], "PLIER" [26], "synapser" [34], "tximport" [35], "immundeconv" [35], and "viper" [32]. All data analyzed in this article are stored on the NF Data Portal [24] (http://nfdataportal.org) with analyses stored at http://synapse.org/nf1landscape. To recapitulate the analysis from these data, all relevant code can be found at https://github.com/Sage-Bionetworks/NF_LandscapePaper_2019.

2.2. Sequencing Data Collection and Processing

Gene expression data were collected from four independent studies and processed via a workflow at https://github.com/Sage-Bionetworks/rare-disease-workflows/tree/master/rna-seq-workflow to be stored on the NF Data Portal (Table 1). Specifically, raw fastq files were downloaded from Synapse and transcripts were quantified using the Salmon pseudo-alignment tool [36] with Gencode V29 transcriptome. Links to specific datasets and the access teams required to download them can be found using Synapse ID *syn21221980*.

Table 1. Description of the gene expression datasets used in the present article.

Dataset Name	Synapse Project Name	Synapse Table Name	Synapse Access Team
WashU Biobank	Preclinical NF1-MPNST Platform Development (*syn11638893*)	WashU Biobank RNA-seq data	WUSTL MPNST PDX Data Access
JHU Biobank [37]	A Nerve Sheath Tumor Bank from Patients with NF1 (*syn4939902*)	Biobank RNASeq Data	JHU Biobank Data Access
cNF Patient Data [38]	Cutaneous Neurofibroma Data Resource (*syn4984604*)	cNF RNASeq Counts	CTF cNF Resource Data Access Group
CBTTC Data [39]	Children's Brain Tumor Tissue Consortium (*syn20629666*)	CBTTC RNASeq Counts	CBTTC Data Access Group

Genomic variant data were collected from exome-Seq [37] or whole-genome sequencing [38]. Variant call format (VCF) files were processed using "vcf2maf" (https://github.com/mskcc/vcf2maf) according to the workflow located at https://github.com/Sage-Bionetworks/rare-disease-workflows/tree/master/gene-variant-workflow and then uploaded to the NF Data Portal. A list of datasets and the access teams required to download them can be found in Table 2 or syn21266269.

Table 2. Description of the genomic variant datasets used in the present article.

Dataset Name	Assay	Synapse Table Name	Synapse Access Team	Synapse Project
JHU Biobank Exome-Seq Data	exomeSeq	Biobank ExomeSeq Data	JHU Biobank Data Access	A Nerve Sheath Tumor Bank from Patients with NF1
cNF WGS Data	wholeGenomeSeq	cNF WGS Harmonized Data	CTF cNF Resource Data Access Group	Cutaneous Neurofibroma Data Resource

2.3. Latent Variable Calculation and Selection

We analyzed transcriptomic data from NF in the context of latent variables from MultiPLIER, a machine learning resource designed to aid in rare disease analyses [27]. Raw transcriptomic data from NF were retrieved from the NF Data Portal, reprocessed, and stored in Synapse as described above. Salmon output files (quant.sf) files were imported into an R session and converted to HUGO gene names using the "tximport" [35] and "org.Hs.eg.db" [40] packages.

MultiPLIER reuses models that were trained on large public compendia. We retrieved a model [41] that was previously trained on the recount2 RNA-seq dataset [25,42] and then used it to assess the expression of latent variables in the pan-NF dataset. We retrieved code for this analysis from the public repository for MultiPLIER [27] (https://github.com/greenelab/multi-plier). To project the NF data into the MultiPLIER model, we used the GetNewDataB function. This analytical approach is described in more detail in a machine learning training module (https://github.com/AlexsLemonade/training-modules) produced by the Alex's Lemonade Stand Foundation's Childhood Cancer Data Lab under an open source license.

In addition, to ensure orthogonality in the final set of latent variables [43], we calculated the pairwise Pearson correlation of all latent variables by comparing the gene loadings (Figure S1), and identified non-self-correlations greater than 0.5. We eliminated one of each of these highly intercorrelated (Pearson correlation > 0.5) latent variables. This process resulted in a final set of 962 latent variables for further analysis. Code for our implementation of this material is available on GitHub at https://github.com/Sage-Bionetworks/NF_LandscapePaper_2019.

We used the R function prcomp to compute the principal components for Figure 1A,C both using genes (Figure 1A) and latent variables (Figure 1C).

2.4. Generation of Ensemble of Random Forests for Feature Selection

To select gene expression patterns of interest, we used ensembles of random forests to sufficiently resample our modestly sized dataset. We compared random forest models built using gene expression data as well as latent variables.

2.4.1. Algorithm Implementation

The main algorithm was implemented using the "caret" and "randomforest" packages in R [44,45]. Figure S2 outlines the steps involved in the generation of the ensemble of random forests. Briefly, the full dataset was first split into 80% *model* set and 20% *independent test* set. The function createDataPartition was used to create balanced splits of the data according to the tumor type. We tuned two parameters to the random forest algorithm, *mtrys*, and *ntrees*, using an iterative approach, evaluating *mtrys* values of 1 to 100 and *ntrees* values of 250, 500, 1000, and 2000. We selected the optimal values ($mtry = 51$, $ntrees = 1000$) using fivefold cross-validation, using latent variables as input features. We then split the *model* training set further to generate 500 samples of *training* data (75%) and hold-out *test* data (25%) (balanced splits randomly sampled without replacement). Each of these *training* and *test* datasets were used to train separate random forests to obtain a distribution of F1 scores and feature importance scores

($n = 500$). Given our noisy dataset with limited sample size, the distribution of feature and F1 scores enabled estimation of confidence intervals for the feature importance as well as model performance.

2.4.2. Feature Selection

The importance of each feature was estimated using raw importance scores that measure the change in correctly classified "class" due to random permutation of the values for the feature. To select the top features for a specific class (i.e., tumor type), we calculated the median *importance score* of each feature from the distribution of *raw importance scores* generated through 500 iterations of random forests. The top 40 features were selected according to the mean decrease of the Gini index. The union of top features from all classes was then used as a restricted feature set to train another 500 iterations of random forests as described above. However, each new forest trained with the restricted feature-set was tested using the *independent test set* to examine the performance of the model on a completely unseen dataset. For each class, the median F1 scores of the new ensemble of forests were compared to the previous ensemble of random forests. Improvement of median F1 scores for each class in the final ensemble of random forests compared to the earlier one suggested that the selected features from each class were sufficiently informative for their classification. This subset of features was then selected for downstream analyses.

2.5. Immune Subtype Prediction

To understand the relative immune infiltration across the nerve sheath tumors studied, we used two tumor deconvolution methods: CIBERSORT and MCP-counter, as implemented through the "immunedeconv" R package [29–31]. Analysis is located at https://github.com/Sage-Bionetworks/NF_LandscapePaper_2019 and results were uploaded to a Synapse table (syn21177277) that includes the tumor-specific immune cell scores for both algorithms as well as associated tumor metadata.

2.6. MetaVIPER

We applied the metaVIPER algorithm [32] to infer protein regulatory activity based on the tissue gene expression profiles. This algorithm builds transcriptional regulatory networks across the cancer genome atlas (TCGA) [32,46] and uses these to build consensus predictions for a sample of other origin. The resulting analysis is uploaded to Synapse and stored at syn21259610 along with tumor-specific metadata.

2.7. VIPER Correlation Clustering and Drug Enrichment Analysis

A heatmap and subclusters of latent variables that had similar VIPER protein predictions were generated using the "pheatmap" R package [47]. We observed five clear clusters of latent variables; these clusters were defined using the R cutree function to isolate the five clusters and their contents. We then calculated the mean correlation of each VIPER protein within each cluster to generate a consensus protein activity prediction for each latent variable cluster. Then, we used gene set enrichment analysis (via the "clusterProfiler" R package [48]) to assess whether drug targets were enriched in the five consensus protein lists. Drug-wise target lists were obtained from the Drug Target Explorer database [49]. Significant enrichment was defined as any positively enriched drug (i.e., a VIPER protein positively correlated with the latent variable cluster) with a Benjamini–Hochberg corrected p-value < 0.05. Results were plotted using "ggplot" and "enrichplot" packages [33,50]. To plot the LV cluster expression by tumor type, we calculated the mean expression of all latent variables for each cluster and for each tumor sample.

3. Results

We collected mRNA sequencing, exome sequencing, and whole genome sequencing data from previously published or publicly available resources as depicted in Table 3. We applied a combination of methods to identify biological mechanisms of interest in the samples that had patient-derived transcriptomic data. Briefly, we employed a transfer learning-inspired approach to group transcripts into latent variables (LVs), and then selected the LVs that best separate out tumors by tumor type using an ensemble of random forest models. We evaluated the selected LVs for patterns of immune cell gene expression, protein activity, and gene variants.

Table 3. Summary of individuals and samples for the NF1 nerve sheath tumors used in this study. All samples have gene expression data and a subset have genomic data derived from whole-exome sequencing or whole genome sequencing. Some neurofibromas did not have more specific pathologic subtyping information available, and therefore were classified as "undefined neurofibromas" or NFs.

Tumor Type	Individuals	Samples	# with Genomic Variant Data
Cutaneous Neurofibroma (cNF)	11	33	23
MPNST	13	13	1
Undefined Neurofibroma (NF)	12	12	11
Plexiform Neurofibroma (pNF)	19	19	5

3.1. Pan-NF Transcriptomic Analysis Identified Most Variable Latent Variables in NF1

To measure the diversity of the nerve sheath tumors at the transcriptomic level, we re-processed RNA-seq data from three published datasets [37–39] and one unpublished dataset. Despite having four types of nerve sheath tumors (cNFs, pNFs, NFs, and MPNSTs) across four datasets, we observed confounding batch effects (Figure 1A) as some tumor types (e.g., cNF and NF) were derived from separate studies. These batch effects, together with a lack of transcriptomic data from normal tissue from NF1 patients, precluded us from carrying out any meaningful differential gene expression analysis and motivated us to seek alternate approaches to pathway identification.

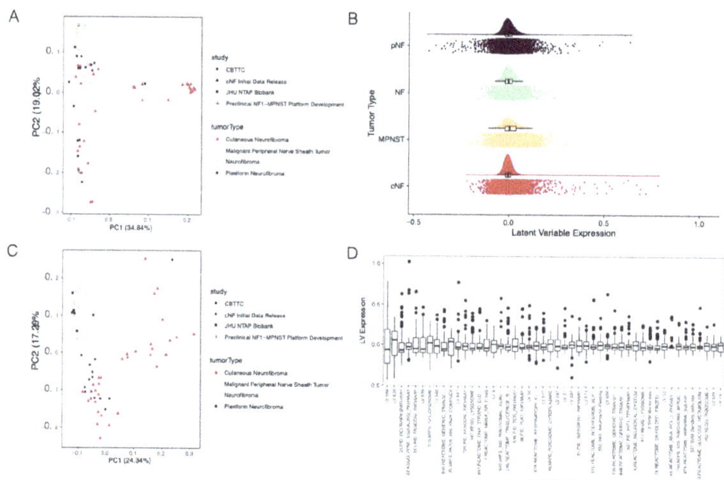

Figure 1. Transfer learning reduced dimensionality and added additional context to gene expression datasets. (**A**) Principal components analysis (PCA) of gene expression data indicated that counts-level data may have been batch confounded. (**B**) The relative distributions of latent variable expression across the four tumor types using a density plot indicated that the majority of latent variables (LVs) had an expression value near 0 and that the four tumor types had similar latent variable expression distributions. (**C**) PCA of LVs indicated that batch effects, although reduced, may still have existed in the LV data (**D**) A look at the 5% most variable LVs across the cohort of gene expression data indicated that the latent variables represented a wide swath of biological processes, as well as some LVs that had no clear association to a defined biological pathway.

To reduce the dimensionality of the RNA-seq data and increase the biological interpretability, we applied MultiPLIER, a transfer learning approach trained on the recount2 dataset, an independent dataset comprising thousands of gene expression experiments. This analysis quantified the expression of 962 latent variables (LVs) in all 77 samples using their gene expression data (Figure 1B, Figure S1, Table S1). We anticipated that as long as technical confounders in our dataset were independent of the technical confounders in the recount2 dataset, we would be able to find some latent variables that were independent of batch effects. We observed that expression of the latent variables was uniformly distributed across the four tumor types (Figure 1B). Furthermore, principal components analysis suggested that this method reduced the size of the previously observed batch effects (Figure 1A,C) as the within-cluster distances were significantly reduced (Figure S3, p-value = 2.1×10^{-182}, Wilcoxon test).

We then sought to identify which LVs characterize the differences between individual tumors. To find these, we examined the LVs with the 5% largest standard deviation across all LVs. Many of the resulting latent variables (Figure 1D) were found to be associated with known biological mechanisms including metabolic, immune, and transcription-related signatures. However, this analysis did not explore the association between tumor types and latent variables. Therefore, we performed additional analyses to (1) identify the gene expression features that best define each tumor type and (2) elucidate the biological mechanisms underlying the most important latent variables.

3.2. Ensemble of Random Forests Identified Latent Variables That Robustly Describe Individual Tumor Types

To identify these expression features, we used random forest models, an approach that is known for its ability to characterize complex decision spaces better than basic clustering approaches [28]. Random forests achieve this by building decision trees that account for both the values of the features and conditional logic at each "branch" of the tree. Although random forest models are generally used to classify samples, in this study we leveraged the capacity of random forest models to identify features in the data that provide meaningful information. Therefore, in this study, our focus was to identify important features rather than building the most accurate classifier.

We generated a random forest using NF-specific latent variable scores (Table S1) as described in the Materials and Methods section and depicted in Figure S2. Given the limited number of samples in our dataset, particularly for some tumor types, we anticipated high variability in the performance of our supervised learning model. To obtain a distribution of model performance, we generated 500 training sets (using stratified sampling from the full dataset without replacement), each training their own random forest. We used the distribution of model accuracy measurements and feature importance scores from these forests to estimate overall model performance and feature importance (Figure 2A). We then compared the class-specific accuracy measurements (median accuracy scores: cNF = 0.97, MPNST = 0.67, NF = 0.67, pNF = 0.57) (Figure 2A) to those from forests built with only the top 40 LVs from each class (a total of 98 LVs) from the first forest ensemble. Using the top 40 set of LVs, we were able to improve model performance for each tumor type on an independent test set (median accuracy scores: cNF = 1.00, MPNST = 0.86, NF = 0.86, pNF = 0.75) (Figure 2B). For each tumor type, reducing the feature set to these 98 LVs significantly improved the median accuracy scores for the models (Mood's median test p-values: cNF < 2.2×10^{-16}, MPNST = 3.789×10^{-11}, NF = 7.297×10^{-09}, pNF < 2.2×10^{-16}). This suggested that the 98 selected LVs with high importance scores for each tumor type were sufficient for characterizing the specific tumor types (Figure 2C).

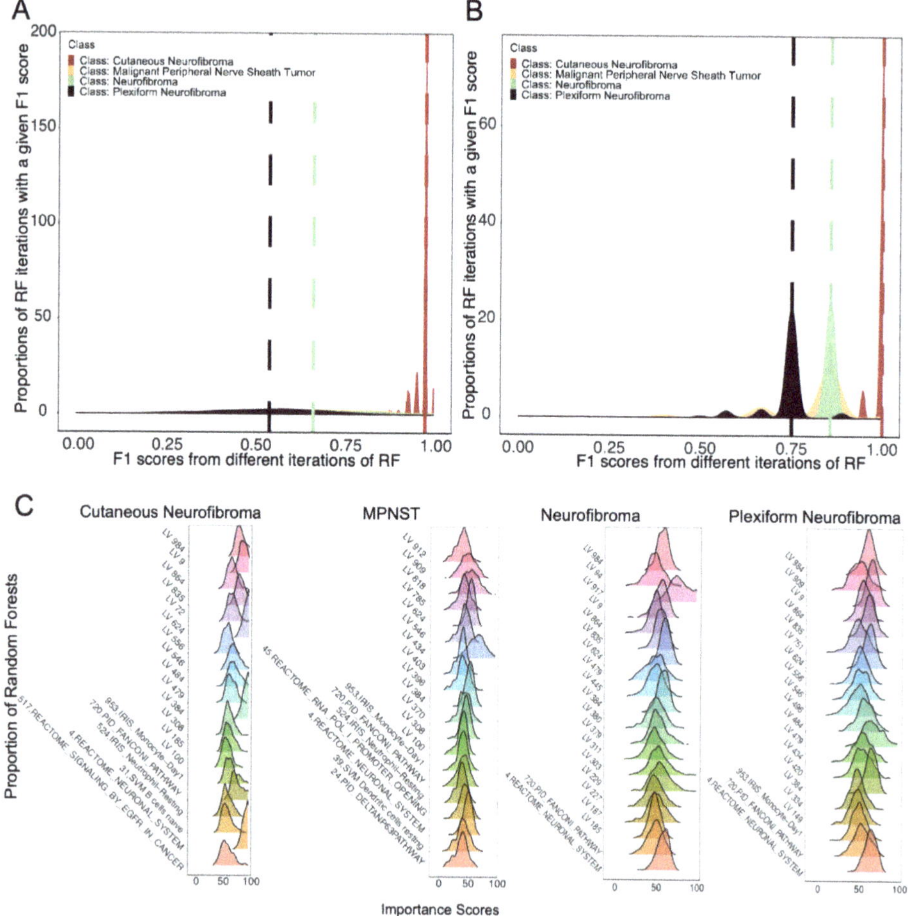

Figure 2. An ensemble of random forests selected the most important latent variables for classifying different tumor types in NF1. (**A**) Density plot showing the distribution of F1 scores of 500 iterations of independent random forest models using all latent variables. (**B**) Density plot showing the distribution of F1 scores of 500 iterations of independent random forest models trained using only the top 40 features with high importance scores for each class obtained from models included in (**A**). (**C**) Ridgeplots of top 20 latent variables selected by the random forest for each tumor type and their importance scores for each class that were selected for later analyses.

3.3. Selected Latent Variables Represented Distinct Biology of Nerve Sheath Tumor Types

To further probe the biology underpinning the distinction between NF1 tumor types, we focused on the 98 latent variables selected by the ensemble of random forests, depicted in Figure 3 and listed in Table S2. We evaluated the tumor-wise LV expression as well as the contributing genes (loadings) for each LV. For example, some latent variables that were selected by the random forest as relevant to predict all four tumor types (Figure 3A) showed differences in expression between various tumor types (dot plots in Figure 3Biii,Ciii), whereas others were less distinct (Figure 3Bi,Ci). By investigating the loadings of each LV, we tried to map them to known biological pathways. For each LV, the loadings of the constituent genes denoted the contribution of that gene to the particular LV. These gene lists associated with the selected LVs can be used to find testable candidates for each tumor type for downstream analysis. For example, one of the four latent variables predictive of all four tumor types was enriched in genes associated with neuronal signaling (Figure 3Bii), suggesting that this measurement is required for the model to distinguish various tumors as the presence of neuronal tissue likely varies across tumor types. Other functionally enriched latent variables, such as those depicted in Figure 3Bi,iii, implicate other biological pathways in tumor growth. However, many of the selected latent variables still remain uncharacterized (few examples shown in Figure 3Ci–iii).

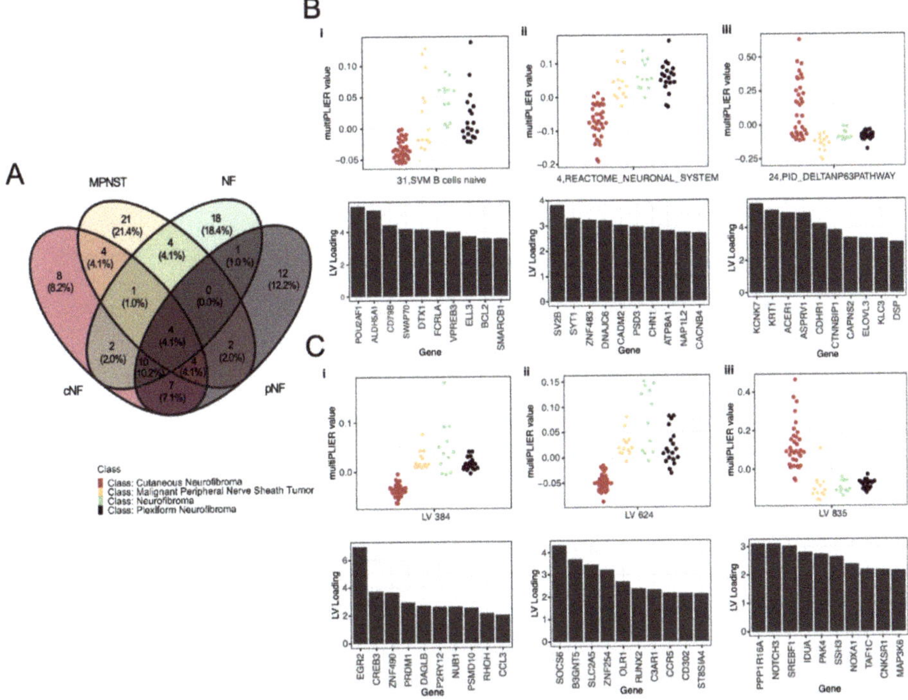

Figure 3. Selected latent variables (LVs) represented gene combinations unique to each tumor type. (**A**) Venn diagram showing the distribution of the top 40 LVs from each tumor type. (**B,C**) Total values of the LVs as measured by multiPLIER across samples are represented in the dot-plots, where color of the dots represents the tumor type ("Class" label colors described in the lower left). Loading values for the top 10 genes for each LV are represented in bar-plots below. The higher the loading, the greater impact that the gene expression had on the total multiPLIER value. (**B,i–iii**) Genes constituting the latent variables associated with known cell signaling pathways. (**C,i–iii**) Genes constituting the uncharacterized latent variables.

3.4. LV Scores May Be Attributed to Specific Gene Variants for Specific Tumor Types

To characterize the 73 LVs with no associated pathway information, we focused on the 40 samples from our dataset (Table 3) that had matched gene variant data (WGS or exome-Seq) to assess if there were any genes that, when mutated, caused a significant change in LV expression. The results of this calculation are found in Table S2.

We identified 22 latent variables that were significantly (Benjamini–Hochberg adjusted $p < 0.01$) associated with single gene variants (Table S3). These latent variables, along with the genes whose variants are associated with altered expression, are depicted in Figure 4A. This approach failed to identify any genes that were mutated across multiple tumor types. In the list of genes in which variants were associated with changes in latent variable expression, we identified nine genes with variants in cNFs and two genes with variants in one neurofibroma sample. An example of how these variants are associated with the expression of a latent variable is depicted in Figure 4B. Mutations in the nine cNF-variant genes are associated with lower expression of LV 851. Because most of these variants occur in cNF samples but not in the other tumor types, it is not surprising that this latent variable is down-regulated across all cNFs (Figure 4C). Figure 4D shows that the MAP1B gene, which has been reported to play a role in glioblastoma [51] but has not been studied in NF1-linked tumors, is a major contributor towards this LV.

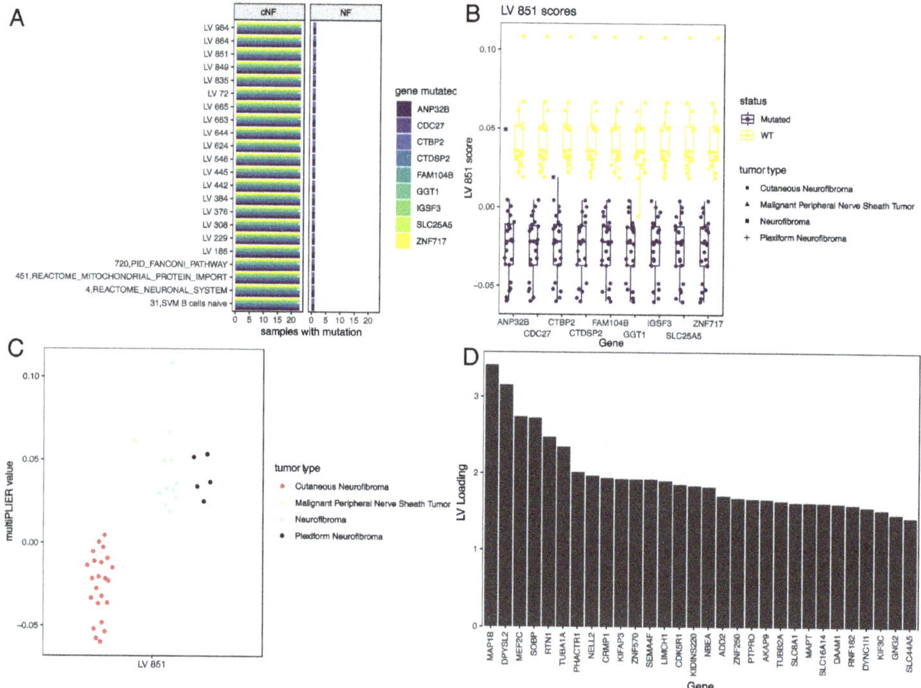

Figure 4. Some genes significantly distinguished expression of latent variables. (**A**) Latent variables (*y*-axis) whose values are significantly altered by mutations in specific genes. (**B**) MultiPLIER value of LV 851 across tumor samples. (**C**) MultiPLIER value of LV 851 across all samples. (**D**) Loading values of the top 20 genes that comprise LV 851.

3.5. Selected Latent Variables Represented Specific Immune Cell Types in the Tumor Microenvironment

Given the limited representation of different NF1 tumor types in our genomic variant dataset, we used alternate gene expression metrics to assess the biological underpinnings of the 73 uncharacterized latent variables. We performed tumor immune cell deconvolution analysis [29–31] to identify potential immune infiltration signatures present in the individual tumors using CIBERSORT and MCP-counter (Table S4). Specifically, CIBERSORT deconvolution indicated the presence of activated mast cells and M2 macrophages in all tumor types (Figure 5A). The analysis further suggested that all of the tumors have a population of resting CD4$^+$ memory T cells. The results from MCP-counter, depicted in Figure 5B, show a complementary view of the tumor cell types due to the slightly different categorization of cell types. Specifically, we found the presence of cancer-associated fibroblasts across all tumors that were not captured in CIBERSORT.

We then used the immune scores from CIBERSORT and MCP-counter to probe some of the latent variables selected by the random forest model to better understand their role in NF1 tumor biology. We first searched for tumor deconvolution scores that were correlated with latent variable expression across all tumors. Figure 5C,E show specific LVs where the immune scores of the samples highly correlated with a predicted immune cell type. LV 546, for example (Figure 5C), was found to have a high correlation of activated mast cells in cNF samples. Figure 5E shows that LV 540 has immune scores strongly correlated with T cells in all NF1 tumor types. Complete results of immune scores across all tumors can be found in Table S4. Together, these results suggest that the selected LVs capture signatures from the tumor microenvironment in NF1 samples.

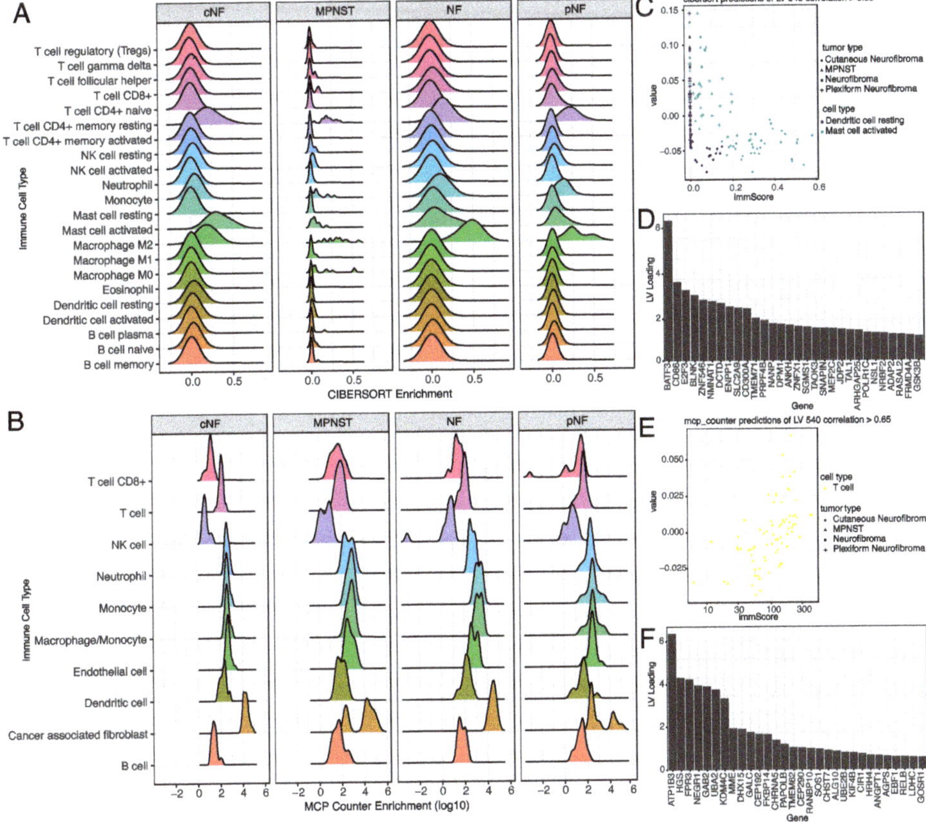

Figure 5. Various immune cell signatures correlated to specific LVs that differentiate tumor types in NF1. (**A**) CIBERSORT deconvolution of bulk nerve sheath tumor expression data predicted the presence of activated mast cells and M2 macrophages and resting CD4$^+$ memory T cells in all of the tested tumor types. (**B**) MCP-counter based deconvolution of bulk nerve sheath tumor expression data predicted the presence of cancer-associated fibroblasts across all tumor types, and diversity in T cell population across tumor types. (**C**) Correlation of CIBERSORT immune score (*x*-axis) with expression of latent variable 546 highlighted the increased presence of activated mast cells and resting dendritic cells in cNFs (circles). (**D**) Top 20 gene loadings of LV 546. (**E**) Correlation of MCP-counter score of Tell infiltration (*x*-axis) with LV 540. (**F**) Top 20 gene loadings of LV 540.

3.6. Selected Latent Variables Captured Protein Regulatory Networks in NF1 Tumors

Of the 98 latent variables selected by the random forest, 55 could not be characterized via enrichment of the gene loadings (Figure 3), mutational patterns (Figure 4), or immune subtypes (Figure 5). To characterize them, we applied the metaVIPER algorithm [32] to identify specific protein activity measurements that correlated with latent variable expression. This algorithm leverages previously published regulatory network information [52] to infer protein activity in each sample to assign numerical scores of activity across 6168 proteins for each of the 77 samples (Table S5). We then measured the correlation of these scores and latent variable scores (Table S6, Figure 6A) to identify functional aspects of the latent variables.

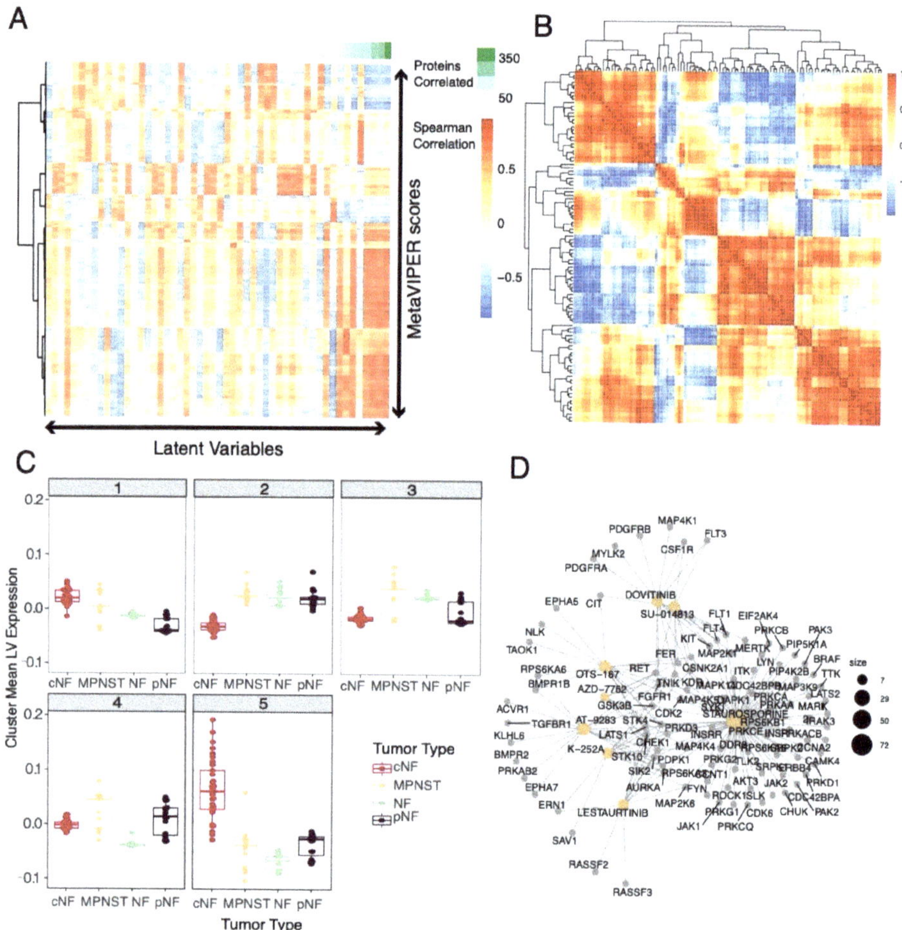

Figure 6. Integration of protein activity information with LVs can identify candidate drug targets for different NF1 tumor types. (**A**) A heatmap of correlation scores of known proteins with regulatory networks (or regulons) that are represented in the characterized and uncharacterized LVs selected above. The green bar across the top depicts how many protein activity scores had a Spearman correlation greater than 0.65. (**B**) Clustering of the LV-correlated VIPER proteins highlighted five clusters of latent variables with similar VIPER protein predictions, suggesting that these five clusters may have functional overlap. (**C**) Mean LV expression within the clusters highlighted differential expression within the clusters across tumor types. Tumor type is indicated by colors on the right. (**D**) Drug set enrichment analysis of the average VIPER protein correlation of cluster 2 identified some drugs and preclinical molecules that are enriched with targets in this cluster.

As seen in previous work [43], we observed that multiple latent variables exhibit similar correlation patterns with active proteins (Figure 6A). We clustered the correlation scores to see if we could group the latent variables with similar protein activity predictions (Figure 6B). In clustering the latent variable–protein activity scores, we observed five distinct clusters of latent variables with similar predicted protein activities (Figure 6B). Aggregation of each cluster of latent variables (mean expression within each cluster) demonstrated that these functional clusters were differentially expressed in the different NF1 tumor types (Figure 6C).

We then assessed the druggability of each of these five clusters by taking the average correlation of each protein within the cluster and performing gene set enrichment analysis against a database of small molecules with known biological targets [49]. This enabled the identification of drugs and drug-like compounds that are significantly enriched for targets in each cluster (Table S7). For example, cluster 2, which is expressed in pNF, NF, and MPNST more than it is expressed in cNF, has correlated VIPER proteins that are enriched for both clinically approved drugs (dovitinib) and drug-like small molecules (Figure 6D). Furthermore, we found that cluster 3 was enriched for compounds that affect cell cycle progression (e.g., dinaciclib, abemaciclib), whereas clusters 1 and 5 were enriched for compounds such as CUDC-101 (7-[4-(3-ethynylanilino)-7-methoxyquinazolin-6-yl]oxy-N-hydroxyheptanamide) and analogs that inhibit histone deacetylases (Figure S4).

4. Discussion

NF1 is the most common of all neurofibromatosis syndromes and is caused by the loss of function of *NF1* gene (a known tumor suppressor) due to mutation or deletion. However, NF1 patients show a great deal of phenotypic heterogeneity [1]. Identification of candidate cellular signaling pathways that differentiate between various tumor types is key for understanding the biology underlying such phenotypic diversity, as well as predicting progression of tumor types towards malignancy. In this study, we integrated various *in silico* resources and analytical techniques to identify candidate genes or pathways unique to different tumor types in an attempt to generate testable hypotheses. By capturing complex gene expression patterns using latent variables (LVs), we identified combinations of LVs that were important to classify tumor types. An ensemble of random forests was then used to select 98 latent variables that were important and sufficient for identifying and classifying the various tumor types with reasonably high accuracy (Figure 2). The selected LVs were then subjected to downstream analyses using different data modalities to gain insight into the composition and relevance of the LVs in the context of NF1 (Figures 3–6). The selected latent variables that correlated with known pathways using tumor deconvolution methods confirmed the presence of previously described tumor microenvironment components. Investigation into previously uncharacterized LVs in the context of NF1 suggested the presence of (a) candidate genes for targeted experiments, (b) previously known as well as unknown tumor microenvironment components, and (c) candidate tissue-specific protein regulatory networks for future drug screening experiments.

To interpret the results of these analyses, we first evaluated the gene loadings of the latent variables that were associated with one or more tumor types. We found that two of the top LVs with lower expression in cNF but higher expression in other tumor types, LV 384 and LV 624 (Figures 3C and 6), had ties to known Schwann cells and NF1 tumor biology, as well as presence of immune cells in the tumor microenvironment. Specifically, the *EGR2* gene, one of the major components of LV 384, has been implicated in diseases associated with the myelin sheath such as Charcot–Marie–Tooth Disease (another disease of the Schwann cell) and is thought to play a role in pathways associated with myelination [53,54]. Indeed, clinical case studies of patients with concurrent NF1-induced tumors and Charcot–Marie–Tooth disease have been reported in the literature, suggesting a possible overlap in underlying biology [55–57]. Although the role of *EGR2* in Schwann cell differentiation is still under active investigation [58], *EGR2*-driven pathways have been found to be significantly downregulated in Lats1/2-deficient Schwann cell-based MPNST models [59]. Similarly, the SOCS6 gene, the top component of LV 624, is known to be directly involved in immune signaling via repression of cytokines [60] and has also been found to be a selective tumor suppressor [61]. Its upregulation in the more malignant NF tumors (Figure 3C) suggests that this gene plays a distinct role in NF-related tumors. Additionally, the *RUNX2* gene (a major component of LV 624), has been shown to drive neurofibromagenesis by repression of the *PMP22* gene encoded myelin protein (a Schwann cell component) [62]. The enrichment of such genes in the selected LVs that differentiate nerve sheath tumors from cNFs showcase the importance of Schwann cell biology in these tumors and serves as a proof of principle for our analyses. Furthermore, EGR2 expression in immune cells has been shown to be important for activation of M1/M2 macrophages [63] as well as regulation of CD4$^+$ T cells [64]. As discussed later in greater detail, our downstream analysis, using all the selected LVs and tumor deconvolution techniques, identified the enrichment of M2 macrophage and CD4$^+$ T cell markers as tumor microenvironment components in our samples (Figure 5A).

Alternatively, we also evaluated LVs that are uniquely associated with specific tumor types. For example, LV 24 was found to be an important feature for classifying MPNST tumor samples but not the other benign tumor samples (Figure 3Biii). LV 24 was found to be significantly associated with the ΔNp63 pathway (FDR < 0.05), a pathway with implications in determining malignancy and poor survival in various subtypes of pancreatic and squamous cell carcinomas [65]. ΔNp63 signaling in the central nervous system is believed to play a role in neural precursor cell survival and neural stem cell dynamics [66,67], but its role in formation or maintenance of malignant NF1 tumors such as MPNSTs is relatively unexplored. Overall, the evidence surrounding EGR2, SOCS6, RUNX2, and ΔNp63 pathway suggest that LV 384, LV 624, and LV 24 may be promising candidates for experimental follow-up.

Beyond looking at individual genes that comprise the latent variables, we employed orthogonal algorithms that measured tumor immune activity and regulatory protein activity to identify specific signatures represented by the latent variables. Tumor deconvolution methods (CIBERSORT, MCP-counter) confirmed previously described observations such as the presence of mast cells in NF1 model systems and patient tumors [68–74]. They also suggested the presence of T cells in human tumor samples. This has previously been described in mouse models of NF1 tumors [75]. In humans, systemic T cell burden has been found to correlate with NF1 nerve sheath tumor progression; T cell presence has also been observed in NF1 gliomas [10,76].

Through the metaVIPER protein regulatory activity predictions, we were able to identify putative therapeutic candidates for further evaluation. Specifically, we found two clusters of latent variables expressed in NF, pNF, and MPNST that had regulatory proteins enriched in targets of dovitinib and drug-like small molecules that inhibit receptor tyrosine kinases, as well as cyclin-dependent kinase (CDK) inhibitors such as abemaciclib or dinaciclib. These findings are consistent with previous studies that identify CDK inhibitors as useful in models of *NF1*-deficient or RAS-dysregulated tumors [77–80]. Because latent variables that are correlated with histone deacetylases (HDACs) were found to be expressed most highly in cNFs (cluster 5, Figure 6C, Figure S4C) and MPNSTs (cluster 1, Figures 4A and 6C), compounds such as CUDC-101 and analogs could be potential candidates for treating cNF and MPNST. Indeed, HDAC inhibitors were previously found to be efficacious in in-vitro and in-vivo models of MPNSTs [77,81]. Thus, our results further suggest that this therapeutic approach might be feasible in both MPNSTs and cNFs. Drugs found in other clusters such as dovitinib and lestaurinib in cluster 2, which is expressed most in MPNSTs, pNFs, and NFs (Figure 6C,D), may also merit further study.

Although the present *in-silico* study brings forth various candidate genes and pathways for further follow up, it also presents a few limitations that could be mitigated with future studies, particularly for tumor types with limited samples such as MPNSTs and neurofibromas. Most notably, analyzing genetic variants across tumor types failed to identify relevant variant signatures (Figure 4). This highlights the challenges in variant analyses using samples with limited class representation and motivates our focus on transcriptional signatures. Additional genomic and transcriptomic data from the same biobanks or additional tumor datasets will improve our ability to identify recurrent genetic markers of tumor type. Furthermore, additional data from tumor-adjacent normal tissue would greatly add value to additional analyses on the basis of differential gene expression. Such differential expression analyses were not possible within the scope of this work since these data are not currently available. Additionally, future studies comparing genomic signatures identified here to other publicly available tumor expression and variant data (e.g., the Cancer Genome Atlas or the International Cancer Genome Consortium [19]) may identify genomic similarities between peripheral nerve sheath tumors and other tumors.

5. Conclusions

In conclusion, this work proposed a short list of testable hypotheses involving specific biological signatures for NF1-deficient nerve sheath tumors. Verification of these mechanisms in in-vitro and in-vivo models of nerve sheath tumors as well as human NF1 nerve sheath tumor tissue needs active and extensive experimental work. Although we analyzed tumor datasets from four different studies, the addition of other neurofibromatosis-driven tumor datasets will greatly aid identification of commonalities or critical differences to inform therapeutic decisions across the family of neurofibromatoses. This study, together with future work, may guide the repositioning of clinically approved drugs in the context of NF1.

Supplementary Materials: The following are available online at http://www.mdpi.com/2073-4425/11/2/226/s1. Figure S1: (a) A density plot indicating the pairwise Pearson correlation of LVs (not including self-self-pairs). A threshold of 0.5 was selected to eliminate non-orthogonal latent variables. Figure S2: Schematic showing the generation of the ensemble of random forests. (A) Stratified split of original dataset into "model" set and "independent test" set. (B) Generation of first ensemble of random forests using training set and test set generated from the "model" set for all 962 latent variables. (i) Example density plot showing distribution of model accuracy (F1) scores for each class. (ii) Example plot showing distribution of feature importance scores from the 500 random forests. (C) Generation of the second ensemble of random forests using restricted feature set of only 98 latent variables. The newly generated models were tested on the "independent test" set. (i) Example plot showing the distribution of model accuracy scores from the second ensemble of random forests. Figure S3: A boxplot representing the decrease in pairwise distances between all tumor samples of the same tumor type from Figure 1A (green, based on gene clustering) to 1C (orange, based on LV clustering). Figure S4: Small molecule-target networks showed enrichment of LV-dependent target classes. (A) Cluster 1 of LV-correlated VIPER proteins was enriched for HDAC inhibitors. (B) Cluster 3 of LV-correlated VIPER proteins was enriched for kinase inhibitors, particularly inhibitors of kinases responsible for cell cycle progression (CDKs). (C) Cluster 5 of LV-correlated VIPER proteins was enriched for HDAC inhibitors. Table S1: Pan-NF1 MultiPLIER results across RNA-seq data for 77 nerve sheath tumors. Table S2: Significance of latent variable status being correlated with gene mutation status. Table S3: Summary of latent variables selected by random forest model and their correlated genes and signatures. Table S4: Tumor deconvolution scores across NF1 patient samples. Table S5: metaVIPER protein scores across all NF1 patient samples. Table S6: Spearman correlation scores of RF-selected latent variables with metaVIPER scores. Table S7: Drug set enrichment analysis of latent variables.

Author Contributions: Conceptualization, R.J.A., J.B., S.J.C.G.; methodology, R.J.A., J.B., J.N.T., C.S.G., A.B., J.G., S.J.C.G.; software, R.J.A., J.B., J.N.T., C.S.G., A.B., J.G., S.J.C.G.; formal analysis, R.J.A., J.B., A.B., S.J.C.G.; investigation, R.J.A., J.B., S.J.C.G.; resources, X.Z., C.I.M., A.H., C.A.P.; writing—original draft preparation, R.J.A., J.B., S.J.C.G.; writing—review and editing, J.B., R.J.A., C.S.G., A.B., J.G., and S.J.C.G.; supervision, S.J.C.G., C.S.G., J.O.B., J.G.; project administration, S.J.C.G.; funding acquisition, S.J.C.G., J.G. All authors have read and agreed to the published version of the manuscript.

Funding: This work was supported with funds from the Neurofibromatosis Therapeutic Acceleration Program.

Acknowledgments: We would like to acknowledge our colleagues at Sage Bionetworks, Matthew Wall and Michael Mason, for many helpful conversations.

Conflicts of Interest: The authors have no conflicts of interest to declare, with the exception of Angela Hirbe and Christine A. Pratilas, who are serving as a co-editor for this issue of Genes. Both Angela Hirbe and Christine A. Pratilas have abstained from any editorial duties relating to this manuscript.

References

1. Friedman, J.M. Epidemiology of neurofibromatosis type 1. *Am. J. Med. Genet.* **1999**, *89*, 1–6. [CrossRef]
2. Evans, D.G.; Howard, E.; Giblin, C.; Clancy, T.; Spencer, H.; Huson, S.M.; Lalloo, F. Birth incidence and prevalence of tumor-prone syndromes: Estimates from a UK family genetic register service. *Am. J. Med. Genet. A* **2010**, *152A*, 327–332. [CrossRef] [PubMed]
3. Allaway, R.J.; Gosline, S.J.C.; La Rosa, S.; Knight, P.; Bakker, A.; Guinney, J.; Le, L.Q. Cutaneous neurofibromas in the genomics era: Current understanding and open questions. *Br. J. Cancer* **2018**, *118*, 1539–1548. [CrossRef] [PubMed]

4. Plotkin, S.R.; Bredella, M.A.; Cai, W.; Kassarjian, A.; Harris, G.J.; Esparza, S.; Merker, V.L.; Munn, L.L.; Muzikansky, A.; Askenazi, M.; et al. Quantitative Assessment of Whole-Body Tumor Burden in Adult Patients with Neurofibromatosis. *PLoS ONE* **2012**, *7*, e35711. [CrossRef]

5. Stucky, C.-C.H.; Johnson, K.N.; Gray, R.J.; Pockaj, B.A.; Ocal, I.T.; Rose, P.S.; Wasif, N. Malignant Peripheral Nerve Sheath Tumors (MPNST): The Mayo Clinic Experience. *Ann. Surg. Oncol.* **2012**, *19*, 878–885. [CrossRef]

6. Yuan, Z.; Xu, L.; Zhao, Z.; Xu, S.; Zhang, X.; Liu, T.; Zhang, S.; Yu, S. Clinicopathological features and prognosis of malignant peripheral nerve sheath tumor: A retrospective study of 159 cases from 1999 to 2016. *Oncotarget* **2017**, *8*, 104785–104795. [CrossRef]

7. Peacock, J.D.; Pridgeon, M.G.; Tovar, E.A.; Essenburg, C.J.; Bowman, M.; Madaj, Z.; Koeman, J.; Boguslawski, E.A.; Grit, J.; Dodd, R.D.; et al. Genomic Status of MET Potentiates Sensitivity to MET and MEK Inhibition in NF1-Related Malignant Peripheral Nerve Sheath Tumors. *Cancer Res.* **2018**, *78*, 3672–3687. [CrossRef]

8. Brohl, A.S.; Kahen, E.; Yoder, S.J.; Teer, J.K.; Reed, D.R. The genomic landscape of malignant peripheral nerve sheath tumors: Diverse drivers of Ras pathway activation. *Sci. Rep.* **2017**, *7*, 14992. [CrossRef]

9. Lee, W.; Teckie, S.; Wiesner, T.; Ran, L.; Prieto Granada, C.N.; Lin, M.; Zhu, S.; Cao, Z.; Liang, Y.; Sboner, A.; et al. PRC2 is recurrently inactivated through EED or SUZ12 loss in malignant peripheral nerve sheath tumors. *Nat. Genet.* **2014**, *46*, 1227–1232. [CrossRef]

10. D'Angelo, F.; Ceccarelli, M.; Tala; Garofano, L.; Zhang, J.; Frattini, V.; Caruso, F.P.; Lewis, G.; Alfaro, K.D.; Bauchet, L.; et al. The molecular landscape of glioma in patients with Neurofibromatosis 1. *Nat. Med.* **2019**, *25*, 176–187.

11. De Raedt, T.; Beert, E.; Pasmant, E.; Luscan, A.; Brems, H.; Ortonne, N.; Helin, K.; Hornick, J.L.; Mautner, V.; Kehrer-Sawatzki, H.; et al. PRC2 loss amplifies Ras-driven transcription and confers sensitivity to BRD4-based therapies. *Nature* **2014**, *514*, 247–251. [CrossRef] [PubMed]

12. Patel, A.V.; Eaves, D.; Jessen, W.J.; Rizvi, T.A.; Ecsedy, J.A.; Qian, M.G.; Aronow, B.J.; Perentesis, J.P.; Serra, E.; Cripe, T.P.; et al. Ras-Driven Transcriptome Analysis Identifies Aurora Kinase A as a Potential Malignant Peripheral Nerve Sheath Tumor Therapeutic Target. *Clin. Cancer Res.* **2012**, *18*, 5020–5030. [CrossRef] [PubMed]

13. Jessen, W.J.; Miller, S.J.; Jousma, E.; Wu, J.; Rizvi, T.A.; Brundage, M.E.; Eaves, D.; Widemann, B.; Kim, M.-O.; Dombi, E.; et al. MEK inhibition exhibits efficacy in human and mouse neurofibromatosis tumors. *J. Clin. Investig.* **2013**, *123*, 340–347. [CrossRef] [PubMed]

14. Choi, K.; Komurov, K.; Fletcher, J.S.; Jousma, E.; Cancelas, J.A.; Wu, J.; Ratner, N. An inflammatory gene signature distinguishes neurofibroma Schwann cells and macrophages from cells in the normal peripheral nervous system. *Sci. Rep.* **2017**, *7*, 43315. [CrossRef]

15. Pemov, A.; Li, H.; Patidar, R.; Hansen, N.F.; Sindiri, S.; Hartley, S.W.; Wei, J.S.; Elkahloun, A.; Chandrasekharappa, S.C.; Boland, J.F.; et al. The primacy of NF1 loss as the driver of tumorigenesis in neurofibromatosis type 1-associated plexiform neurofibromas. *Oncogene* **2017**, *36*, 3168–3177. [CrossRef]

16. Pemov, A.; Hansen, N.F.; Sindiri, S.; Patidar, R.; Higham, C.S.; Dombi, E.; Miettinen, M.M.; Fetsch, P.; Brems, H.; Chandrasekharappa, S.; et al. Low mutation burden and frequent loss of CDKN2A/B and SMARCA2, but not PRC2, define pre-malignant neurofibromatosis type 1-associated atypical neurofibromas. *Neuro-Oncology* **2019**, *21*, 981–992. [CrossRef]

17. Thomas, L.; Kluwe, L.; Chuzhanova, N.; Mautner, V.; Upadhyaya, M. Analysis of NF1 somatic mutations in cutaneous neurofibromas from patients with high tumor burden. *Neurogenetics* **2010**, *11*, 391–400. [CrossRef]

18. Weinstein, J.N.; Collisson, E.A.; Mills, G.B.; Shaw, K.M.; Ozenberger, B.A.; Ellrott, K.; Shmulevich, I.; Sander, C.; Stuart, J.M. The Cancer Genome Atlas Pan-Cancer Analysis Project. *Nat. Genet.* **2013**, *45*, 1113–1120. [CrossRef]

19. International network of cancer genome projects. *Nature* **2010**, *464*, 993–998. [CrossRef]

20. Guinney, J.; Dienstmann, R.; Wang, X.; de Reyniès, A.; Schlicker, A.; Soneson, C.; Marisa, L.; Roepman, P.; Nyamundanda, G.; Angelino, P.; et al. The Consensus Molecular Subtypes of Colorectal Cancer. *Nat. Med.* **2015**, *21*, 1350–1356. [CrossRef]

21. Ali, M.; Aittokallio, T. Machine learning and feature selection for drug response prediction in precision oncology applications. *Biophys. Rev.* **2019**, *11*, 31–39. [CrossRef] [PubMed]

22. Costello, J.C.; Heiser, L.M.; Georgii, E.; Gönen, M.; Menden, M.P.; Wang, N.J.; Bansal, M.; Ammad-ud-din, M.; Hintsanen, P.; Khan, S.A.; et al. A community effort to assess and improve drug sensitivity prediction algorithms. *Nat. Biotechnol.* **2014**, *32*, 1202–1212. [CrossRef] [PubMed]

23. Way, G.P.; Sanchez-Vega, F.; La, K.; Armenia, J.; Chatila, W.K.; Luna, A.; Sander, C.; Cherniack, A.D.; Mina, M.; Ciriello, G.; et al. Machine Learning Detects Pan-cancer Ras Pathway Activation in The Cancer Genome Atlas. *Cell Rep.* **2018**, *23*, 172–180.e3. [CrossRef] [PubMed]

24. Allaway, R.J.; La Rosa, S.; Verma, S.; Mangravite, L.; Guinney, J.; Blakeley, J.; Bakker, A.; Gosline, S.J.C. Engaging a community to enable disease-centric data sharing with the NF Data Portal. *Sci. Data* **2019**, *6*, 319. [CrossRef] [PubMed]

25. Collado-Torres, L.; Nellore, A.; Kammers, K.; Ellis, S.E.; Taub, M.A.; Hansen, K.D.; Jaffe, A.E.; Langmead, B.; Leek, J.T. Reproducible RNA-seq analysis using recount2. *Nat. Biotechnol.* **2017**, *35*, 319–321. [CrossRef]

26. Mao, W.; Zaslavsky, E.; Hartmann, B.M.; Sealfon, S.C.; Chikina, M. Pathway-level information extractor (PLIER) for gene expression data. *Nat. Methods* **2019**, *16*, 607–610. [CrossRef]

27. Taroni, J.N.; Grayson, P.C.; Hu, Q.; Eddy, S.; Kretzler, M.; Merkel, P.A.; Greene, C.S. MultiPLIER: A Transfer Learning Framework for Transcriptomics Reveals Systemic Features of Rare Disease. *Cell Syst.* **2019**, *8*, 380–394. [CrossRef]

28. Breiman, L. Random Forests. *Mach. Learn.* **2001**, *45*, 5–32. [CrossRef]

29. Newman, A.M.; Liu, C.L.; Green, M.R.; Gentles, A.J.; Feng, W.; Xu, Y.; Hoang, C.D.; Diehn, M.; Alizadeh, A.A. Robust enumeration of cell subsets from tissue expression profiles. *Nat. Methods* **2015**, *12*, 453–457. [CrossRef]

30. Becht, E.; Giraldo, N.A.; Lacroix, L.; Buttard, B.; Elarouci, N.; Petitprez, F.; Selves, J.; Laurent-Puig, P.; Sautès-Fridman, C.; Fridman, W.H.; et al. Estimating the population abundance of tissue-infiltrating immune and stromal cell populations using gene expression. *Genome Biol.* **2016**, *17*, 218. [CrossRef]

31. Sturm, G.; Finotello, F.; Petitprez, F.; Zhang, J.D.; Baumbach, J.; Fridman, W.H.; List, M.; Aneichyk, T. Comprehensive evaluation of transcriptome-based cell-type quantification methods for immuno-oncology. *Bioinformatics* **2019**, *35*, i436–i445. [CrossRef] [PubMed]

32. Ding, H.; Douglass, E.F.; Sonabend, A.M.; Mela, A.; Bose, S.; Gonzalez, C.; Canoll, P.D.; Sims, P.A.; Alvarez, M.J.; Califano, A. Quantitative assessment of protein activity in orphan tissues and single cells using the metaVIPER algorithm. *Nat. Commun.* **2018**, *9*, 1471. [CrossRef] [PubMed]

33. Wickham, H.; Averick, M.; Bryan, J.; Chang, W.; McGowan, L.; François, R.; Grolemund, G.; Hayes, A.; Henry, L.; Hester, J.; et al. Welcome to the Tidyverse. *J. Open Source Softw.* **2019**, *4*, 1686. [CrossRef]

34. Hoff, B.; Ladia, K. Synapser: R Language Bindings for Synapse API. 2019. Available online: https://r-docs.synapse.org/ (accessed on 19 February 2020).

35. Soneson, C.; Love, M.I.; Robinson, M.D. Differential analyses for RNA-seq: Transcript-level estimates improve gene-level inferences. *F1000Research* **2016**, *4*, 1521. [CrossRef]

36. Patro, R.; Duggal, G.; Love, M.I.; Irizarry, R.A.; Kingsford, C. Salmon: Fast and bias-aware quantification of transcript expression using dual-phase inference. *Nat. Methods* **2017**, *14*, 417–419. [CrossRef]

37. Pollard, K.; Banerjee, J.; Doan, X.; Wang, J.; Guo, X.; Allaway, R.; Langmead, S.; Slobogean, B.; Meyer, C.F.; Loeb, D.M.; et al. A clinically and genomically annotated nerve sheath tumor biospecimen repository. *bioRxiv* **2019**. [CrossRef]

38. Gosline, S.J.C.; Weinberg, H.; Knight, P.; Yu, T.; Guo, X.; Prasad, N.; Jones, A.; Shrestha, S.; Boone, B.; Levy, S.E.; et al. A high-throughput molecular data resource for cutaneous neurofibromas. *Sci. Data* **2017**, *4*, 170045. [CrossRef]

39. Ijaz, H.; Koptyra, M.; Gaonkar, K.S.; Rokita, J.L.; Baubet, V.P.; Tauhid, L.; Zhu, Y.; Brown, M.; Lopez, G.; Zhang, B.; et al. Pediatric high-grade glioma resources from the Children's Brain Tumor Tissue Consortium. *Neuro-Oncology* **2020**, *22*, 163–165. [CrossRef]

40. Carlson, M. *org.Hs.eg.db: Genome Wide Annotation for Human*; R Package Version 3.8.2; R Development Core Team: Vienna, Austria, 2019.

41. MultiPLIER Fileset 2019. Available online: https://doi.org/10.6084/m9.figshare.6982919.v2 (accessed on 19 February 2020).

42. Collado-Torres, L.; Nellore, A.; Jaffe, A.E.; Taub, M.A.; Kammers, K.; Ellis, S.E.; Hansen, K.D.; Langmead, B.; Leek, J.T. Recount: Explore and Download Data from the Recount Project; Bioconductor Version: Release (3.10). 2019. Available online: https://rdrr.io/bioc/recount/ (accessed on 19 February 2020).

43. Tan, J.; Huyck, M.; Hu, D.; Zelaya, R.A.; Hogan, D.A.; Greene, C.S. ADAGE signature analysis: Differential expression analysis with data-defined gene sets. *BMC Bioinform.* **2017**, *18*, 512. [CrossRef]

44. Liaw, A.; Wiener, M. Classification and Regression by randomForest. *R News* **2002**, *2*, 5.

45. Kuhn, M. Building Predictive Models in R Using the caret Package. *J. Stat. Softw.* **2008**, *28*, 1–26. [CrossRef]

46. Giorgi, F.M. aracne.networks: ARACNe-inferred gene networks from TCGA tumor datasets. *R Package Vers.* **2018**.

47. Kolde, R. pheatmap: Pretty Heatmaps. *R Package Vers.* **2019**, *61*, 617.

48. Yu, G.; Wang, L.-G.; Han, Y.; He, Q.-Y. clusterProfiler: An R Package for Comparing Biological Themes Among Gene Clusters. *OMICS J. Integr. Biol.* **2012**, *16*, 284–287. [CrossRef] [PubMed]

49. Allaway, R.J.; La Rosa, S.; Guinney, J.; Gosline, S.J.C. Probing the chemical-biological relationship space with the Drug Target Explorer. *J. Cheminform.* **2018**, *10*, 41. [CrossRef]

50. Yu, G. enrichplot: Visualization of Functional Enrichment Result. *R Package Vers.* **2019**, *112*.

51. Laks, D.R.; Oses-Prieto, J.A.; Alvarado, A.G.; Nakashima, J.; Chand, S.; Azzam, D.B.; Gholkar, A.A.; Sperry, J.; Ludwig, K.; Condro, M.C.; et al. A molecular cascade modulates MAP1B and confers resistance to mTOR inhibition in human glioblastoma. *Neuro-Oncology* **2018**, *20*, 764–775. [CrossRef]

52. Margolin, A.A.; Nemenman, I.; Basso, K.; Wiggins, C.; Stolovitzky, G.; Dalla Favera, R.; Califano, A. ARACNE: An algorithm for the reconstruction of gene regulatory networks in a mammalian cellular context. *BMC Bioinform.* **2006**, *7* (Suppl. 1), S7. [CrossRef]

53. Sevilla, T.; Sivera, R.; Martínez-Rubio, D.; Lupo, V.; Chumillas, M.J.; Calpena, E.; Dopazo, J.; Vílchez, J.J.; Palau, F.; Espinós, C. The EGR2 gene is involved in axonal Charcot–Marie–Tooth disease. *Eur. J. Neurol.* **2015**, *22*, 1548–1555. [CrossRef]

54. Warner, L.E.; Mancias, P.; Butler, I.J.; McDonald, C.M.; Keppen, L.; Koob, K.G.; Lupski, J.R. Mutations in the early growth response 2 (EGR2) gene are associated with hereditary myelinopathies. *Nat. Genet.* **1998**, *18*, 382–384. [CrossRef]

55. Roos, K.L.; Pascuzzi, R.M.; Dunn, D.W. Neurofibromatosis, Charcot-Marie-Tooth disease, or both? *Neurofibromatosis* **1989**, *2*, 238–243.

56. Lancaster, E.; Elman, L.B.; Scherer, S.S. A patient with Neurofibromatosis type 1 and Charcot-Marie-Tooth Disease type 1B. *Muscle Nerve* **2010**, *41*, 555–558. [CrossRef] [PubMed]

57. Lupski, J.R.; Pentao, L.; Williams, L.L.; Patel, P.I. Stable inheritance of the CMT1A DNA duplication in two patients with CMT1 and NF1. *Am. J. Med. Genet.* **1993**, *45*, 92–96. [CrossRef]

58. Lin, H.-P.; Oksuz, I.; Svaren, J.; Awatramani, R. Egr2-dependent microRNA-138 is dispensable for peripheral nerve myelination. *Sci. Rep.* **2018**. [CrossRef] [PubMed]

59. Wu, J.; Williams, J.P.; Rizvi, T.A.; Kordich, J.J.; Witte, D.; Meijer, D.; Stemmer-Rachamimov, A.O.; Cancelas, J.A.; Ratner, N. Plexiform and Dermal Neurofibromas and Pigmentation Are Caused by Nf1 Loss in Desert Hedgehog-Expressing Cells. *Cancer Cell* **2008**, *13*, 105–116. [CrossRef] [PubMed]

60. Kabir, N.N.; Sun, J.; Rönnstrand, L.; Kazi, J.U. SOCS6 is a selective suppressor of receptor tyrosine kinase signaling. *Tumor Biol.* **2014**, *35*, 10581–10589. [CrossRef] [PubMed]

61. Yuan, D.; Wang, W.; Su, J.; Zhang, Y.; Luan, B.; Rao, H.; Cheng, T.; Zhang, W.; Xiao, S.; Zhang, M.; et al. SOCS6 Functions as a Tumor Suppressor by Inducing Apoptosis and Inhibiting Angiogenesis in Human Prostate Cancer. Available online: http://www.eurekaselect.com/158762/article (accessed on 10 February 2020).

62. Hall, A.; Choi, K.; Liu, W.; Rose, J.; Zhao, C.; Yu, Y.; Na, Y.; Cai, Y.; Coover, R.A.; Lin, Y.; et al. RUNX represses Pmp22 to drive neurofibromagenesis. *Sci. Adv.* **2019**, *5*, eaau8389. [CrossRef]

63. Veremeyko, T.; Yung, A.W.Y.; Anthony, D.C.; Strekalova, T.; Ponomarev, E.D. Early Growth Response Gene-2 Is Essential for M1 and M2 Macrophage Activation and Plasticity by Modulation of the Transcription Factor CEBPβ. *Front. Immunol.* **2018**, *9*, 2515. [CrossRef]

64. Okamura, T.; Yamamoto, K.; Fujio, K. Early Growth Response Gene 2-Expressing CD4+LAG3+ Regulatory T Cells: The Therapeutic Potential for Treating Autoimmune Diseases. *Front. Immunol.* **2018**, *9*, 340. [CrossRef]

65. Hamdan, F.H.; Johnsen, S.A. DeltaNp63-dependent super enhancers define molecular identity in pancreatic cancer by an interconnected transcription factor network. *Proc. Natl. Acad. Sci. USA* **2018**, *115*, E12343–E12352. [CrossRef]

66. Cancino, G.I.; Yiu, A.P.; Fatt, M.P.; Dugani, C.B.; Flores, E.R.; Frankland, P.W.; Josselyn, S.A.; Miller, F.D.; Kaplan, D.R. p63 Regulates Adult Neural Precursor and Newly Born Neuron Survival to Control Hippocampal-Dependent Behavior. *J. Neurosci.* **2013**, *33*, 12569–12585. [CrossRef]

67. Packard, A.; Schnittke, N.; Romano, R.-A.; Sinha, S.; Schwob, J.E. ΔNp63 Regulates Stem Cell Dynamics in the Mammalian Olfactory Epithelium. *J. Neurosci.* **2011**, *31*, 8748–8759. [CrossRef] [PubMed]

68. Carr, N.J.; Warren, A.Y. Mast cell numbers in melanocytic naevi and cutaneous neurofibromas. *J. Clin. Pathol.* **1993**, *46*, 86–87. [CrossRef] [PubMed]

69. Zhu, Y.; Ghosh, P.; Charnay, P.; Burns, D.K.; Parada, L.F. Neurofibromas in NF1: Schwann Cell Origin and Role of Tumor Environment. *Science* **2002**, *296*, 920–922. [CrossRef] [PubMed]

70. Isaacson, P. Mast cells in benign nerve sheath tumours. *J. Pathol.* **1976**, *119*, 193–196. [CrossRef] [PubMed]

71. Tucker, T.; Riccardi, V.M.; Sutcliffe, M.; Vielkind, J.; Wechsler, J.; Wolkenstein, P.; Friedman, J.M. Different patterns of mast cells distinguish diffuse from encapsulated neurofibromas in patients with neurofibromatosis 1. *J. Histochem. Cytochem.* **2011**, *59*, 584–590. [CrossRef]

72. Greggio: Les Cellules Granuleuses (Mastzellen) Dans-Google Scholar. Available online: https://scholar.google.com/scholar_lookup?journal=Arch.+Med.+Exp.&title=Les+cellules+granuleuses+(Mastzellen)+dans+les+tissus+normaux+et+dans+certaines+maladies+chirurgicales&author=H+Greggio&volume=23&publication_year=1911&pages=323-375& (accessed on 13 January 2020).

73. Chen, Z.; Mo, J.; Brosseau, J.-P.; Shipman, T.; Wang, Y.; Liao, C.-P.; Cooper, J.M.; Allaway, R.J.; Gosline, S.J.C.; Guinney, J.; et al. Spatiotemporal Loss of NF1 in Schwann Cell Lineage Leads to Different Types of Cutaneous Neurofibroma Susceptible to Modification by the Hippo Pathway. *Cancer Discov.* **2019**, *9*, 114–129. [CrossRef]

74. Nürnberger, M.; Moll, I. Semiquantitative aspects of mast cells in normal skin and in neurofibromas of neurofibromatosis types 1 and 5. *Dermatology* **1994**, *188*, 296–299. [CrossRef]

75. Brosseau, J.-P.; Liao, C.-P.; Wang, Y.; Ramani, V.; Vandergriff, T.; Lee, M.; Patel, A.; Ariizumi, K.; Le, L.Q. NF1 heterozygosity fosters de novo tumorigenesis but impairs malignant transformation. *Nat. Commun.* **2018**, *9*, 1–11. [CrossRef]

76. Farschtschi, S.; Park, S.-J.; Sawitzki, B.; Oh, S.-J.; Kluwe, L.; Mautner, V.F.; Kurtz, A. Effector T cell subclasses associate with tumor burden in neurofibromatosis type 1 patients. *Cancer Immunol. Immunother.* **2016**, *65*, 1113–1121. [CrossRef]

77. Kahen, E.J.; Brohl, A.; Yu, D.; Welch, D.; Cubitt, C.L.; Lee, J.K.; Chen, Y.; Yoder, S.J.; Teer, J.K.; Zhang, Y.O.; et al. Neurofibromin level directs RAS pathway signaling and mediates sensitivity to targeted agents in malignant peripheral nerve sheath tumors. *Oncotarget* **2018**, *9*, 22571–22585. [CrossRef]

78. Allaway, R.J.; Fischer, D.A.; de Abreu, F.B.; Gardner, T.B.; Gordon, S.R.; Barth, R.J.; Colacchio, T.A.; Wood, M.; Kacsoh, B.Z.; Bouley, S.J.; et al. Genomic characterization of patient-derived xenograft models established from fine needle aspirate biopsies of a primary pancreatic ductal adenocarcinoma and from patient-matched metastatic sites. *Oncotarget* **2016**, *7*, 17087–17102. [CrossRef] [PubMed]

79. Discovery of a Small Molecule Targeting IRA2 Deletion in Budding Yeast and Neurofibromin Loss in Malignant Peripheral Nerve Sheath Tumor Cells | Molecular Cancer Therapeutics. Available online: https://mct.aacrjournals.org/content/10/9/1740.figures-only (accessed on 13 January 2020).

80. Danilov, A.V.; Hu, S.; Orr, B.; Godek, K.; Mustachio, L.M.; Sekula, D.; Liu, X.; Kawakami, M.; Johnson, F.M.; Compton, D.A.; et al. Dinaciclib Induces Anaphase Catastrophe in Lung Cancer Cells via Inhibition of Cyclin-Dependent Kinases 1 and 2. Mol. *Cancer Ther.* **2016**, *15*, 2758–2766. [CrossRef] [PubMed]

81. Malone, C.F.; Emerson, C.; Ingraham, R.; Barbosa, W.; Guerra, S.; Yoon, H.; Liu, L.L.; Michor, F.; Haigis, M.; Macleod, K.F.; et al. mTOR and HDAC Inhibitors Converge on the TXNIP/Thioredoxin Pathway to Cause Catastrophic Oxidative Stress and Regression of RAS-Driven Tumors. *Cancer Discov.* **2017**, *7*, 1450–1463. [CrossRef] [PubMed]

Review

From Genes to -Omics: The Evolving Molecular Landscape of Malignant Peripheral Nerve Sheath Tumor

Kathryn M. Lemberg [1,2], **Jiawan Wang** [1,2] **and Christine A. Pratilas** [1,2,*]

[1] Sidney Kimmel Comprehensive Cancer Center at Johns Hopkins, Baltimore, 401 N Broadway, Baltimore, MD 21231, USA; klember1@jhmi.edu (K.M.L.); jwang255@jhmi.edu (J.W.)
[2] Johns Hopkins University School of Medicine, Baltimore, 733 N Broadway, Baltimore, MD 21205, USA
*   Correspondence: cpratil1@jhmi.edu

Received: 15 May 2020; Accepted: 17 June 2020; Published: 24 June 2020

Abstract: Malignant peripheral nerve sheath tumors (MPNST) are rare, aggressive soft tissue sarcomas that occur with significantly increased incidence in people with the neuro-genetic syndrome neurofibromatosis type I (NF1). These complex karyotype sarcomas are often difficult to resect completely due to the involvement of neurovascular bundles, and are relatively chemotherapy- and radiation-insensitive. The lifetime risk of developing MPNST in the NF1 population has led to great efforts to characterize the genetic changes that drive the development of these tumors and identify mutations that may be used for diagnostic or therapeutic purposes. Advancements in genetic sequencing and genomic technologies have greatly enhanced researchers' abilities to broadly and deeply investigate aberrations in human MPNST genomes. Here, we review genetic sequencing efforts in human MPNST samples over the past three decades. Particularly for NF1-associated MPNST, these overall sequencing efforts have converged on a set of four common genetic changes that occur in most MPNST, including mutations in neurofibromin 1 (*NF1*), *CDKN2A*, *TP53*, and members of the polycomb repressor complex 2 (PRC2). However, broader genomic studies have also identified recurrent but less prevalent genetic variants in human MPNST that also contribute to the molecular landscape of MPNST and may inform further research. Future studies to further define the molecular landscape of human MPNST should focus on collaborative efforts across multiple institutions in order to maximize information gathered from large numbers of well-annotated MPNST patient samples, both in the NF1 and the sporadic MPNST populations.

Keywords: MPNST; NF1; genomics

1. Clinical Overview of MPNST

MPNST are aggressive soft tissue sarcomas originating from Schwann cells in the peripheral nervous system [1,2]. Half of MPNST occur in patients with the cancer predisposition syndrome NF1, caused by germline loss of function (LOF) of one copy of the tumor suppressor gene *NF1*. In patients with NF1, most MPNST arise from within plexiform neurofibromas (pNF), which are pre-malignant tumors of the peripheral nerve [3–5]. pNF can, themselves, be a major source of disfigurement or dysfunction. MPNST can also occur sporadically or following radiation treatment in the general population, although the incidence of the latter is substantially lower. MPNST carry a high risk of sarcoma-specific death; in the absence of complete surgical resection with wide negative margins, the five-year event-free survival is ~30% [6,7]. Conventional chemotherapy and radiation often do not improve patient outcomes [8].

2. Germline Loss of NF1: Correlations with NF1 Phenotype

NF1 is one of the most common monogenic inherited syndromes with an incidence of approximately 1:3000 live births [9]. This neurocutaneous syndrome is characterized by several hallmark skin findings (café au lait macules, axillary freckling, cutaneous neurofibromas), may involve additional organ systems (including CNS, musculoskeletal, and vascular manifestations) [10], and predisposes patients to an increased risk of malignancy, with an estimated lifetime cancer risk ~60% [11]. One of the hallmark lesions in NF1 patients is the pNF, a complex lesion that grows along major nerve bundles. While benign, pNF can result in significant anatomic, functional, cosmetic, and psychological effects [12]. In patients with NF1, MPNST may arise within existing pNF and are often accompanied by rapid growth, increased pain, or other nervous system deficits. Studies correlating the pathologic changes and genetic alterations in the peripheral nerves of NF1 patients or model organisms, along the spectrum from healthy to pNF to atypical neurofibroma (ANF) to MPNST, have aided in understanding the roles of specific genetic mutations in MPNST tumorigenesis [13].

NF1 syndrome is characterized by a wide variation in phenotypic expression which partially reflects the large number of mutations in the *NF1* gene that have been identified in people with the condition [14–16]. The *NF1* gene was originally cloned nearly three decades ago [17]. It is a large gene, approximately 350 kb in length, located on human chromosome 17q11.2. There may be multiple splice variants [18] but the primary gene product codes for the NF1 protein of 2818 amino acids, which acts as a GTPase-activating protein (GAP) for RAS oncogenes [19–22]. Loss of *NF1*, therefore, leads to constitutive activation of RAS signaling [23,24] (Figure 1), likely accounting for the pro-tumor phenotype observed in patients with NF1 [25].

Historically, the diagnosis of neurofibromatosis was based on clinical symptoms and physical findings, without a requirement for clinical genetic testing [26]. With the advances in detailed sequencing efforts, however, disruption of a copy of *NF1* in the germline may be identified in the majority of patients with NF1 [27]. Mutational analysis demonstrates a very high rate of mutations occurring in *NF1*, as evidenced by the fact that approximately 50% of cases of NF1 appear to be *de novo*. To date, hundreds of mutations associated with the syndrome have been characterized [9,14]. Identification of the specific LOF mutation in patients can be helpful for testing family members, particularly offspring of those affected, and for counseling patients about syndrome-specific risks.

Genotype–phenotype correlations associated with specific germline *NF1* alterations have been observed in a limited number of cases. Two examples are associated with limited risk for MPNST. A small in-frame deletion (c.2970_2972del(p.Met992del)) leading to loss of a methionine in the cysteine-serine rich domain (CSRD) of *NF1* is associated with suppression of cutaneous neurofibroma (cNF) and clinically apparent pNF formation, though these individuals have an increased risk for learning disabilities (48%) and brain tumors (~5%) [28,29]. Several missense mutations affecting arginine 1809 (e.g., p.Arg1809Cys) have also been characterized in multiple unrelated families. These patients have a high prevalence of developmental delay and learning disabilities as well as short stature and pulmonic stenosis, but few cutaneous or plexiform neurofibromas, and low risk of malignancy [30].

By contrast, two other *NF1* genotypes have been strongly associated with a higher risk of MPNST. Microdeletion of a 1.4 Mbp segment of chromosome 17 due to homologous recombination within duplication regions of the chromosome leads to deletion of 14 functional genes [31,32]. Individuals with the microdeletion syndrome (approximately 5% of NF1 cases) tend to present with a more severe NF1 phenotype [33], including dysmorphic features, developmental delay, intellectual disability, increased number of neurofibromas, and a two-fold higher lifetime risk for MPNST (16–26%, compared to approximately 8–13% risk in the general NF1 population) [34,35]. Missense mutations in NF1 protein codons 844–848 (including Leu844, Cys845, Ala846, Leu847, and Gly848; located in the CSRD) occur in ~0.8% of studied NF1 cases and are also reported as a risk factor for severe phenotypic presentation. These patients have higher numbers of clinically apparent major pNF, symptomatic spinal neurofibromas, optic pathway gliomas, and skeletal abnormalities, and up to 10% develop malignancy, including MPNST [36].

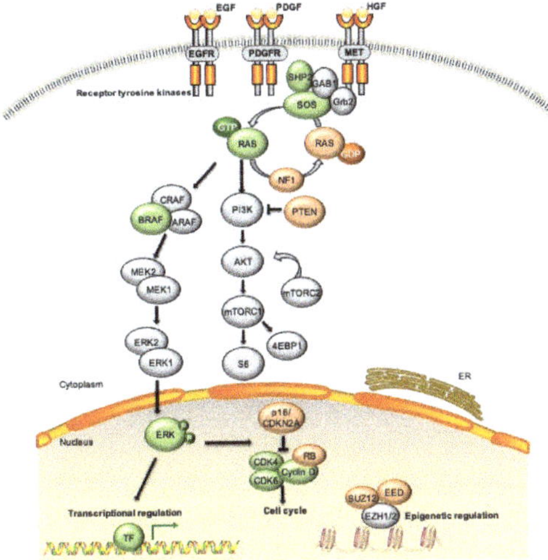

Figure 1. Signaling pathways altered due to genetic changes observed in malignant peripheral nerve sheath tumors (MPNST). The most common alterations in MPNST are loss of function of multiple tumor suppressors including NF1, p16/CDKN2A, TP53, and SUZ12/EED. Loss of *NF1*, as well as epigenetic changes due to loss of PRC2 components, leads to increased signaling through the RAS/RAF/MEK and PI3K/AKT pathways. Additional molecular events observed in subsets of MPNST include mutations in *BRAF*, amplification of *EGFR* or *MET* receptor tyrosine kinases (RTKs), and changes to chromatin structure through mutations in alpha thalassemia/mental retardation syndrome X (ATRX) and other epigenetic modifiers. EGF/EGFR = epidermal growth factor/receptor; PDGF/PDGFR = platelet derived growth factor/receptor; HGF = hepatocyte growth factor; ERK = extracellular signal regulated kinase; CDK = cyclin dependent kinase; RB = retinoblastoma; TF = transcription factor; ER = endoplasmic reticulum.

3. Sequencing Efforts in Human MPNST Samples: Improvements in Technology with Variability in Study Design

A collated summary of human MPNST sequencing efforts over the past two decades is shown in Table 1. MPNST have complex karyotypes with multiple chromosomal losses and gains and structural anomalies; a single recurrent translocation for diagnostic purposes has not been defined for MPNST as it has for some other mesenchymal tumors [37]. Expanded knowledge of MPNST gene alterations originated in the era of targeted gene evaluation using sequencing specific to the *NF1* locus or a small number of related genes. More recent studies have employed whole exome, whole genome, or targeted next-generation sequencing (NGS) on discovery cohorts for MPNST, with follow up studies performed by targeted gene sequencing in validation cohorts. Whole exome sequencing (WES) efforts have also been performed on patient tumors with paired neurofibroma or blood samples in a minority of cases. Individual studies vary with respect to how much additional clinical information is available (e.g., clinical background, treatment effect, comparison to neurofibroma or blood leukocytes). Some studies include sporadic and radiation-associated cases, while others focus purely on NF1-associated MPNST. In addition, in several studies multiple MPNST samples are derived from the same patient or fragments of the same tumor. These differences in study design, sample collection and annotation, and data analysis likely account for some of the differences and depth of discovery in genomic alterations across the literature. Taken together, however, a clear picture emerges of several characteristic alterations (i.e., *CDKN2A*, genes encoding PRC2 components) involved in evolution of benign nerve sheath tumor to MPNST. Less frequent alterations (i.e., *BRAF*, *MET*) identified in smaller subsets also merit additional attention in follow up evaluations, particularly as new diagnostic and treatment strategies for these tumors are being developed.

Table 1. Genomic sequencing studies for most common genetic alterations in human MPNST. All MPNST or neurofibromas in study reported under *n*. Reported sequencing results given as in reference (cases or percentages) for human MPNST specimens.

Ref.	Study Author Year	Description	*n* Total MPNST (*n* NF1 Associated) *n* (Other Specimen Types)	NF1	CDKN2A	TP53	EED	SUZ12	Notes
[38]	Mantripragada, 2008	Targeted seq, aCGH	35 (35) 16 pNF 8 cNF	71%	39%	17%	NR	NR	
[39]	Verdijk, 2010	Targeted seq	88 (26)	NR	ND	17/72	ND	ND	36% of *TP53* mutations detected were from NF1 patients
[40]	Yang, 2011	aCGH	51 (16)	~30%	65%	~30%	NR	NR	
[41]	DeRaedt, 2014	Targeted seq, aCGH	51 (51)	51/51	NR	NR	15/51	32/51	
[42]	Zhang, 2014	WGS (5), WES (3), Targeted seq (42)	50 (39) 11 (paired neurofibroma)	22/50	1/8	1/8	1/50	16/50	
[43]	Lee, 2014	WES (15), SNP, targeted (37)	52 (27) 7 neurofibromas	45/52	42/52	23/52	19/52	25/52	RNAseq analysis of MPNST with PRC2 loss vs. intact PRC2 demonstrates enrichment of genes associated with development and morphogenesis
[44]	Sohier, 2017	Exome seq, aCGH	8 (8) 1 pNF 7 cNF	8/8	5/8	1/8	2/8	7/8	No *TP53* point mutations identified
[45]	Brohl, 2017	WES + SNP	5 (4) + 7 TCGA cases (6)	11/12	7/12	6/12	4/12	5/12	5/12 MPNST contain somatic Ras-pathway activating mutation
[46]	Zehir, 2017	IMPACT NGS	11	2/11	6/11	NR	1/11	2/11	Data accessible through cBioPortal

Table 1. *Cont.*

Ref.	Study Author Year	Description	*n* Total MPNST (*n* NF1 Associated) *n* (Other Specimen Types)	*NF1*	*CDKN2A*	*TP53*	*EED*	*SUZ12*	Notes
[47]	Kaplan, 2018	Foundation Medicine NGS 2014–2016	186 (clinical data NR)	102 of 186	57% overall (71% *NF1*- altered, 80% *BRAF* altered, 34% non-*NF1*/ non-*BRAF* altered)	32% of *NF1* 14% of non-*NF1*	8% of *NF1*-altered, 13% of *BRAF*-altered, 3% of non–*NF1*/ non–*BRAF*-altered	20% of *NF1*-altered, 13% of *BRAF*-altered, 9% of non–*NF1*/ non–*BRAF*-altered	Data reported as % of NF1/BRAF cohorts rather than absolute numbers
[48]	Pemov, 2019	NF1 deep sequencing (4); WES (3); CNV (28)	31 (4) 16 ANF	4/4; 10/28 (Loss, CNV)	4/4; 20/28 (Loss, CNV)	0/3 (WES); 10/28 (Loss, CNV)	1/3 (WES); 10/28 (Loss, CNV)	1/3 (WES); 9/28 (Loss, CNV)	RNAseq reported for ANF and 4 MPNST
[49]	Pollard, 2020	WES	1 (1) 7 pNF 13 cNF	1	0/1	0/1	0/1	1/1	RNAseq on cNF, pNF, and MPNST samples from 23 patients

aCGH = array comparative genomic hybridization; WGS = whole genome sequencing; WES = whole exome sequencing; SNP = single nucleotide polymorphism; NGS = next generation sequencing; CNV = copy number variation; NR = not reported; ND = not determined; cNF = cutaneous neurofibroma; pNF = plexiform neurofibroma; ANF = atypical neurofibroma.

4. Somatic NF1 Mutations in Tumors Including MPNST

Consistent with its role as a classical tumor suppressor gene, loss of heterozygosity (LOH) or "second-hit" somatic mutations in the inherited wild-type *NF1* allele have been detected in a variety of tumors in patients with NF1, including pheochromocytomas [50], breast cancer [51], and hematologic malignancies [52]. Somatic LOH analysis using PCR markers performed on the *NF1* locus in dermal neurofibromas identified deletions in a subset of tumors in several early studies [53,54]; those cases known to be familial were analyzed further and shown to have deletions in the non-germline allele, demonstrating that somatic inactivation of *NF1* occurs in these benign lesions. Several studies have compared germline and somatic *NF1* mutations in MPNST. In a single study which investigated 34 MPNST from 27 NF1 patients, germline mutations were identified by lymphocyte DNA in 22 cases—these included one large 1.4 Mbp genomic deletion, one two-exon deletion, and smaller mutations (missense, nonsense, frameshift, and splicing anomalies) in the remainder [55]. In the same cohort, somatic *NF1* mutations were identified in 31 out of 34 MPNST samples—of these, 28 (91% tumors) were large genomic deletions that partially or entirely deleted the *NF1* gene. The authors speculate that in some cases somatic *NF1* mutations arise upon aberrant intrachromosomal recombination of the *NF1* gene during mitosis. Similarly, another report screened 47 MPNST from patients with or without NF1 syndrome (n = 25 and 22 cases, respectively). Of the somatic *NF1* mutations identified (n = 10/25 NF1-associated and 9/22 sporadic), approximately 55–60% involved large genomic copy number changes (i.e., deletions) in both NF1 and sporadic MPNST [32]. By contrast, in MPNST analyzed from NF1 patients with the 1.4 Mbp germline *NF1* microdeletion, the *NF1* somatic hit is typically a small (e.g., missense) mutation [31]. Interestingly, in a single patient with clinical NF1 syndrome who developed asynchronous cNF, a primary breast tumor, and later gluteal MPNST, WES revealed three distinct *NF1* somatic mutations compared to the germline mutation noted in the blood [51].

5. Acquired Mutations during Transformation from pNF

5.1. Loss of CDKN2A/B: Correlations with the pNF to ANF Transition

NF1 LOH is considered to be an initiating event in pNF formation as confirmed in several animal models [56]. Several additional mutations are necessary for malignant transformation. ANF (now re-classified as atypical neurofibromatous neoplasms of uncertain biological potential, ANNUBP) are precursor lesions to NF1-associated MPNST, representing an intermediate step from the malignant transformation of pNF into MPNST [57–59]. Alterations to chromosome 9q have been observed in a high proportion of ANF and MPNST [48,60]; one study noted deletion at 9p21.3, identified in 94% (15/16) of ANF and in 70% (16/23) of high-grade MPNST but not in pNF [57]. This locus encompasses several candidate tumor suppressors, including *CDKN2A/B*. *CDKN2A* encodes two gene products each the result of differential splicing: p16^{ink4a} (a negative regulator of CDK4 and CDK6 cyclin dependent kinases) and p19^{Arf}, a negative regulator of the *TP53* E3 ligase MDM2. Several early studies on human NF1-associated MPNST specimens identified deletions within the short arm of chromosome 9, in the region of *CDKN2A*, as well as low expression of p19, while these were not detected in neurofibroma samples [61,62]. A more recent study identified frequent somatic deletions of *CDKN2A/B* (69%) and *SMARCA2* (42%), apart from recurrent *NF1* somatic mutations (81%), in 16 ANF [48]. These studies indicate that *CDKN2A/B* deletion is the first step in the progression of pNF toward ANF and eventually MPNST.

5.2. LOH and Mutation in the Tumor Suppressor TP53: Not Universal in Human MPNST

Copy number variation and mutations in the tumor suppressor gene *TP53* have been identified in some cases of NF1-associated MPNST. Early studies on small subsets of NF1-associated neurofibrosarcomas identified deletions on chromosome 17 outside of the *NF1* locus [63,64], which included the coding region for *TP53*. Screening for *TP53* inactivation in a panel of 20 MPNST identified LOH in over half of the tumors tested [55]. The first genetically-engineered mouse (GEM) model for MPNST made use of LOH of both *NF1* and *TP53* from mouse chromosome 11 as the tumor initiating event [65]. Numerous subsequent studies have focused on identifying the true incidence of *TP53* mutation in human MPNST; from compiled data on 25 studies including 114 MPNST (both NF1-associated and sporadic), *TP53* mutations were observed in 14% of MPNST, with LOH in 39% of cases (Table 1) [39]. WES of NF1 tumor samples from a single patient with pNF, MPNST, and metastatic sites also identified loss of one copy of *TP53* in the MPNST and metastatic lesion, but not the primary pNF [66]. Genetic changes in *TP53* are thus present in some MPNST but not necessary for all cases of pNF malignant transformation.

5.3. Loss of PRC2 or H3K27me3: Recurrently and Specifically Occurs in MPNST

Components of the epigenetic regulatory PRC2are recurrently and specifically inactivated in MPNST (Table 1). Chi and colleagues identified genomic alterations in *EED* (37%, or 19/52) and *SUZ12* (48% or 25/52) in MPNST, alongside frequent somatic alterations in *CDKN2A* (81%, 42/52) and *NF1* (87%, 45/52) [43]. Bettegowda and colleagues simultaneously reported PRC2 loss via *EED* (2%, 1/50) and *SUZ12* (32%, 16/50) mutations in 50 MPNST [42]. De Raedt et al. similarly reported alterations in *EED* in 29% (15/51) and *SUZ12* in 63% (32/51) of NF1-associated MPNST [41]. PRC2-component loss in MPNST is associated with complete loss of histone H3 trimethylation at lysine 27 (H3K27me3) and increased level of H3K27 acetylation (H3K27Ac), which can serve as biomarkers to improve upon the accuracy of the diagnosis of MPNST [41,43]. *SUZ12* loss potentiates the effects of NF1 loss by amplifying RAS-driven transcription through effects on chromatin that triggers an epigenetic switch [41]. Further detail on the role and function of PRC2 elements in MPNST is found in the review article by Zhang et al. dedicated to this topic, also included in this Special Issue on *Genomics and Models of Nerve Sheath Tumors* [67]. Collectively, the highly recurrent and specific inactivation of PRC2 components, *NF1*, and *CDKN2A/B* posits their critical and potentially cooperative roles in MPNST pathogenesis.

6. Less Common Recurrent Variants Identified with Modern Sequencing Investigations of MPNST

MPNST demonstrate complex genomic imbalances and chromosomal aberrations [58,59]. In addition to the common deletions of tumor suppressor genes *NF1*, *CDKN2A*, *TP53* and LOF in the PRC2 genes *EED* and *SUZ12*, several other recurrent genomic events have been identified in NF1-associated and sporadic MPNST. Significant findings from these studies are highlighted in Table 2 and described below.

Table 2. Less frequent genomic alterations identified in MPNST.

Gene	Description	*n*	NF1	Altered	Details	Study	Ref.
BRAF	Targeted seq	18	NR	0	18 MPNST out of 1,320 nervous system tumors	Schindler, 2011	[68]
	Targeted seq	47	25	1	1/1 N581S	Bottillo, 2009	[32]
	Targeted seq	24	NR	0	0/24 MPNST with BRAF exon 15 mutation	Je, 2012	[69]
	Foundation NGS	186	102	10	5 of 10 BRAF V600E; 9 of 10 pathogenic; 1 of 10 VUS 47% with alteration in >/=1 non-NF1/non-BRAF gene in the RAS/RAF pathway (*ERBB2, ERBB3, ERBB4, KRAS, MET, HRAS, MAP2K1, MAP2K2, NRAS*). 7% with alteration in RTK (e.g., *KIT/PDGFR/FGFR1*)—some likely pathogenic 70% with alteration in DNA repair genes (*ATM, BARD1, BRCA1/2, FANCx, PBRM1, CHEK2, MSH2, MSH3, MSH6, NBN, PBRM1, POLE, RAD51, RAD51C*)	Kaplan, 2018	[47]
MET	WES	1	1	1 (amplified)	Single patient longitudinal sampling (pre/post treatment, recurrence, mets) Copy number alterations in *HGF, EGFR*	Peacock, 2018	[70]
	Targeted seq, aCGH	35		25% (amplified)	*HGF, EGR, PDGFRA* amplifications in 25-29% samples	Mantripragada, 2008	[38]
EGFR	aCGH	51		37% (19/51)	At least one EGFR pathway gene was altered in 84% of samples, including *GRB2, HRAS, MAPK1, STAT1*, and others.	Du, 2013	[71]
	Targeted gene sequencing	37	29/37	28% gain	Direct sequencing of *EGFR* exons 18–24	Holtkamp, 2008	[72]
	Targeted gene sequencing and FISH	27	14 of 25 pts	14 of 23 (copy number gain)	Direct sequencing of *EGFR* exons 18–21	Perrone, 2009	[73]
IGF1R	aCGH	51	16	24% (amplified)	>/= 1 gene in IGFR1 pathway altered in 82% cases	Yang, 2011	[40]
AURKA	SNP array, qPCR	13	NR	8	1/8 neurofibromas also with *AURKA* locus copy number increase	Patel, 2012	[74]

Table 2. *Cont.*

Gene	Description	n	NF1	Altered	Details	Study	Ref.
TYK2		7	7	2	Tyrosine kinase 2, activates STAT signaling and promotes cancer cell survival	Hirbe, 2017	[75]
ATRX	NGS clinical genomic profiling	7	7	NR		Hirbe, 2017	[75]
	NGS	4	4	2	Of 3 ALT-positive MPNST, 2 had *ATRX* mutations. One ALT positive MPNST had *RECQL4* variant.	Rodriguez, 2019	[76]
	WGS, WES	8	5	1	Additional chromatin organization-related genes: *EZH2*, *CHD4*, and *AEBP2* mutations $n = 1$ tumor each; *RBBP7* mutation in $n = 2$ tumors.	Zhang, 2014	[42]
KDM2B	Exome seq, aCGH	8	8	1	Jumonji histone lysine demethylase; identified in single patient MPNST lacking SUZ12 or EED mutation	Sohier, 2017	[44]
LATS2	aCGH	51	16	NR (copy number loss in ~25%)	Copy number gains and losses in HIPPO effector loci (*TAZ*, *CTGF*, *BIRC5*) and HIPPO inhibitory loci (*LATS2*, *AMOTL2*) graphically illustrated. Same dataset as Yang et al. Clin Can Res 2011.	Wu, 2018	[77]
HMMR/RHAMM	aCGH	35	71%	46%	Deletions in hyaluronan binding protein may affect signaling through ERK or AURAK	Mantripragada, 2008	[38]

NGS = next generation sequencing; aCGH = array comparative genomic hybridization; WGS = whole genome sequencing; WES = whole exome sequencing; SNP = single nucleotide polymorphism; NR = not reported; ND = not determined.

6.1. BRAF Mutation: An Alternate Mechanism for Activation of RAS Signaling

In addition to loss of *NF1* and PRC2 function, *BRAF* mutations are reported as an alternate mechanism for aberrant activation of RAS signaling in MPNST, albeit at a lower frequency (ranging from 0–9.7%) [32,47,69,78,79], and occurring more commonly in sporadic than NF1-associated cases [78]. Strongly activating kinase mutations (BRAF V600E) occurred in five out of ten *BRAF*-mutant *NF1*-wild type MPNST (*n* = 84; Table 2) [47]. *BRAF* amplification has also been described, with a frequency of 31% in another study cohort consisting of 51 MPNST [40]. Brohl et al. suggest that the relative strength of RAS-activating mutations may determine whether *BRAF* and *NF1* mutations (or *NRAS/KRAS* and *NF1*) co-occur and thereby serve together to result in ERK signaling hyperactivation [45].

6.2. EGFR, MET and Other Receptor Tyrosine Kinases: Frequent Copy Number Gains in MPNST

A variety of oncogenic receptor tyrosine kinases (RTK) are frequently altered in MPNST. In MPNST, alterations in RTK usually take the form of amplification, rather than single nucleotide variations that result in constitutively activated kinases (Table 2). Several early aCGH studies revealed amplifications of *HGF*, *MET*, *EGFR*, *PDGFRA*, and *IGF1R* in approximately 25% to 40% of analyzed MPNST [38,40]. These studies and others [71,72,80] suggest a putative role of these genes and their respective biological pathways in the initiation and/or progression of MPNST.

Notably, *HGF* and its receptor *MET*, co-located at chromosome 7q, are highly expressed in a relatively large panel of human MPNST samples, and increased phospho-MET expression level directly correlates with shorter MPNST patient survival [81]. A single patient study revealed progressive amplifications of *HGF*, *MET* and *EGFR* in a patient with MPNST harboring early *NF1* and *TP53* loss, using longitudinal genomic analysis from pNF, to MPNST, to metastatic recurrence. These studies further justify investigation of the role of RTK signaling, in particular HGF/MET, on the progression of MPNST.

6.3. AURKA Amplification

Dramatic upregulation (7.9-fold) of *AURKA* (the gene encoding aurora kinase A) was observed through RAS-driven transcriptome analysis on a GEM model and 14 human MPNST samples compared with normal nerves. Further analysis using SNP-array and qPCR confirmed copy number gains in the *AURKA* locus in eight out of 13 primary MPNST and five out of five MPNST cell lines but not neurofibromas [74]. Reducing the expression and activity of Aurora kinase using shRNA knockdown and a kinase inhibitor MLN8237, respectively, inhibits MPNST cell survival in vitro and in vivo, and supports the role of aurora kinase as a rational therapeutic target for MPNST [82].

6.4. Tyrosine Kinase 2 Overexpression in MPNST

NGS on a set of seven NF1-associated MPNST identified a predicted pathogenic mutation in tyrosine kinase 2 (*TYK2*) in two out of seven tumors [75]. TYK2 P1104A mutated tumors demonstrated strong immunoreactivity, whereas TYK2 wild type tumors were not immunoreactive. Strong TYK2 expression as assayed by immunohistochemical staining was observed in 63% of MPNST in an independent tissue set, while only 11% of pNF samples stained for TYK2. Ablation of TYK2 expression in human and murine MPNST cells resulted in increased cell death in vitro and decreased tumor growth in a murine model [83]. The example of *TYK2* suggests the role that sequencing efforts can play in development of novel markers of MPNST biology.

6.5. ATRX Mutation and Evidence for Alternative Lengthening of Telomeres

In addition to the role of PRC2 in MPNST chromatin regulation, the chromatin regulator ATRX (Alpha Thalassemia/Mental Retardation Syndrome X) has been identified as mutated in a subset of MPNST [75]. Loss of ATRX function is involved in alternative lengthening of telomeres (ALT), a telomerase-independent means of telomere maintenance which prevents tumor cell senescence and promotes tumorigenesis. Subsequent studies on a larger subset of MPNST identified decreased nuclear expression of ATRX and demonstrated a correlation between aberrant ATRX expression and decreased overall survival in NF1-associated MPNST [84]. In a separate study a small subset ($n = 3$) of NF1-associated MPNST that were ALT-positive were analyzed by NGS and found to have *ATRX* mutations in two out of three cases [76]. While this study did not identify inferior overall survival (OS) for ALT-positive MPNST compared to those with normal telomere length, short telomeres were significantly correlated with improved OS.

6.6. Beyond SUZ12: Less Common Variant Mutations in Other Chromatin Modifying Genes

In addition to loss of *SUZ12* and *EED*, several studies have demonstrated additional alterations in PRC2 components or associated chromatin modifying genes. Sohier and colleagues detected a novel sequence change in the histone lysine demethylase *KDM2B* by WES (c3376C > T) in one out of eight human MPNST. This change is thought to potentially impact protein function; in an additional set of 14 tumors assayed by qPCR, KDM2B expression was reduced [44]. Whole genome and whole exome sequencing on an additional subset of NF1-associated MPNST identified mutations in additional chromatin associated genes including *CHD4*, *AEBP2*, *EPC1*, and *EZH2*, particularly in tumors with intact *SUZ12* [42].

6.7. Evidence for Alterations in the HIPPO Pathway in a Subset of MPNST and Schwann Cell Derived Tumors

Several studies have found evidence for alterations in the HIPPO–YAP pathway in MPNST. Analysis of aCGH from 51 MPNST samples [40] revealed an increase in the copy number of HIPPO effector gene loci, including *TAZ*, *CTGF* and *BIRC5* and a loss of HIPPO inhibitory gene loci, such as *LATS2* and *AMOTL2* [77]. In agreement with these findings, transcriptome sequencing of human MPNST samples from two additional patient cohorts revealed elevated YAP-activated gene expression in MPNST relative to normal nerves and NF1-associated neurofibromas [85,86]. Genomic alterations in the HIPPO pathway appear to occur in additional NF1 patient tissues including somatic mutations in seven of 33 cNF described in a recent study and as germline mutations (e.g., missense, frameshift and occasionally insertion) in seven of nine NF1 patients from the same dataset [87]. Together these studies validate the role of HIPPO pathway in neurofibroma biology and as a driver of MPNST tumorigenesis [77,87].

7. Beyond Genomics: The State of Understanding MPNST Transcriptomes, Proteomes, Epigenomes, and Metabolomes

In addition to the genomic alterations described above, these and other studies on human MPNST have revealed downstream effects on MPNST gene product expression and signaling. These investigations have confirmed or supplemented the genomic data by assaying downstream pathway effects in human MPNST. Several studies have broadly analyzed gene expression in human MPNST samples using microarray or RNAseq approaches [48,88]; these data can be examined in relation to known genetic changes to generate additional hypotheses for effects on downstream signaling pathways. Recent work compared gene expression in multiple functional pathways across pNF, ANF/ANNUBP, and MPNST and found that some ANNUBP share signaling pathway characteristics that more closely resemble pNF (e.g., ERK/MAPK) and others (e.g., AKT/mTOR) are more similar to MPNST [88]. Phospho-proteome arrays may be used to investigate kinase signaling in relation to various genomic alterations or therapeutic interventions in MPNST; to date this has primarily been used in MPNST cell lines or animal models (see article by Grit et al. in this Special Issue on Genomics and Models of Nerve Sheath Tumors) [89]. Methylation analysis on MPNST has revealed overall decreased histone and DNA methylation [90], and has also revealed how methylation changes in MPNST can affect expression of other tumor suppressor genes (e.g., *PTEN*) in MPNST [91]. Parallel methylation analysis and proteomic analysis on a set of nine MPNST samples characterized the relationship between PRC2 LOF on histone and DNA modification and consequent gene product expression. This work found that PRC2 loss was associated with increased pro-growth and immune evasion protein expression [92]. To date global metabolomics profiling has not been reported on human MPNST specimens; several recent efforts have examined metabolic shifts in animal models of MPNST in response to preclinical therapeutic interventions [93–95].

8. Translating Molecular Landscape of MPNST into Improved Therapies for Patients

One overarching goal of improved molecular characterization of MPNST is to translate genomic discoveries into improved treatments for this classically chemo-refractory tumor. As a result of improved understanding of MPNST genomic variants, several targeted therapies have been trialed in preclinical MPNST models. For example, the MET-specific tyrosine kinase inhibitor capmatinib has shown promise, particularly in combination with the MEK inhibitor trametinib, in an NF1-MET driven MPNST GEM model [70]. BRAF mutant MPNST may also respond to targeted therapy; one case report described a dramatic response to the RAF inhibitor vemurafenib in a patient with sporadic metastatic MPNST harboring the BRAF V600E mutation [96]. Efforts to target histone acetylation in a preclinical MPNST model with loss of SUZ12 shrank tumors when combined with MEK inhibition [41], while other DNA methyltransferase inhibitors appear to affect immune surveillance of MPNST [92]. It is likely that in the near future MPNST clinical trials will incorporate therapies inhibiting components of the epigenetic machinery.

9. Conclusions

Significant research efforts over the past three decades have significantly advanced the state of knowledge of the genetic landscape of human MPNST. Particularly in NF1-associated MPNST, it is generally accepted that alterations in *NF1*, *CDKN2A*, *TP53*, and *SUZ12* are involved in tumor progression from benign to malignant tumors. However, less frequent alterations in genes with complementary function have been described in subsets of tumors, and additional tumor-driving mutations may be present in sporadic or recurrent/metastatic tumor samples. Future genomic studies should aim to incorporate as many well-annotated samples as feasible and clearly report on differences between NF1-associated and sporadic MPNST subtypes. Exciting future work will also incorporate additional technologies to improve our understanding of the downstream consequences of genomic alterations for MPNST biology and aid in development of improved treatments.

Author Contributions: All authors researched the published literature, conceived, wrote, and edited the manuscript. All authors have read and agreed to the published version of the manuscript.

Funding: K.M.L. is supported by a Young Investigator Award from CureSearch for Children's Cancer. C.A.P. receives research funding from the Neurofibromatosis Therapeutic Acceleration Program (NTAP).

Conflicts of Interest: The authors have no conflicts of interest to declare, with the exception of Christine A. Pratilas, who served as a co-editor for this special issue of Genes. Christine Pratilas has abstained from editorial duties relating to this manuscript. Christine A. Pratilas is a paid consultant for Genentech/ Roche.

References

1. Carroll, S.L. Molecular mechanisms promoting the pathogenesis of Schwann cell neoplasms. *Acta Neuropathol.* **2012**, *123*, 321–348. [CrossRef] [PubMed]
2. Zhu, Y.; Ghosh, P.; Charnay, P.; Burns, D.K.; Parada, L.F. Neurofibromas in NF1: Schwann cell origin and role of tumor environment. *Science* **2002**, *296*, 920–922. [CrossRef] [PubMed]
3. Ferner, R.E.; Gutmann, D.H. International consensus statement on malignant peripheral nerve sheath tumors in neurofibromatosis. *Cancer Res.* **2002**, *62*, 1573–1577.
4. Evans, D.G.; Baser, M.E.; McGaughran, J.; Sharif, S.; Howard, E.; Moran, A. Malignant peripheral nerve sheath tumours in neurofibromatosis 1. *J. Med. Genet.* **2002**, *39*, 311–314. [CrossRef] [PubMed]
5. McCaughan, J.A.; Holloway, S.M.; Davidson, R.; Lam, W.W. Further evidence of the increased risk for malignant peripheral nerve sheath tumour from a Scottish cohort of patients with neurofibromatosis type 1. *J. Med. Genet.* **2007**, *44*, 463–466. [CrossRef] [PubMed]

6. Kattan, M.W.; Leung, D.H.; Brennan, M.F. Postoperative nomogram for 12-year sarcoma-specific death. *J. Clin. Oncol.* **2002**, *20*, 791–796. [CrossRef]

7. Fletcher, C.D.; McKee, P.H. Sarcomas—A clinicopathological guide with particular reference to cutaneous manifestation. II. Malignant nerve sheath tumour, leiomyosarcoma and rhabdomyosarcoma. *Clin. Exp. Dermatol* **1985**, *10*, 201–216. [CrossRef]

8. Higham, C.S.; Steinberg, S.M.; Dombi, E.; Perry, A.; Helman, L.J.; Schuetze, S.M.; Ludwig, J.A.; Staddon, A.; Milhem, M.M.; Rushing, D.; et al. SARC006: Phase II Trial of Chemotherapy in Sporadic and Neurofibromatosis Type 1 Associated Chemotherapy-Naive Malignant Peripheral Nerve Sheath Tumors. *Sarcoma* **2017**, *2017*, 8685638. [CrossRef]

9. Gutmann, D.H.; Ferner, R.E.; Listernick, R.H.; Korf, B.R.; Wolters, P.L.; Johnson, K.J. Neurofibromatosis type 1. *Nat. Rev. Dis. Primers* **2017**, *3*, 17004. [CrossRef]

10. Ly, K.I.; Blakeley, J.O. The Diagnosis and Management of Neurofibromatosis Type 1. *Med. Clin. N. Am.* **2019**, *103*, 1035–1054. [CrossRef]

11. Uusitalo, E.; Rantanen, M.; Kallionpää, R.A.; Poyhonen, M.; Leppavirta, J.; Yla-Outinen, H.; Riccardi, V.M.; Pukkala, E.; Pitkaniemi, J.; Peltonen, S.; et al. Distinctive Cancer Associations in Patients with Neurofibromatosis Type 1. *J. Clin. Oncol.* **2016**, *34*, 1978–1986. [CrossRef] [PubMed]

12. Gross, A.M.; Singh, G.; Akshintala, S.; Baldwin, A.; Dombi, E.; Ukwuani, S.; Goodwin, A.; Liewehr, D.J.; Steinberg, S.M.; Widemann, B.C. Association of Plexiform Neurofibroma Volume Changes and Development of Clinical Morbidities in Neurofibromatosis 1. *Neuro. Oncol.* **2018**, *12*, 1643–1651. [CrossRef] [PubMed]

13. Kim, A.; Stewart, D.R.; Reilly, K.M.; Viskochil, D.; Miettinen, M.M.; Widemann, B.C. Malignant Peripheral Nerve Sheath Tumors State of the Science: Leveraging Clinical and Biological Insights into Effective Therapies. *Sarcoma* **2017**, *2017*, 7429697. [CrossRef] [PubMed]

14. Shen, M.H.; Harper, P.S.; Upadhyaya, M. Molecular genetics of neurofibromatosis type 1 (NF1). *J. Med. Genet.* **1996**, *33*, 2–17. [CrossRef]

15. Jett, K.; Friedman, J.M. Clinical and genetic aspects of neurofibromatosis 1. *Genet. Med.* **2010**, *12*, 1–11. [CrossRef]

16. Anderson, J.L.; Gutmann, D.H. Neurofibromatosis type 1. *Handb. Clin. Neurol.* **2015**, *132*, 75–86.

17. Wallace, M.R.; Marchuk, D.A.; Andersen, L.B.; Letcher, R.; Odeh, H.M.; Saulino, A.M.; Fountain, J.W.; Brereton, A.; Nicholson, J.; Mitchell, A.L.; et al. Type 1 neurofibromatosis gene: Identification of a large transcript disrupted in three NF1 patients. *Science* **1990**, *249*, 181–186. [CrossRef]

18. Andersen, L.B.; Ballester, R.; Marchuk, D.A.; Chang, E.; Gutmann, D.H.; Saulino, A.M.; Camonis, J.; Wigler, M.; Collins, F.S. A conserved alternative splice in the von Recklinghausen neurofibromatosis (NF1) gene produces two neurofibromin isoforms, both of which have GTPase-activating protein activity. *Mol. Cell Biol.* **1993**, *13*, 487–495. [CrossRef]

19. Gutmann, D.H.; Wood, D.L.; Collins, F.S. Identification of the neurofibromatosis type 1 gene product. *Proc. Natl. Acad. Sci. USA* **1991**, *88*, 9658–9662. [CrossRef]

20. DeClue, J.E.; Cohen, B.D.; Lowy, D.R. Identification and characterization of the neurofibromatosis type 1 protein product. *Proc. Natl. Acad. Sci. USA* **1991**, *88*, 9914–9918. [CrossRef]

21. Martin, G.A.; Viskochil, D.; Bollag, G.; McCabe, P.C.; Crosier, W.J.; Haubruck, H.; Conroy, L.; Clark, R.; O'Connell, P.; Cawthon, R.M.; et al. The GAP-related domain of the neurofibromatosis type 1 gene product interacts with ras p21. *Cell* **1990**, *63*, 843–849. [CrossRef]

22. Ballester, R.; Marchuk, D.; Boguski, M.; Letcher, R.; Wigler, M.; Collins, F. The NF1 locus encodes a protein functionally related to mammalian GAP and yeast IRA proteins. *Cell* **1990**, *63*, 851–859. [CrossRef]

23. Basu, T.N.; Gutmann, D.H.; Fletcher, J.A.; Glover, T.W.; Collins, F.S.; Downward, J. Aberrant regulation of ras proteins in malignant tumour cells from type 1 neurofibromatosis patients. *Nature* **1992**, *356*, 713–715. [CrossRef] [PubMed]

24. Bollag, G.; Clapp, D.W.; Shih, S.; Adler, F.; Zhang, Y.Y.; Thompson, P.; Lange, B.J.; Freedman, M.H.; McCormick, F.; Jacks, T.; et al. Loss of NF1 results in activation of the Ras signaling pathway and leads to aberrant growth in haematopoietic cells. *Nat. Genet.* **1996**, *12*, 144–148. [CrossRef] [PubMed]

25. Cichowski, K.; Jacks, T. NF1 tumor suppressor gene function: Narrowing the GAP. *Cell* **2001**, *104*, 593–604. [CrossRef]

26. Rosser, T.; Packer, R.J. Neurofibromas in children with neurofibromatosis 1. *J. Child Neurol.* **2002**, *17*, 585–651. [CrossRef]

27. Messiaen, L.M.; Callens, T.; Mortier, G.; Beysen, D.; Vandenbroucke, I.; Van Roy, N.; Speleman, F.; Paepe, A.D. Exhaustive mutation analysis of the NF1 gene allows identification of 95% of mutations and reveals a high frequency of unusual splicing defects. *Hum. Mutat.* **2000**, *15*, 541–555. [CrossRef]

28. Koczkowska, M.; Callens, T.; Gomes, A.; Sharp, A.; Chen, Y.; Hicks, A.D.; Aylsworth, A.S.; Azizi, A.A.; Basel, D.G.; Bellus, G.; et al. Expanding the clinical phenotype of individuals with a 3-bp in-frame deletion of the NF1 gene (c.2970_2972del): An update of genotype-phenotype correlation. *Genet. Med.* **2019**, *21*, 867–876. [CrossRef]

29. Upadhyaya, M.; Huson, S.M.; Davies, M.; Thomas, N.; Chuzhanova, N.; Giovannini, S.; Evans, D.G.; Howard, E.; Kerr, B.; Griffiths, S.; et al. An absence of cutaneous neurofibromas associated with a 3-bp inframe deletion in exon 17 of the NF1 gene (c.2970-2972 delAAT): Evidence of a clinically significant NF1 genotype-phenotype correlation. *Am. J. Hum. Genet.* **2007**, *80*, 140–151. [CrossRef]

30. Rojnueangnit, K.; Xie, J.; Gomes, A.; Sharp, A.; Callens, T.; Chen, Y.; Liu, Y.; Cochran, M.; Abbott, M.A.; Atkin, J.; et al. High Incidence of Noonan Syndrome Features Including Short Stature and Pulmonic Stenosis in Patients carrying NF1 Missense Mutations Affecting p.Arg1809: Genotype-Phenotype Correlation. *Hum. Mutat.* **2015**, *36*, 1052–1063. [CrossRef]

31. De Raedt, T.; Maertens, O.; Chmaram, M.; Brems, H.; Heyns, I.; Sciot, R.; Majounie, E.; Upadhyaya, M.; De Schepper, S.; Speleman, F.; et al. Somatic loss of wild type NF1 allele in neurofibromas: Comparison of NF1 microdeletion and non-microdeletion patients. *Genes Chromosomes Cancer* **2006**, *45*, 893–904. [CrossRef] [PubMed]

32. Bottillo, I.; Ahlquist, T.; Brekke, H.; Danielsen, S.A.; van den Berg, E.; Mertens, F.; Lothe, R.A.; Dallapiccola, B. Germline and somatic NF1 mutations in sporadic and NF1-associated malignant peripheral nerve sheath tumours. *J. Pathol.* **2009**, *217*, 693–701. [CrossRef] [PubMed]

33. Pasmant, E.; Sabbagh, A.; Spurlock, G.; Laurendeau, I.; Grillo, E.; Hamel, M.J.; Martin, L.; Barbarot, S.; Leheup, B.; Rodriguez, D.; et al. NF1 microdeletions in neurofibromatosis type 1: From genotype to phenotype. *Hum. Mutat.* **2010**, *31*, E1506–E1518. [CrossRef] [PubMed]

34. Kehrer-Sawatzki, H.; Mautner, V.F.; Cooper, D.N. Emerging genotype-phenotype relationships in patients with large NF1 deletions. *Hum. Genet.* **2017**, *136*, 349–376. [CrossRef]

35. De Raedt, T.; Brems, H.; Wolkenstein, P.; Vidaud, D.; Pilotti, S.; Perrone, F.; Mautner, V.; Frahm, S.; Sciot, R.; Legius, E. Elevated risk for MPNST in NF1 microdeletion patients. *Am. J. Hum. Genet.* **2003**, *72*, 1288–1292. [CrossRef]

36. Koczkowska, M.; Chen, Y.; Callens, T.; Gomes, A.; Sharp, A.; Johnson, S.; Hsiao, M.C.; Chen, Z.; Balasubramanian, M.; Barnett, C.P.; et al. Genotype-Phenotype Correlation in NF1: Evidence for a More Severe Phenotype Associated with Missense Mutations Affecting NF1 Codons 844–848. *Am. J. Hum. Genet.* **2018**, *102*, 69–87. [CrossRef]

37. Bridge, R.S., Jr.; Bridge, J.A.; Neff, J.R.; Naumann, S.; Althof, P.; Bruch, L.A. Recurrent chromosomal imbalances and structurally abnormal breakpoints within complex karyotypes of malignant peripheral nerve sheath tumour and malignant triton tumour: A cytogenetic and molecular cytogenetic study. *J. Clin. Pathol.* **2004**, *57*, 1172–1178. [CrossRef]

38. Mantripragada, K.K.; Spurlock, G.; Kluwe, L.; Chuzhanova, N.; Ferner, R.E.; Frayling, I.M.; Dumanski, J.P.; Guha, A.; Mautner, V.; Upadhyaya, M. High-resolution DNA copy number profiling of malignant peripheral nerve sheath tumors using targeted microarray-based comparative genomic hybridization. *Clin. Cancer Res.* **2008**, *14*, 1015–1024. [CrossRef]

39. Verdijk, R.M.; den Bakker, M.A.; Dubbink, H.J.; Hop, W.C.; Dinjens, W.N.; Kros, J.M. TP53 mutation analysis of malignant peripheral nerve sheath tumors. *J. Neuropathol. Exp. Neurol.* **2010**, *69*, 16–26. [CrossRef]

40. Yang, J.; Ylipää, A.; Sun, Y.; Zheng, H.; Chen, K.; Nykter, M.; Trent, J.; Ratner, N.; Lev, D.C.; Zhang, W. Genomic and molecular characterization of malignant peripheral nerve sheath tumor identifies the IGF1R pathway as a primary target for treatment. *Clin. Cancer Res.* **2011**, *17*, 7563–7573. [CrossRef]

41. De Raedt, T.; Beert, E.; Pasmant, E.; Luscan, A.; Brems, H.; Ortonne, N.; Helin, K.; Hornick, J.L.; Mautner, V.; Kehrer-Sawatzki, H.; et al. PRC2 loss amplifies Ras-driven transcription and confers sensitivity to BRD4-based therapies. *Nature* **2014**, *514*, 247–251. [CrossRef] [PubMed]

42. Zhang, M.; Wang, Y.; Jones, S.; Sausen, M.; McMahon, K.; Sharma, R.; Wang, Q.; Belzberg, A.J.; Chaichana, K.; Gallia, G.L.; et al. Somatic mutations of SUZ12 in malignant peripheral nerve sheath tumors. *Nat. Genet.* **2014**, *46*, 1170–1172. [CrossRef]

43. Lee, W.; Teckie, S.; Wiesner, T.; Prieto Granada, C.N.; Lin, M.; Zhu, S.; Cao, Z.; Liang, Y.; Sboner, A.; Tap, W.D.; et al. PRC2 is recurrently inactivated through EED or SUZ12 loss in malignant peripheral nerve sheath tumors. *Nat. Genet.* **2014**, *46*, 1227–1232. [CrossRef] [PubMed]

44. Sohier, P.; Luscan, A.; Lloyd, A.; Ashelford, K.; Laurendeau, I.; Briand-Suleau, A.; Vidaud, D.; Ortonne, N.; Pasmant, E.; Upadhyaya, M. Confirmation of mutation landscape of NF1-associated malignant peripheral nerve sheath tumors. *Genes Chromosomes Cancer* **2017**, *56*, 421–426. [CrossRef] [PubMed]

45. Brohl, A.S.; Kahen, E.; Yoder, S.J.; Teer, J.K.; Reed, D.R. The genomic landscape of malignant peripheral nerve sheath tumors: Diverse drivers of Ras pathway activation. *Sci. Rep.* **2017**, *7*, 14992. [CrossRef] [PubMed]

46. Zehir, A.; Benayed, R.; Shah, R.H.; Syed, A.; Middha, S.; Kim, H.R.; Srinivasan, P.; Gao, J.; Chakravarty, D.; Devlin, S.M.; et al. Mutational landscape of metastatic cancer revealed from prospective clinical sequencing of 10,000 patients. *Nat. Med.* **2017**, *23*, 703–713. [CrossRef]

47. Kaplan, H.G.; Rostad, S.; Ross, J.S.; Ali, S.M.; Millis, S.Z. Genomic Profiling in Patients with Malignant Peripheral Nerve Sheath Tumors Reveals Multiple Pathways with Targetable Mutations. *J. Natl. Compr. Cancer Netw.* **2018**, *16*, 967–974. [CrossRef]

48. Pemov, A.; Hansen, N.F.; Sindiri, S.; Patidar, R.; Higham, C.S.; Dombi, E.; Miettinen, M.M.; Fetsch, P.; Brems, H.; Chandrasekharappa, S.; et al. Low mutation burden and frequent loss of CDKN2A/B and SMARCA2, but not PRC2, define pre-malignant neurofibromatosis type 1-associated atypical neurofibromas. *Neuro. Oncol.* **2019**, *21*, 981–992. [CrossRef]

49. Pollard, K.; Banerjee, J.; Doan, X.; Wang, J.; Guo, X.; Allaway, R.; Langmead, S.; Slobogean, B.; Meyer, C.F.; Loeb, D.M.; et al. A clinically and genomically annotated nerve sheath tumor biospecimen repository. *Sci. Data* **2020**, *7*, 184. [CrossRef]

50. Xu, W.; Mulligan, L.M.; Ponder, M.A.; Liu, L.; Smith, B.A.; Mathew, C.G.; Ponder, B.A. Loss of NF1 alleles in phaeochromocytomas from patients with type I neurofibromatosis. *Genes Chromosomes Cancer* **1992**, *4*, 337–342. [CrossRef]

51. McPherson, J.R.; Ong, C.K.; Ng, C.C.; Rajasegaran, V.; Heng, H.L.; Yu, W.S.; Tan, B.K.; Madhukumar, P.; Teo, M.C.; Ngeow, J.; et al. Whole-exome sequencing of breast cancer, malignant peripheral nerve sheath tumor and neurofibroma from a patient with neurofibromatosis type 1. *Cancer Med.* **2015**, *4*, 1871–1878. [CrossRef] [PubMed]

52. Shannon, K.M.; O'Connell, P.; Martin, G.A.; Paderanga, D.; Olson, K.; Dinndorf, P.; McCormick, F. Loss of the normal NF1 allele from the bone marrow of children with type 1 neurofibromatosis and malignant myeloid disorders. *N. Engl. J. Med.* **1994**, *330*, 597–601. [CrossRef] [PubMed]

53. Colman, S.D.; Williams, C.A.; Wallace, M.R. Benign neurofibromas in type 1 neurofibromatosis (NF1) show somatic deletions of the *NF1* gene. *Nat. Genet.* **1995**, *11*, 90–92. [CrossRef]

54. Serra, E.; Puig, S.; Otero, D.; Gaona, A.; Kruyer, H.; Ars, E.; Estivill, X.; Lázaro, C. Confirmation of a double-hit model for the *NF1* gene in benign neurofibromas. *Am. J. Hum. Genet.* **1997**, *61*, 512–519. [CrossRef]

55. Upadhyaya, M.; Kluwe, L.; Spurlock, G.; Monem, B.; Majounie, E.; Mantripragada, K.; Ruggieri, M.; Chuzhanova, N.; Evans, D.G.; Ferner, R.; et al. Germline and somatic *NF1* gene mutation spectrum in NF1-associated malignant peripheral nerve sheath tumors (MPNSTs). *Hum. Mutat.* **2008**, *29*, 74–82. [CrossRef] [PubMed]

56. Yang, F.C.; Ingram, D.A.; Chen, S.; Zhu, Y.; Yuan, J.; Li, X.; Yang, X.; Knowles, S.; Horn, W.; Li, Y.; et al. Nf1-dependent tumors require a microenvironment containing Nf1+/− and c-kit-dependent bone marrow. *Cell* **2008**, *135*, 437–448. [CrossRef]

57. Beert, E.; Brems, H.; Daniëls, B.; De Wever, I.; Van Calenbergh, F.; Schoenaers, J.; Debiec-Rychter, M.; Gevaert, O.; De Raedt, T.; Van Den Bruel, A.; et al. Atypical neurofibromas in neurofibromatosis type 1 are premalignant tumors. *Genes Chromosomes Cancer* **2011**, *50*, 1021–1032. [CrossRef]

58. Higham, C.S.; Dombi, E.; Rogiers, A.; Bhaumik, S.; Pans, S.; Connor, S.; Miettinen, M.; Sciot, R.; Tirabosco, R.; Brems, H.; et al. The characteristics of 76 atypical neurofibromas as precursors to neurofibromatosis 1 associated malignant peripheral nerve sheath tumors. *Neuro. Oncol.* **2018**, *20*, 818–825. [CrossRef]

59. Miettinen, M.M.; Antonescu, C.R.; Fletcher, C.D.M.; Bhaumik, S.; Pans, S.; Connor, S.; Miettinen, M.; Sciot, R.; Tirabosco, R.; Brems, H.; et al. Histopathologic evaluation of atypical neurofibromatous tumors and their transformation into malignant peripheral nerve sheath tumor in patients with neurofibromatosis 1-a consensus overview. *Hum. Pathol.* **2017**, *67*, 1–10. [CrossRef]

60. Nielsen, G.P.; Stemmer-Rachamimov, A.O.; Ino, Y.; Moller, M.B.; Rosenberg, A.E.; Louis, D.N. Malignant transformation of neurofibromas in neurofibromatosis 1 is associated with CDKN2A/p16 inactivation. *Am. J. Pathol.* **1999**, *155*, 1879–1884. [CrossRef]

61. Kourea, H.P.; Orlow, I.; Scheithauer, B.W.; Cordon-Cardo, C.; Woodruff, J.M. Deletions of the INK4A gene occur in malignant peripheral nerve sheath tumors but not in neurofibromas. *Am. J. Pathol.* **1999**, *155*, 1855–1860. [CrossRef]

62. Berner, J.M.; Sørlie, T.; Mertens, F.; Henriksen, J.; Saeter, G.; Mandahl, N.; Brøgger, A.; Myklebost, O.; Lothe, R.A. Chromosome band 9p21 is frequently altered in malignant peripheral nerve sheath tumors: Studies of CDKN2A and other genes of the pRB pathway. *Genes Chromosomes Cancer* **1999**, *26*, 151–160. [CrossRef]

63. Menon, A.G.; Anderson, K.M.; Riccardi, V.M.; Chung, R.Y.; Whaley, J.M.; Yandell, D.W.; Farmer, G.E.; Freiman, R.N.; Lee, J.K.; Li, F.P.; et al. Chromosome 17p deletions and p53 gene mutations associated with the formation of malignant neurofibrosarcomas in von Recklinghausen neurofibromatosis. *Proc. Natl. Acad. Sci. USA* **1990**, *87*, 5435–5439. [CrossRef] [PubMed]

64. Legius, E.; Dierick, H.; Wu, R.; Hall, B.K.; Marynen, P.; Cassiman, J.J.; Glover, T.W. TP53 mutations are frequent in malignant NF1 tumors. *Genes Chromosomes Cancer* **1994**, *10*, 250–255. [CrossRef] [PubMed]

65. Cichowski, K.; Shih, T.S.; Schmitt, E.; Santiago, S.; Reilly, K.; McLaughlin, M.E.; Bronson, R.T.; Jacks, T. Mouse models of tumor development in neurofibromatosis type 1. *Science* **1999**, *286*, 2172–2176. [CrossRef]

66. Hirbe, A.C.; Dahiya, S.; Miller, C.A.; Li, T.; Fulton, R.S.; Zhang, X.; McDonald, S.; DeSchryver, K.; Duncavage, E.J.; Walrath, J.; et al. Whole Exome Sequencing Reveals the Order of Genetic Changes during Malignant Transformation and Metastasis in a Single Patient with NF1-plexiform Neurofibroma. *Clin. Cancer Res.* **2015**, *21*, 4201–4211. [CrossRef]

67. Zhang, X.; Murray, B.; Mo, G.; Shern, J.F. The Role of Polycomb Repressive Complex in Malignant Peripheral Nerve Sheath Tumor. *Genes* **2020**, *11*, 287. [CrossRef]

68. Schindler, G.; Capper, D.; Meyer, J.; Janzarik, W.; Omran, H.; Herold-Mende, C.; Schmieder, K.; Wesseling, P.; Mawrin, C.; Hasselblatt, M.; et al. Analysis of BRAF V600E mutation in 1,320 nervous system tumors reveals high mutation frequencies in pleomorphic xanthoastrocytoma, ganglioglioma and extra-cerebellar pilocytic astrocytoma. *Acta Neuropathol.* **2011**, *121*, 397–405. [CrossRef]

69. Je, E.M.; An, C.H.; Yoo, N.J.; Lee, S.H. Mutational analysis of PIK3CA, JAK2, BRAF, FOXL2, IDH1, AKT1 and EZH2 oncogenes in sarcomas. *APMIS* **2012**, *120*, 635–639. [CrossRef]

70. Peacock, J.D.; Pridgeon, M.G.; Tovar, E.A.; Essenburg, C.J.; Bowman, M.; Madaj, Z.; Koeman, J.; Boguslawski, E.A.; Grit, J.; Dodd, R.D.; et al. Genomic Status of MET Potentiates Sensitivity to MET and MEK Inhibition in NF1-Related Malignant Peripheral Nerve Sheath Tumors. *Cancer Res.* **2018**, *78*, 3672–3687. [CrossRef]

71. Du, X.; Yang, J.; Ylipää, A.; Zhu, Z. Genomic amplification and high expression of EGFR are key targetable oncogenic events in malignant peripheral nerve sheath tumor. *J. Hematol. Oncol.* **2013**, *6*, 93. [CrossRef] [PubMed]

72. Holtkamp, N.; Malzer, E.; Zietsch, J.; Okuducu, A.F.; Mucha, J.; Mawrin, C.; Mautner, V.F.; Schildhaus, H.U.; von Deimling, A. EGFR and erbB2 in malignant peripheral nerve sheath tumors and implications for targeted therapy. *Neuro. Oncol.* **2008**, *10*, 946–957. [CrossRef] [PubMed]

73. Perrone, F.; Da Riva, L.; Orsenigo, M.; Losa, M.; Jocollè, G.; Millefanti, C.; Pastore, E.; Gronchi, A.; Pierotti, M.A.; Pilotti, S. PDGFRA, PDGFRB, EGFR, and downstream signaling activation in malignant peripheral nerve sheath tumor. *Neuro. Oncol.* **2009**, *11*, 725–736. [CrossRef] [PubMed]

74. Patel, A.V.; Da Riva, L.; Orsenigo, M.; Rizvi, T.A.; Ecsedy, J.A.; Qian, M.G.; Aronow, B.J.; Perentesis, J.P.; Serra, E.; Cripe, T.P.; et al. Ras-driven transcriptome analysis identifies aurora kinase A as a potential malignant peripheral nerve sheath tumor therapeutic target. *Clin. Cancer Res.* **2012**, *18*, 5020–5030. [CrossRef] [PubMed]

75. Hirbe, A.C.; Kaushal, M.; Sharma, M.K.; Dahiya, S.; Pekmezci, M.; Perry, A.; Gutmann, D.H. Clinical genomic profiling identifies TYK2 mutation and overexpression in patients with neurofibromatosis type 1-associated malignant peripheral nerve sheath tumors. *Cancer* **2017**, *123*, 1194–1201. [CrossRef]

76. Rodriguez, F.J.; Graham, M.K.; Brosnan-Cashman, J.A.; Cashman, J.A.; Barber, J.R.; Davis, C.; Vizcaino, M.A.; Palsgrove, D.N.; Giannini, C.; Pekmezci, M.; et al. Telomere alterations in neurofibromatosis type 1-associated solid tumors. *Acta Neuropathol. Commun.* **2019**, *7*, 139. [CrossRef]

77. Wu, L.M.N.; Deng, Y.; Wang, J.; Zhao, C.; Wang, J.; Rao, R.; Xu, L.; Zhou, W.; Choi, K.; Rizvi, T.A.; et al. Programming of Schwann Cells by Lats1/2-TAZ/YAP Signaling Drives Malignant Peripheral Nerve Sheath Tumorigenesis. *Cancer Cell* **2018**, *33*, 292–308.e7. [CrossRef]

78. Hirbe, A.C.; Pekmezci, M.; Dahiya, S.; Apicelli, A.J.; Van Tine, B.A.; Perry, A.; Gutmann, D.H. BRAFV600E mutation in sporadic and neurofibromatosis type 1-related malignant peripheral nerve sheath tumors. *Neuro Oncol.* **2014**, *16*, 466–467. [CrossRef]

79. Serrano, C.; Simonetti, S.; Hernández-Losa, J.; Valverde, C.; Carrato, C.; Bagué, S.; Orellana, R.; Somoza, R.; Moliné, T.; Carles, J.; et al. BRAF V600E and KRAS G12S mutations in peripheral nerve sheath tumours. *Histopathology* **2013**, *62*, 499–504. [CrossRef]

80. DeClue, J.E.; Heffelfinger, S.; Benvenuto, G.; Ling, B.; Li, S.; Rui, W.; Vass, W.C.; Viskochil, D.; Ratner, N. Epidermal growth factor receptor expression in neurofibromatosis type 1-related tumors and NF1 animal models. *J. Clin. Investig.* **2000**, *105*, 1233–1241. [CrossRef]

81. Torres, K.E.; Zhu, Q.S.; Bill, K.; Lopez, G.; Ghadimi, M.P.; Xie, X.; Young, E.D.; Liu, J.; Nguyen, T.; Bolshakov, S.; et al. Activated MET is a molecular prognosticator and potential therapeutic target for malignant peripheral nerve sheath tumors. *Clin. Cancer Res.* **2011**, *17*, 3943–3955. [CrossRef] [PubMed]

82. Mohan, P.; Castellsague, J.; Jiang, J.; Allen, K.; Chen, H.; Nemirovsky, O.; Spyra, M.; Hu, K.; Kluwe, L.; Pujana, M.A.; et al. Genomic imbalance of HMMR/RHAMM regulates the sensitivity and response of malignant peripheral nerve sheath tumour cells to aurora kinase inhibition. *Oncotarget* **2013**, *4*, 80–93. [CrossRef] [PubMed]

83. Qin, W.; Godec, A.; Zhang, X.; Zhu, C.; Shao, J.; Tao, Y.; Bu, X.; Hirbe, A.C. TYK2 promotes malignant peripheral nerve sheath tumor progression through inhibition of cell death. *Cancer Med.* **2019**, *8*, 5232–5241. [CrossRef]

84. Lu, H.C.; Eulo, V.; Apicelli, A.J.; Pekmezci, M.; Tao, Y.; Luo, J.; Hirbe, A.C.; Dahiya, S. Aberrant ATRX protein expression is associated with poor overall survival in NF1-MPNST. *Oncotarget* **2018**, *9*, 23018–23028. [CrossRef]

85. Jessen, W.J.; Miller, S.J.; Jousma, E.; Wu, J.; Rizvi, T.A.; Brundage, M.E.; Eaves, D.; Widemann, B.; Kim, M.O.; Dombi, E.; et al. MEK inhibition exhibits efficacy in human and mouse neurofibromatosis tumors. *J. Clin. Investig.* **2013**, *123*, 340–347. [CrossRef]

86. Kolberg, M.; Høland, M.; Lind, G.E.; Agesen, T.H.; Skotheim, R.I.; Hall, K.S.; Mandahl, N.; Smeland, S.; Mertens, F.; Davidson, B.; et al. Protein expression of BIRC5, TK1, and TOP2A in malignant peripheral nerve sheath tumours—A prognostic test after surgical resection. *Mol. Oncol.* **2015**, *9*, 1129–1139. [CrossRef] [PubMed]

87. Chen, Z.; Mo, J.; Brosseau, J.P.; Shipman, T.; Wang, Y.; Liao, C.P.; Cooper, J.M.; Allaway, R.J.; Gosline, S.; Guinney, J.; et al. Spatiotemporal Loss of NF1 in Schwann Cell Lineage Leads to Different Types of Cutaneous Neurofibroma Susceptible to Modification by the Hippo Pathway. *Cancer Discov.* **2019**, *9*, 114–129. [CrossRef]

88. Kohlmeyer, J.L.; Kaemmer, C.A.; Pulliam, C.; Maharjan, C.K.; Samayoa, A.M.; Major, H.J.; Cornick, K.E.; Knepper-Adrian, V.; Khanna, R.; Sieren, J.C.; et al. RABL6A is an essential driver of MPNSTs that negatively regulates the RB1 pathway and sensitizes tumor cells to CDK4/6 inhibitors. *Clin. Cancer Res.* **2020**, *26*, 2997–3011. [CrossRef]

89. Brossier, N.M.; Prechtl, A.M.; Longo, J.F.; Barnes, S.; Wilson, L.S.; Byer, S.J.; Brosius, S.N.; Carroll, S.L. Classic Ras Proteins Promote Proliferation and Survival via Distinct Phosphoproteome Alterations in Neurofibromin-Null Malignant Peripheral Nerve Sheath Tumor Cells. *J. Neuropathol. Exp. Neurol.* **2015**, *74*, 568–586. [CrossRef]

90. Nix, J.S.; Haffner, M.C.; Ahsan, S.; Hicks, J.; De Marzo, A.M.; Blakeley, J.; Raabe, E.H.; Rodriguez, F.J. Malignant Peripheral Nerve Sheath Tumors Show Decreased Global DNA Methylation. *J. Neuropathol. Exp. Neurol.* **2018**, *77*, 958–963. [CrossRef]

91. Bradtmoller, M.; Hartmann, C.; Zietsch, J.; Jäschke, S.; Mautner, V.F.; Kurtz, A.; Park, S.J.; Baier, M.; Harder, A.; Reuss, D.; et al. Impaired Pten expression in human malignant peripheral nerve sheath tumours. *PLoS ONE* **2012**, *7*, e47595. [CrossRef] [PubMed]

92. Wojcik, J.B.; Marchione, D.M.; Sidoli, S.; Djedid, A.; Lisby, A.; Majewski, J.; Garcia, B.A. Epigenomic Reordering Induced by Polycomb Loss Drives Oncogenesis but Leads to Therapeutic Vulnerabilities in Malignant Peripheral Nerve Sheath Tumors. *Cancer Res.* **2019**, *79*, 3205–3219. [CrossRef] [PubMed]

93. Malone, C.F.; Fromm, J.A.; Maertens, O.; DeRaedt, T.; Ingraham, R.; Cichowski, K. Defining key signaling nodes and therapeutic biomarkers in NF1-mutant cancers. *Cancer Discov.* **2014**, *4*, 1062–1073. [CrossRef] [PubMed]

94. Sheikh, T.N.; Patwardhan, P.P.; Cremers, S.; Schwartz, G.K. Targeted inhibition of glutaminase as a potential new approach for the treatment of NF1 associated soft tissue malignancies. *Oncotarget* **2017**, *8*, 94054–94068. [CrossRef]

95. Lemberg, K.M.; Zhao, L.; Wu, Y.; Veeravalli, V.; Alt, J.; Aguilar, J.; Dash, R.P.; Lam, J.; Tenora, L.; Rodriguez, C.; et al. The novel glutamine antagonist prodrug JHU395 has antitumor activity in malignant peripheral nerve sheath tumor. *Mol. Cancer Ther.* **2019**, *19*, 397–408. [CrossRef]

96. Kaplan, H.G. Vemurafenib treatment of BRAF V600E-mutated malignant peripheral nerve sheath tumor. *J. Natl. Compr. Cancer Netw.* **2013**, *11*, 1466–1470. [CrossRef]

 *genes*

Review

New Model Systems and the Development of Targeted Therapies for the Treatment of Neurofibromatosis Type 1-Associated Malignant Peripheral Nerve Sheath Tumors

Kyle B. Williams [1,2] and **David A. Largaespada** [1,2,*]

1   Department of Pediatrics, University of Minnesota, Minneapolis, MN 55455, USA; kbwillia@umn.edu
2   Masonic Cancer Center, University of Minnesota, Minneapolis, MN 55455, USA
*   Correspondence: larga002@umn.edu; Tel.: +1-612-626-4979; Fax: +1-612-625-4648

Received: 2 April 2020; Accepted: 26 April 2020; Published: 28 April 2020

Abstract: Neurofibromatosis Type 1 (NF1) is a common genetic disorder and cancer predisposition syndrome (1:3000 births) caused by mutations in the tumor suppressor gene *NF1*. *NF1* encodes neurofibromin, a negative regulator of the Ras signaling pathway. Individuals with NF1 often develop benign tumors of the peripheral nervous system (neurofibromas), originating from the Schwann cell linage, some of which progress further to malignant peripheral nerve sheath tumors (MPNSTs). Treatment options for neurofibromas and MPNSTs are extremely limited, relying largely on surgical resection and cytotoxic chemotherapy. Identification of novel therapeutic targets in both benign neurofibromas and MPNSTs is critical for improved patient outcomes and quality of life. Recent clinical trials conducted in patients with NF1 for the treatment of symptomatic plexiform neurofibromas using inhibitors of the mitogen-activated protein kinase (MEK) have shown very promising results. However, MEK inhibitors do not work in all patients and have significant side effects. In addition, preliminary evidence suggests single agent use of MEK inhibitors for MPNST treatment will fail. Here, we describe the preclinical efforts that led to the identification of MEK inhibitors as promising therapeutics for the treatment of NF1-related neoplasia and possible reasons they lack single agent efficacy in the treatment of MPNSTs. In addition, we describe work to find targets other than MEK for treatment of MPNST. These have come from studies of RAS biochemistry, in vitro drug screening, forward genetic screens for Schwann cell tumors, and synthetic lethal screens in cells with oncogenic *RAS* gene mutations. Lastly, we discuss new approaches to exploit drug screening and synthetic lethality with *NF1* loss of function mutations in human Schwann cells using CRISPR/Cas9 technology.

Keywords: malignant peripheral nerve sheath tumors; plexiform neurofibromas; Schwann cells; neurofibromatosis type 1 syndrome; neurofibromin 1; genetically engineered mouse models

---

1. Neurofibromatosis Type 1 Syndrome and Associated Peripheral Nerve Sheath Tumors

Neurofibromatosis type 1 (NF1) syndrome is a common, autosomal dominant genetic disease that causes a predisposition to several kinds of tumors, especially a spectrum of benign and malignant forms of peripheral nerve sheath tumors (PNSTs). Patients with NF1 have inherited one mutant copy of the *NF1* gene, encoding the Ras GTPase activating protein neurofibromin, and tumors develop after somatic cell loss of the remaining wild type *NF1* allele. Benign Schwann cell PNSTs in patients with NF1 called plexiform neurofibromas (PNs) are common and problematic, occurring in roughly 60% of patients [1]. PNs have limited treatment options and can cause significant pain and morbidity. These PNs are composed of a complex mixture of cell types, but the neoplastic component is derived from a Schwann cell lineage cell, which has undergone loss of heterozygosity (LOH) of the *NF1* locus, with retention of the mutant allele [2]. Thus, these PN cells have no functional copies of *NF1* and do not produce any functional neurofibromin protein. Other reactive cell types within the PN, several of which are thought to help initiate and drive PN growth, include perineural and CD34+ fibroblasts, endothelial cells, neurons, and various cells of hematopoietic origin including mast cells, macrophages, and T cells [2–4]. PNs can affect any peripheral nerve, are thought to be congenital, and often grow aggressively during childhood [3]. A feared complication of the PNs is malignant transformation.

A newly recognized type of tumor along the spectrum of neurofibroma to malignant peripheral nerve sheath tumors (MPNST) is called atypical neurofibromatosis neoplasms of uncertain biological potential (ANNUBP) [5]. ANNUBP have at least two of three features not common in PNs, including loss of neurofibroma architecture, high cellularity, and high mitotic activity [5,6]. ANNUBPs are very important because they may well be premalignant tumors and an important transition step to MPNST. They often show loss of nuclear p16INK4A protein expression with variable loss of S100 and SOX10 expression, which are also common findings in MPNSTs [5]. ANNUBP have frequent *CDKN2A/CDKN2B* gene copy number loss [6,7].

MPNSTs are aggressive soft tissue sarcomas thought to be derived from PN Schwann cells. MPNSTs can occur in any nerve and do not respond to current therapies. In fact, MPNSTs are the most common cause of death of patients with NF1 [1]. It is estimated that roughly half of all MPNST patients have NF1, the other half of MPNSTs occur sporadically in patients without any obvious cancer predisposition syndrome [8]. As might be expected, sporadic MPNST occurs more commonly in older patients compared to patients with NF1 syndrome, many of whom develop MPNSTs in adolescence or young adulthood. While disputed, some data suggests that MPNSTs developing in the context of NF1 syndrome have worse clinical outcomes [9].

2. Molecular Genetics of the *NF1* Gene Product and MPNST

As mentioned above, *NF1* encodes a large GTPase activating protein (GAP) called neurofibromin. GAPs increase the intrinsic GTPase activity of small GTPases, such as the Ras superfamily of proteins. Neurofibromin has GAP activity for several Ras proteins including HRAS, KRAS, NRAS, RRAS, and perhaps others [10]. Patients with NF1 are heterozygous for *NF1* gene mutations, but the benign and malignant tumors that develop in these patients are caused, in part, by somatic cell loss of the remaining wildtype *NF1* allele [11]. Preclinical models suggest, many NF1-associated tumors have increased and prolonged RAS activation and MEK/ERK signaling after stimulation [1,12]. However, monotherapy often leads to emergence of drug resistance, and work from our lab indicates MEK inhibition can synergize with other therapeutics, such as mTOR inhibitors [13]. It is unclear if MEK inhibition will be useful in MPNST treatment, but preclinical data suggests that it may [12]. It is likely, however, that other recurrent genetic changes in MPNSTs make them relatively resistant to treatment with a MEK inhibitor alone [14–17]. Moreover, these other genetic abnormalities may be associated with additional drug sensitivities, which might be usefully exploited along with MEK inhibition.

MPNSTs also have frequent loss of either *CDKN2A* or *TP53* [18–20]. In addition, loss of function mutations in the Polycomb repressive complex 2 (PRC2) component genes, such as *SUZ12* or *EED*, are now known to be *very* common in MPNSTs, occurring in ~70% of NF1-associated cases and more than 90% of sporadic cases [21,22]. PRC2 represses many loci by causing histone H3 lysine 27 trimethylation (H3K27me3) in promoter regions [23]. Loss of H3K27me3, known to be caused by mutations in *SUZ12* or *EED*, is an indicator of a bad prognosis in MPNST [24]. These common mutations in chromatin remodeling machinery provide unique avenues of therapeutic targeting. In fact, it has been reported that *SUZ12* mutant MPNSTs are sensitive to bromodomain inhibitors, such as JQ1 [25]. To determine if such therapeutic ideas might have merit in the clinic, it is critical to ensure that the very best model systems for MPNSTs are utilized.

3. MPNST Model Systems

Human MPNST tumor models for testing new therapeutic ideas include genetically engineered mouse models (GEMMs), genetically engineered zebrafish [26–29], and established human MPNST cell lines [30]. There are two main problems with these approaches. The most often used GEMM does not recapitulate all of the genetic features of human MPNSTs, especially NF1 syndrome-associated MPNSTs, which likely develop from a pre-existing PN or ANNUBP. Secondly, there are relatively few established human MPNST cell lines in common usage in the field. We believe that MPNST therapy would benefit from development of better model systems. These should include: (1) better GEMMs, (2) genetically engineered human cells of the correct lineage(s) for an MPNST, and (3) use of primary human MPNSTs grown in vitro and in vivo as patient derived xenografts. Below, each of these three types of models are described in more detail including work already done and published, current efforts, and future innovations.

3.1. Genetically Engineered Mouse Models (GEMM)

Several GEMMs of MPNSTs have been published. While the most relevant incorporate loss of the *Nf1* gene, others have shown that Ras oncogene activation itself can cooperate with *Pten* loss to induce MPNST-like tumors [31]. Others have shown that overexpression of certain growth factor ligands or receptors in Schwann lineage cells can also induce these tumors [32]. These models have been useful for many things, including testing new therapies. However, major current limitations of these models include: few demonstrate metastasis (as is common in human MPNSTs), most do not incorporate loss of PRC2 function, and it is currently unclear if any really represent human MPNSTs at the transcriptomic level or reflect human MPNST molecular subtypes if they exist.

Several investigators have shown that Schwann cell lineage overexpression of growth factor receptors or ligands can promote peripheral nerve sheath tumor development. The ligand neuregulin-1, which binds to the ERBB3 and ERBB4 receptors, when expressed from the P_0 promoter leads to peripheral nerve sheath tumor formation [33,34], which is accelerated to a higher grade by concomitant loss of *Trp53* [35]. Expression of human epidermal growth factor receptor (hEGFR) from a Desert hedgehog (*Dhh*) promoter can induce neurofibromas by itself [36], and cooperate with expression of a dominantly acting $Trp53^{R270H}$ mutant [37], or loss of one or two copies of *Pten* [32]. Human *PTEN* is often deleted in human MPNSTs and is expressed at reduced levels compared to normal human Schwann cells or Schwann cells from benign neurofibromas [38]. Indeed, codeletion of *Nf1* and *Pten* in *Dhh-Cre* positive cells caused rapid high grade peripheral nerve sheath tumors [38]. Another paper also showed that loss of *Pten* potently accelerated high grade MPNST-like tumors in the presence of Schwann cell lineage expression of $Kras^{G12D}$ [31]. Taken together these data strongly suggest that *PTEN*-regulated pathways are major suppressors of progression to MPNST. Which PTEN-regulated processes are most important is not clear, but evidence suggests AKT-driven beta-catenin activation [39], and/or mTOR activation might be critical effectors [40]. *CDKN2A* is a well-established human MPNST tumor suppressor, frequently lost in pre-malignant, atypical neurofibromas [5,41]. When combined with deletion of *Nf1* in the Schwann cell lineage, signatures of senescence are suppressed in resultant neurofibromas, which appear as faster developing, higher grade tumors than produced by deletion of *Nf1* alone [7]. Thus, this work provides a good model of ANNUBP and it suggests that senescence limits neurofibroma progression. Therefore, therapies that target senescent cells, so called senolytics [42], could be useful to prevent peripheral nerve sheath tumor progression.

Two papers report *Sleeping Beauty* (SB) transposon-based insertion mutation screens in *Dhh-Cre* or *Cnp-Cre* positive cells in mice in vivo [43,44]. These studies were done on an $Nf1^{flox/flox}$ [44] or *loxP-STOP-loxP-Trp53*^{R270H} (*LSL-Trp53*^{R270H}) plus *Cnp-hEGFR* mutant backgrounds [43]. The result of SB mutagenesis in *Dhh-Cre* positive cells on the $Nf1^{flox/flox}$ background was to increase tumor multiplicity but not tumor grade, which were all benign neurofibromas. However, SB mutagenesis in *Cnp-Cre* positive cells on a *LSL-Trp53*^{R270H} plus *Cnp-hEGFR* background resulted in accelerated high-grade MPNSTs like tumor development in most mice. While this model is primarily useful for the genes and pathways it revealed as potential drivers of MPNST, tumors made in this way could be used to select for additional tumor phenotypes because the tumors can be allografted and SB transposon mutagenesis is an ongoing process. It may be possible to select for metastatic potential and treatment resistance using this SB model. In any case, the SB screen revealed PTEN regulated signaling (e.g., PI3 kinase), Wnt-beta catenin signaling, RAS-MAPK signaling, Hippo/Yap signaling, Myc activation through FOXR2, SHH signaling, and other pathways as potential drivers of MPNST [43]. Moreover, while inactivating SB transposon insertion mutations in *Eed* and *Suz12* were not recovered in the screen, inactivating *Jarid2* and *Nsd1* insertion mutations were recovered [43]. JARID2 is a critical component of the PCR2 complex required for localization of the PRC2 complex on chromatin, and has been reviewed before [45]. NSD1 is required to prevent histone H3 lysine 27 trimethylation (H3K27me3) from spreading to new regions of chromatin genome-wide, the effect of which is to decrease H3K27me3 in its usual domains [46]. Thus, the SB screen did reveal a role of H3K27me3 function in MPNST suppression.

Perhaps the most well-studied GEMM for MPNST is the so-called *NPCis* mouse, in which loss of function mutations in *Nf1* and *Trp53* are placed in a Cis configuration as both genes reside closely linked on chromosome 11 [47,48]. As in the mouse, these two TSGs are closely linked in the human genome. In *NPCis* mice, therefore, a single loss of heterozygosity event can result in complete elimination of both TSGs. Two different groups reported that *NPCis* mice develop a variety of sarcomas, including MPNST-like tumors, as well as gliomas [48]. Strain-specific effects influence the frequency of MPNST-like tumors, with more developing on a 129/SvJ background than on a C57BL/6J background [49]. The *NPCis* mouse model has been used to test a variety of new therapies [40,50–53]. The major limitation of the *NPCis* model may have to do with the fact that these tumors do not develop through a process that involves progression from benign neurofibroma (with loss of *Nf1* only), to pre-malignant atypical neurofibroma (with loss of *Nf1* and *Cdkn2a*), to MPNST with loss of *Suz12* or *Eed*. However, heterozygous germline loss of function mutations in *Nf1* and *Suz12* have been combined in Cis on chromosome 11 and this resulted in the acceleration of neurofibroma and MPNST-like tumors [25]. Furthermore, heterozygous germline mutations in *Nf1*, *Trp53* and *Suz12* have been combined and the result was acceleration of MPNST-like tumor formation, along with other tumor types including glioma, lymphoma and histiocytic sarcoma [25]. The utility of the MPNST models utilizing the *Suz12* germline mutation is limited by the fact that other types of cancer develop in the mice, the highly stochastic nature of tumor development, and lack of a suitable method for labeling the MPNST cells as they develop so that in vivo imaging might be accomplished. Some of these concerns could be addressed using conditional alleles for *Cdkn2a* or *Trp53* and for *Suz12*. Another approach would be to use somatic cell editing in situ. Indeed, adenoviral delivery of Cre to the sciatic nerves of *Nf1^{flox/flox}; Cdkn2a^{flox/flox}* mice resulted in development of MPNST-like tumors at the injection site [54]. This model was used to demonstrate a positive, driver role for the *Nf1*+/− heterozygous field in promoting MPNST-like tumor development [55]. More recently, it has been shown that adenovirus-mediated somatic cell delivery of single guide RNAs (sgRNA) and Cas9 to the sciatic nerve can produce MPNST-like tumors [56]. The flexibility of this approach is promising since building murine MPNST-like tumors that are truly genetically similar to their human counterparts may require layering many mutations in the correct temporal order. Achieving this remains a major challenge in the field.

3.2. Human MPNST Cell Lines and Patient Derived Xenografts

The MPNST research field has been hampered by a relative dearth of established cell lines [57,58]. Moreover, cell lines that do exist have not been well characterized. Commonly, only a few are used to test new hypotheses. Critically, none are included in large scale whole exome/whole genome sequencing, gene expression profiling, drug, RNAi, or CRISPR/Cas9 screening projects. The field will benefit tremendously from better characterization of existing lines and establishment of more human MPNST cell lines. Similarly, only a few human MPNST patient derived xenografts (PDXs) have been described in the literature [59,60]. Apparently, MPNST PDXs can be fairly readily established in immunodeficient mice and this also has been our experience using NOD-*Rag1^{null.} Il2rg^{null}* (NRG) mice. New efforts to make additional human MPNST cell lines and PDXs are underway and some of these new these resources are available to the research field already (https://www.hopkinsmedicine.org/kimmel_cancer_center/centers/pediatric_oncology/research_and_clinical_trials/pratilas/nf1_biospecimen_repository.html).

3.3. Human Cell-Based Models for MPNST

In theory, engineering relevant mutations into primary human cells of the correct lineage would provide models useful for synthetic lethal genetic screens and drug screens. Models of neurofibroma and MPNST made in this way would also guard against the possibility that mouse cells fail to accurately recapitulate aspects of human cell biology relevant to MPNST development. Making a permanent, non-perishable cell line resource of this kind is difficult using purified, primary human Schwann cells as they have limited proliferative potential in vitro and are difficult to culture without contaminating fibroblasts [61]. For this reason, Dr. Margaret (Peggy) Wallace, Ph.D., at the University of Florida, Gainesville, has pioneered the generation of immortalized human Schwann cells from normal nerves of non-patients with NF1 and patients with NF1, as well as neurofibromas [62]. These cells were immortalized using lentiviral transduction of human *TERT* and murine *Cdk4* transgenes. These immortalized cells have been useful for testing the effects of candidate MPNST oncogenes and tumor suppressor genes on human Schwan transformation [43,63]. The *NF1−/−* plexiform neurofibroma derived Schwann cells provide a useful cell culture model of these benign tumors [62], and provide a logical platform for studying progression to MPNST.

A permanent, non-perishable, and human cell-based model for MPNST modeling would be best accomplished using induced pluripotent stem cells (iPSC). Indeed, *NF1−/−* iPSC have been made from plexiform neurofibroma cells from patients with NF1 [64]. When these *NF1−/−* iPSCs were differentiated to neural crest cells and subsequently Schwann cells, it was found that they had an enhanced proliferation rate, poor myelination ability, and a tendency to form 3D spheres like those from primary neurofibromas compared to Schwann cells made from wild type iPSC. Thus, these Schwann cells made from a *NF1−/−* iPSC represent a valuable model to study and treat plexiform neurofibromas. It remains to be seen whether these iPSC, or differentiated progeny derived from them, can be used to model ANNUBP and MPNST by knocking out *CDKN2A* and *SUZ12*. It is also unclear how the differentiation state of the target cell and order of mutation would affect the outcome of studies like these. In any case, CRISPR/Cas9 based models produced this way are an ideal method to produce isogenic sets of relevant human cells for drug and synthetic genetic lethality screens.

3.4. Synthetic Lethality as A Tool for NF1 Drug Discovery

Synthetic lethality is the genetic incompatibility of the loss of two or more gene products, leading to cell death. However, the deficiency of any of these gene products on their own results in viable cells (Figure 1). The notion of synthetic lethality was first described in work on Drosophila genetics in the 1920's by Calvin B. Bridges and others who noticed parental lines of flies harboring mutations were able to successfully reproduce, yet when these independent mutant lines were crossed it was not possible to obtain viable offspring with a combination phenotype [65]. Using the power of synthetic lethality to identify novel vulnerabilities in a given cancer based on defined genetic backgrounds was first suggested over 20 years ago [66]. The first type of investigation to move to clinical relevance was that of the use of inhibitors of poly(ADP-ribose) polymerase (PARP) to selectively kill BRCA-2 deficient tumors [67]. Now PARP inhibitors such are olaparib are approved for BRACA mutated ovarian cancers [68].

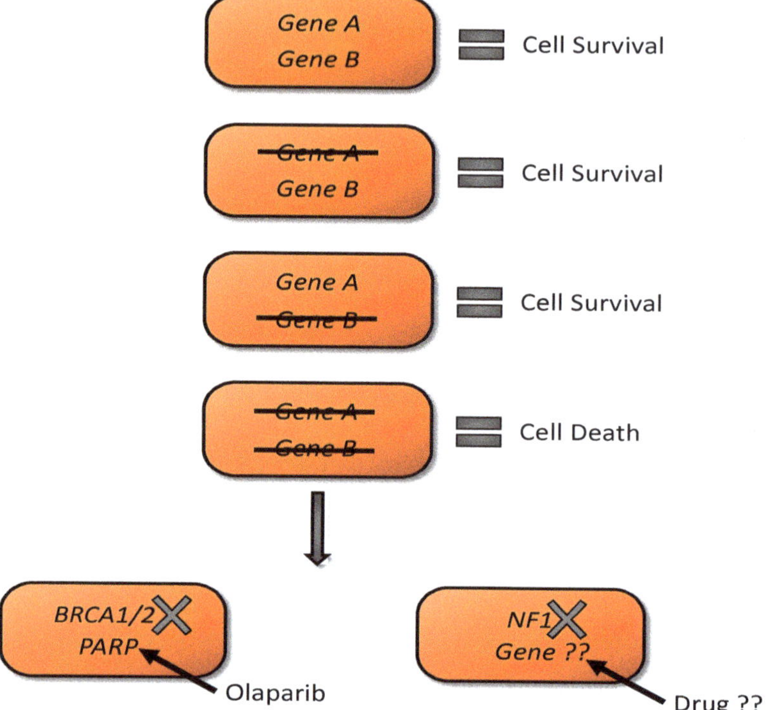

Figure 1. Synthetic lethality if the genetic incompatibility of the loss of two independent gene products results in cell death. Whereas a hypothetical cell can survive with *either* loss of *Gene A* or *Gene B*, it cannot live with loss of both simultaneously. A prime example of this being exploited therapeutically is in *BRCA1/2* mutated cancers and their sensitivity to PARI inhibitors such as olaprarib. Similarly, loss of *NF1* could render human Schwann cells sensitive to pharmacologic intervention with novel agents.

Treatment options for plexiform neurofibromas and malignant peripheral nerve sheath tumors (MPNST) are limited, relying mostly on surgical resection and broad-spectrum chemotherapy. The genetic basis of NF1 syndrome is well suited for using synthetic lethal genetic screens and related approaches to uncover unique variabilities in *NF1* deficient cells, as well as cells closely mimicking the genetics of an MPNST. Loss of the *NF1* gene could uncover sensitivity to loss or impairment of another gene or pathway. All neurofibromas, and derived MPNSTs, harbor *NF1*-deficient Schwann cells, which may have re-wired signaling to reveal unique vulnerabilities. However, the synthetic lethal interactions after loss of *NF1* are likely to be highly context dependent. Ideally, they should be operative in human Schwann cells. The discovery of such interactions would most easily be found using pairs of isogenic, human *NF1*-deficient and *NF1*-proficient Schwann cells. Genome engineering technologies, such as CRISPR/Cas9, allow introduction of clinically relevant mutations into cells of the correct tissue type for the disease being studied. Cells engineered in this manner would be poised to be used for novel drug discovery for not only NF1-associated malignancies, but many other cancers harboring loss of tumor suppressors or with known oncogenic drivers [69].

Efforts have been made to undertake this type of study for the *NF1* deficient cells. A small-scale screening project was conducted using mouse embryonic fibroblasts harboring biallelic loss of *NF1* [70]. This screening effort yielded only modest amounts of selective lethality toward *NF1* null cells and it is unclear how these could translate to the clinic. It would be most useful to undertake these types of screening efforts in cells of the correct cell-type for neurofibromas and MPNST (i.e., human Schwann cells). There are two cell-based platforms that would be amenable to the required genetic manipulation and required screening efforts: 1. Immortalized human Schwann cells. 2. iPSC derived Schwann cells. A number of immortalized human Schwann cell lines have been created that could be used for these types of assays [71]. These do have the limitation of being immortalized with retroviral vectors carrying *hTERT* for example and thus not mirroring the exact genetics of an MPNST cell. Induced pluripotent stem cell-based models are also a powerful tool for these types of studies. For example, protocols have been established to differentiate iPSC cells to the neural crest linage and then further to Schwann cells [64]. iPSC lines harboring *NF1* loss would have the advantage of being a "cleaner" genetic background, but perhaps not as easy to work with on large scale, genome-wide genetic screens. However, both systems could be used for: 1. Medium to large scale small molecule screens looking for drugs that could selectively kill *NF1* null cells. 2. Genome-wide genetic screens looking for other gene products whose loss would result in cell death when combined with *NF1* deficiency (Figure 2). The genetic screens are extremely powerful, as they offer a truly non-biased approach for novel target discovery. Whereas the small molecule screens could yield a potential therapeutic compound faster (particularity if the drug is already approved for another indication).

3.5. Preclinical Development of New Therapies for NF1-Associated MPNST

A clear, standout success has been the identification of MEK inhibitors as effective treatment for symptomatic, inoperable plexiform neurofibromas [72]. Work examining the preclinical effectiveness of a selective pharmacological inhibitor of MEK, PD0325901, was reported in 2013 [12]. This work demonstrated that MAPK signaling suppression was effective in controlling neurofibroma growth in a neurofibromatosis mouse model (*Nf1^{fl/fl}*;*Dhh-Cre*) and an NF1 patient MPNST cell xenograft. PD0325901 treatment increased survival of mice implanted with human MPNST cells, and shrank neurofibromas in more than 80% of the mice enrolled on treatment [73]. This important work clearly demonstrated that Ras/ERK signaling is critical for growth of NF1-associated PNSTs and provided reasoning to initiate clinical trials of MEK inhibitors for the treatment of these tumors in patients with NF1.

Following these promising results with PD0325901 in various in vivo models, investigation of other MEK inhibitors (including trametinib and selumetinib) were performed and advanced to human clinical trials. The most striking of these to date were the responses seen with selumetinib in pediatric patients with NF1 and inoperable plexiform neurofibromas [72]. Again, for this study the preclinical modeling of the MEK inhibitor was tested in the *Nf1^{fl/fl}*; *Dhh-Cre* mouse model. As assessed by volumetric MRI, 67% of mice showed a reduction in tumor volume from baseline. When assessed in the human patient population, the results were marked, with confirmed partial responses in >70% of patients ($\geq$20% decrease in volume from baseline). Moreover, disease progression was not reported in any patient and decreased tumor-related pain and functional impairment was widely reported. Recent Phase II trials also reported reduced tumor growth and in some cases a reduction in PN-associated pain and quality of life [74]. This is a major breakthrough in PN therapy and selumetinib was granted approval by the FDA for treatment of NF1-associated PN in April of 2020.

The success of selumetinib and other MEK inhibitors for NF1-associated PNs and other benign tumors represents a major milestone in therapy for this disease. However, several important questions remain. Selumetinib treatment effects are not complete, as tumors usually shrink by 20% or more, but do not disappear completely and typically regrow after treatment cessation [74]. It would be ideal if robust responses were observed in more patients. Moreover, MEK inhibitors have serious side effects when used long term, including skin rashes, ocular and cardiac toxicities [75]. Finally, it is unclear if MEK inhibition will be useful in the context of the premalignant atypical NF/ANNUBP in which *CDKN2A/2B* gene deletions are present or MPNSTs.

Given the success of MEK inhibition for the control of plexiform neurofibromas and that many MPNST arise from within existing plexiform neurofibromas or ANNUBPs, it would seem likely that patients with MPNST could already be undergoing treatment with a MEK inhibitor. As such, it would be critical to establish any future targeted therapy for MPNST is at the least not antagonistic when used in combination with MEK inhibitors, such as selumetinib. Indeed, it would be desirable to identify novel candidate therapeutics exhibiting synergistic effects against MPNST models when used alongside MEK inhibitors.

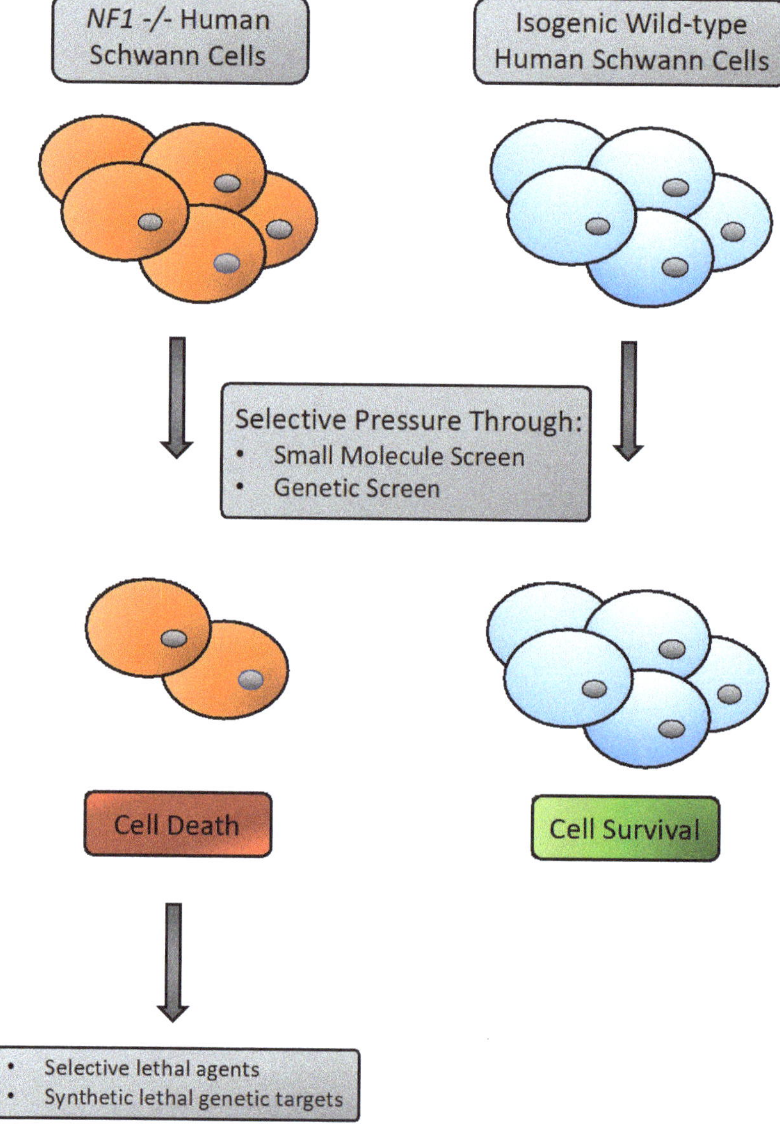

Figure 2. Screening scheme utilizing *NF1*-deficeint immortalized or iPSC derived human Schwann cells. *NF1* −/− cells (of either origin) and their isogenic match parental lines can be used in selective lethal pharmacologic screens or synthetic lethal genetic screens. Compounds found to only kill *NF1*-deficient cells could be then prioritized for further preclinical testing, using in vivo models of MPNST for example. Novel genetic targets identified from genome-wide synthetic lethal genetic screens could provide the basis for additional drug discovery efforts.

3.6. Targeting Ras or Ras-Activated Signaling Pathways

Ras proteins are small, membrane associated GTPases that exist in an inactive guanosine diphosphate (GDP) bound form and an active guanosine triphosphate (GTP) bound form. Biallelic loss of the *NF1* gene, and the encoded protein neurofibromin, has been shown to lead to increased and prolonged Ras activation (i.e., Ras-GTP), including in benign and malignant peripheral nerve sheath tumors [10]. Ras-GTP itself has been considered undruggable as it lacks deep grooves that would fit a small molecule [76]. Ras proteins are post-translationally modified, including by farnesylation, and the enzymes that carry out these effects have been tested for therapeutic effects with some success in treating NF1-associated peripheral nerve sheath tumor cells [77–79]. However, translation into human clinical trials was disappointing [80,81]. Ras-GTP activated pathways have provided an alternate pathway to reverse the effects of *NF1* gene loss. A clear example, is Ras activation of the MAPK pathway, which proceeds from RAS activation to activation of a series of kinases: RAF to MEK to ERK(MAPK) kinase activation. MAPK ultimately phosphorylates many other substrates that can confer a survival or proliferation advantage to RAS activated cells [13]. As described above, a number of MEK inhibitors, notably selumetinib, have shown some clinical activity versus PN and other benign NF1-associated tumors [74,82].

3.7. Combination Signal Pathway Inhibition

Given the incomplete effects of MEK inhibition alone on PN and MPNST, combinations with MEK inhibitors have been explored. MEK inhibitors have been shown to cooperate in vitro and in vivo with inhibition of the MNK kinases, which are active in many MPNSTs and converge on mTOR-dependent eIF4E phosphorylation [53]. Indeed, several papers show that *NF1*-deficient cells are relatively dependent on mTOR signaling and that mTORC1 inhibition synergizes with MEK inhibition [50]. mTORC1 inhibition as a single agent was weakly effective for PN treatment in patients [83]. Better results may be obtained with a direct mTOR kinase inhibitor that inhibits both mTORC1 and mTORC2 [84]. MEK inhibition and mTORC1, or mTORC1 and mTORC2 inhibition are synergistic in vitro and in vivo [13,84], and such a combination may be useful for PNs and MPNST treatment. The mTORC1 inhibitor sirolimus plus selumetinib is currently in a Phase II MPNST trial [NCT03433183; https://www.ctf.org/research/clinical-drug-pipeline]. mTOR inhibition was also combined with other signaling inhibitors including the HSP90 inhibitor ganetespib [NCT02008877], which did not produce responses in MPNST patients [85], despite promising preclinical data in the *NPCis* GEM model [86]. The mTORC1 inhibitor everolimus plus the VEGF inhibitor bevacizumab also failed to show potential in a Phase II MPNST study [87]. The multi-kinase inhibitor PLX3397 in combination with rapamycin inhibited MPNST xenografts in vivo [88], and this has led to a Phase II clinical trial [NCT02584647]. A more recent study revealed that blockade of mTOR with sapanisertib, which inhibits mTORC1 and mTORC2, with histone deacetylase (HDAC) inhibition is selectively toxic to Ras pathway-driven tumors, including human MPNST xenografts and the *NPCis* GEM model, by converging on the TXNIP/thioredoxin pathway [52]. The polo-like kinases (PLKs) have also emerged as targets for MPSNT therapy [89].

3.8. Targeting Cyclin-Dependent Kinases for MPNST

Premalignant atypical PN/ANNUBP have often acquired loss of *CDKN2A* and the *CDKN2B* genes, also a feature of many MPNSTs. These genes encode cyclin dependent kinase inhibitors and their loss is thought to lead to unregulated cyclin/cyclin dependent kinase (CDK) activity, in particular CDK4 or CDK6. Thus, these tumors may become sensitive to CDK4/6 inhibitors like ribociclib or palbociclib, which are approved for breast cancer treatment. Interestingly, high level expression of RABL6A in some MPNSTs promotes growth, in part by inhibition of RB1 protein, and its effects can be blocked by CDK4/6 inhibition [90]. CDK4/6 inhibition using ribociclib plus doxorubicin is in Phase II trial for MPNST now [NCT03009201].

3.9. Sensitivities Associated with Loss of PRC2 Function in MPNST Cells

As described above, progression to MPNST usually involves biallelic loss of the *EED* or *SUZ12* genes, essential components of the polycomb repressor complex 2 (PRC2). Most MPNSTs lack, or have reduced, histone H3 lysine 27 di- and tri-methylation, a chromatin repression mark mediated by EZH2, the enzymatic component of the PRC2 complex [22,24]. Loss of PRC2 activity has been proposed to sensitize MPNST cells to BRD4 inhibition, as BRD4 seems to be involved in activation of Ras pathway-dependent transcription dependent on histone H3 lysine 27 acetylation, which is enhanced following loss of methylation at this same site [25]. Another recent publication strongly suggests that loss of PRC2 in MPNST cells sensitizes them to histone deacetylase (HDAC) and DNA methyltransferase inhibition [91]. If these results are validated in isogenic model systems and proper in vivo preclinical MPNST models, these drugs should be tested in vivo, possibly with the signaling inhibitors described above that leverage sensitivities from loss of *NF1* expression.

3.10. Other Therapeutic Approaches

Several other therapeutic approaches have made it to clinical trials for MPNST in recent years. Excitement for immunotherapies, especially checkpoint blockade and chimeric antigen receptor (CAR) T cell or CAR-T therapy has been building for treatment of MPNST. Indeed, blockade of PD1 [NCT02691026] or PD-1 and CTLA-4 [NCT02834013] are in clinical trials for MPNST. CAR-T cells specific for human EGFR are also being tested clinically in EGFR+ MPSNT [NCT03618381]. The perinuclear compartment disrupter metarrestin [92] is in Phase I trials for MPNST [NCT04222413], although not published specifically in MPNST. Inhibition of NFκB signaling and MPNST preclinical effectiveness was demonstrated using selinexor, a compound that induces IkB nuclear localization [93], and this agent is in Phase I testing for MPNST [NCT03880123]. Oncolytic viral therapy is a relatively new approach for MPNST therapy, but several reports describe promising preclinical data [94–100].

3.11. Comprehensive Pharmacological Profiling of Neurofibromatosis Type 1 Cancer Cell Lines

One attractive method for discovering new therapeutic options for NF1-associated tumors is to perform drug screens using one or more of the model systems described above. Drug screens in animal models are likely to be most predictive of future efficacy in vivo. Indeed, some measure of predictability for drugs to treat PNs in people was obtained using the *P0-Cre; Nf1^{fl}/Nf1^{fl}* or *Dhh-Cre.; Nf1^{fl}/Nf1^{fl}* models [12,87,101]. Unfortunately, these in vivo models are not well suited to testing many drugs and drug combinations. Therefore, screens using cell lines or cell strains have been used instead. In one medium throughput drug screen of about 470 compounds, immortalized *Nf1*-deficient mouse embryo fibroblasts (MEFs) were screened, using wildtype MEFs as a counter screen [70]. This project revealed several drugs with some selectivity for Nf1-deficienct cells. Moreover, validation of screening results using a human NF1-associated MPNST cell line xenograft model showed that nifedipine treatment significantly decreased local tumor growth. The reported data suggest that inhibitors of PP2A, including cantharidins, as well as calcium channel blockers like nifedipine, might be useful in MPNST treatment. A screen of this nature would benefit from the use of human isogenic cells of the correct cell lineage rather than fibroblasts. It is also possible that screening a larger number of drugs would yield additional drug candidates.

3.12. Synthetic Sensitivities Identified by Drug Screening in Nf1-Deficient Mouse Embryo Fibroblasts

In another in vitro drug screening effort, seven NF1-associated human MPNST and one sporadic MPNST cell line were used in a medium throughput screen of 130 drugs predicted to be useful for NF1-associated cancer therapy based on an analysis of the literature [102]. These drugs were based on mechanistic knowledge of neurofibromin and merlin function, as *NF2*-mutant cells were screened also, as well as important cancer pathways and classic chemotherapies. Drugs were found that clearly differentiated *NF1* from *NF2* mutant cells. Raf, MEK, PI3K/mTOR and Pak inhibitors were active versus *NF1*-mutant MPNST cells, while EGFR, GSK3, and AKT inhibitors had almost no activity. These data are consistent with the concept of using combination therapy centered on targeting the RAS-MAPK pathway in NF1-associated tumor cells. Additional drugs can then be layered on to maximize tumor-specific killing by targeting additional parallel vulnerabilities. This concept should be tested in vivo in improved models for NF1-associated peripheral nerve sheath tumors.

4. Summary and Future Perspectives

It has become clear that improved model systems are needed both to better validate current ideas before clinical testing in patients and to discover new MPNST treatment approaches and vulnerabilities. Too many MPNST clinical approaches have stalled at Phase II because they did not meet primary response rate endpoints and/or the effects were modest. The field is in desperate need of a home run. MPNST lack obvious activated oncogene products to target. Therefore, we favor the concept of using carefully engineered isogenic human Schwann lineage cell lines, harboring MPNST-relevant mutations, to screen for mutation-specific drug sensitivities and synthetic lethal genetic interactions to discover new therapies. Human iPSC derived model systems are likely to be especially useful in this regard. New therapeutic drugs and combinations must be better vetted before clinical testing and we favor testing in multiple human MPSNT PDXs and looking for evidence of very profound and long-lasting tumor shrinkage. Because in NF1 we can consider measures to reduce the chance of malignant progression we encourage the development of genetically accurate models in which benign PN, premalignant atypical PN, and MPNST can be studied using human cells and GEMMs. Immunoproficient GEMMs may also play an important role in development of new immune based and oncolytic viral therapies. Many good candidate therapeutic targets for *NF1*, *CDKN2A* and PRC2-deficient MPNST have emerged in recent years. These drugs, perhaps in combination, may also be relevant in tumors that characterize other RASopathies or which show activation of RAS-MAPK signaling after NF1 gene loss in sporadic settings.

Author Contributions: D.A.L. and K.B.W. both conceived of and wrote portions of the manuscript and each reviewed the entire manuscript. All authors have read and agreed to the published version of the manuscript.

Acknowledgments: We thanks Pauline Beckman, for critical review of the manuscript.

Conflicts of Interest: D.A.L. is the co-founder and co-owner of several biotechnology companies including NeoClone Biotechnologies, Inc., Discovery Genomics, Inc. (recently acquired by Immusoft, Inc.), B-MoGen Biotechnologies, Inc. (recently acquired by the biotechne corporation), and Luminary Therapeutics, Inc. He consults for Genentech, Inc., which is funding some of his research. D.A.L. holds equity in and serves as the Chief Scientific Officer of Surrogen, a subsidiary of Recombinetics, a genome-editing company. The business of all these companies is unrelated to the contents of this manuscript. K.B.W. has no conflicts of interest to disclose.

References

1. Gutmann, D.H.; Ferner, R.E.; Listernick, R.H.; Korf, B.R.; Wolters, P.L.; Johnson, K.J. Neurofibromatosis type 1. *Nat. Rev. Dis Primers.* **2017**, *3*, 17004. [CrossRef] [PubMed]

2. Hernandez-Martin, A.; Duat-Rodriguez, A. An Update on Neurofibromatosis Type 1: Not Just Cafe-au-Lait Spots and Freckling. Part II. Other Skin Manifestations Characteristic of NF1. NF1 and Cancer. *Actas Dermosifiliogr.* **2016**, *107*, 465–473. [CrossRef] [PubMed]

3. Longo, J.F.; Weber, S.M.; Turner-Ivey, B.P.; Carroll, S.L. Recent Advances in the Diagnosis and Pathogenesis of Neurofibromatosis Type 1 (NF1)-associated Peripheral Nervous System Neoplasms. *Adv. Anat. Pathol.* **2018**, *25*, 353–368. [CrossRef]

4. Brosseau, J.P.; Liao, C.P.; Wang, Y.; Ramani, V.; Vandergriff, T.; Lee, M.; Patel, A.; Ariizumi, K.; Le, L.Q. NF1 heterozygosity fosters de novo tumorigenesis but impairs malignant transformation. *Nat. Commun.* **2018**, *9*, 5014. [CrossRef]

5. Miettinen, M.M.; Antonescu, C.R.; Fletcher, C.D.M.; Kim, A.; Lazar, A.J.; Quezado, M.M.; Reilly, K.M.; Stemmer-Rachamimov, A.; Stewart, D.R.; Viskochil, D.; et al. Histopathologic evaluation of atypical neurofibromatous tumors and their transformation into malignant peripheral nerve sheath tumor in patients with neurofibromatosis 1-a consensus overview. *Hum. Pathol.* **2017**, *67*, 1–10. [CrossRef] [PubMed]

6. Beert, E.; Brems, H.; Daniels, B.; De Wever, I.; Van Calenbergh, F.; Schoenaers, J.; Debiec-Rychter, M.; Gevaert, O.; De Raedt, T.; Van Den Bruel, A.; et al. Atypical neurofibromas in neurofibromatosis type 1 are premalignant tumors. *Genes Chromosomes Cancer* **2011**, *50*, 1021–1032. [CrossRef]

7. Carrio, M.; Gel, B.; Terribas, E.; Zucchiatti, A.C.; Moline, T.; Rosas, I.; Teule, A.; Ramon, Y.C.S.; Lopez-Gutierrez, J.C.; Blanco, I.; et al. Analysis of intratumor heterogeneity in Neurofibromatosis type 1 plexiform neurofibromas and neurofibromas with atypical features: Correlating histological and genomic findings. *Hum. Mutat.* **2018**, *39*, 1112–1125. [CrossRef]

8. Farid, M.; Demicco, E.G.; Garcia, R.; Ahn, L.; Merola, P.R.; Cioffi, A.; Maki, R.G. Malignant peripheral nerve sheath tumors. *Oncologist* **2014**, *19*, 193–201. [CrossRef]

9. Kolberg, M.; Holand, M.; Agesen, T.H.; Brekke, H.R.; Liestol, K.; Hall, K.S.; Mertens, F.; Picci, P.; Smeland, S.; Lothe, R.A. Survival meta-analyses for >1800 malignant peripheral nerve sheath tumor patients with and without neurofibromatosis type 1. *Neuro. Oncol.* **2013**, *15*, 135–147. [CrossRef]

10. Ratner, N.; Miller, S.J. A RASopathy gene commonly mutated in cancer: The neurofibromatosis type 1 tumour suppressor. *Nat. Rev. Cancer* **2015**, *15*, 290–301. [CrossRef]

11. Laycock-van Spyk, S.; Thomas, N.; Cooper, D.N.; Upadhyaya, M. Neurofibromatosis type 1-associated tumours: Their somatic mutational spectrum and pathogenesis. *Hum. Genom.* **2011**, *5*, 623–690. [CrossRef]

12. Jessen, W.J.; Miller, S.J.; Jousma, E.; Wu, J.; Rizvi, T.A.; Brundage, M.E.; Eaves, D.; Widemann, B.; Kim, M.O.; Dombi, E.; et al. MEK inhibition exhibits efficacy in human and mouse neurofibromatosis tumors. *J. Clin. Invest.* **2013**, *123*, 340–347. [CrossRef]

13. Watson, A.L.; Anderson, L.K.; Greeley, A.D.; Keng, V.W.; Rahrmann, E.P.; Halfond, A.L.; Powell, N.M.; Collins, M.H.; Rizvi, T.; Moertel, C.L.; et al. Co-targeting the MAPK and PI3K/AKT/mTOR pathways in two genetically engineered mouse models of schwann cell tumors reduces tumor grade and multiplicity. *Oncotarget* **2014**, *5*, 1502–1514. [CrossRef]

14. Fischer-Huchzermeyer, S.; Chikobava, L.; Stahn, V.; Zangarini, M.; Berry, P.; Veal, G.J.; Senner, V.; Mautner, V.F.; Harder, A. Testing ATRA and MEK inhibitor PD0325901 effectiveness in a nude mouse model for human MPNST xenografts. *BMC Res. Notes.* **2018**, *11*, 520. [CrossRef]

15. Peacock, J.D.; Cherba, D.; Kampfschulte, K.; Smith, M.K.; Monks, N.R.; Webb, C.P.; Steensma, M. Molecular-guided therapy predictions reveal drug resistance phenotypes and treatment alternatives in malignant peripheral nerve sheath tumors. *J. Transl. Med.* **2013**, *11*, 213. [CrossRef]

16. Peacock, J.D.; Pridgeon, M.G.; Tovar, E.A.; Essenburg, C.J.; Bowman, M.; Madaj, Z.; Koeman, J.; Boguslawski, E.A.; Grit, J.; Dodd, R.D.; et al. Genomic Status of MET Potentiates Sensitivity to MET and MEK Inhibition in NF1-Related Malignant Peripheral Nerve Sheath Tumors. *Cancer Res.* **2018**, *78*, 3672–3687. [CrossRef]

17. Ahsan, S.; Ge, Y.; Tainsky, M.A. Combinatorial therapeutic targeting of BMP2 and MEK-ERK pathways in NF1-associated malignant peripheral nerve sheath tumors. *Oncotarget.* **2016**, *7*, 57171–57185. [CrossRef]

18. Verdijk, R.M.; den Bakker, M.A.; Dubbink, H.J.; Hop, W.C.; Dinjens, W.N.; Kros, J.M. TP53 mutation analysis of malignant peripheral nerve sheath tumors. *J. Neuropathol. Exp. Neurol.* **2010**, *69*, 16–26. [CrossRef]

19. Brohl, A.S.; Kahen, E.; Yoder, S.J.; Teer, J.K.; Reed, D.R. The genomic landscape of malignant peripheral nerve sheath tumors: Diverse drivers of Ras pathway activation. *Sci. Rep.* **2017**, *7*, 14992. [CrossRef]

20. Kaplan, H.G.; Rostad, S.; Ross, J.S.; Ali, S.M.; Millis, S.Z. Genomic Profiling in Patients With Malignant Peripheral Nerve Sheath Tumors Reveals Multiple Pathways With Targetable Mutations. *J. Natl. Compr. Cancer Netw.* **2018**, *16*, 967–974. [CrossRef]

21. Zhang, M.; Wang, Y.; Jones, S.; Sausen, M.; McMahon, K.; Sharma, R.; Wang, Q.; Belzberg, A.J.; Chaichana, K.; Gallia, G.L.; et al. Somatic mutations of SUZ12 in malignant peripheral nerve sheath tumors. *Nat. Genet.* **2014**, *46*, 1170–1172. [CrossRef] [PubMed]

22. Lee, W.; Teckie, S.; Wiesner, T.; Ran, L.; Prieto Granada, C.N.; Lin, M.; Zhu, S.; Cao, Z.; Liang, Y.; Sboner, A.; et al. PRC2 is recurrently inactivated through EED or SUZ12 loss in malignant peripheral nerve sheath tumors. *Nat. Genet.* **2014**, *46*, 1227–1232. [CrossRef] [PubMed]

23. Moritz, L.E.; Trievel, R.C. Structure, mechanism, and regulation of polycomb repressive complex 2. *J. Biol. Chem.* **2018**, *293*, 13805–13814. [CrossRef] [PubMed]

24. Cleven, A.H.; Sannaa, G.A.; Briaire-de Bruijn, I.; Ingram, D.R.; van de Rijn, M.; Rubin, B.P.; de Vries, M.W.; Watson, K.L.; Torres, K.E.; Wang, W.L.; et al. Loss of H3K27 tri-methylation is a diagnostic marker for malignant peripheral nerve sheath tumors and an indicator for an inferior survival. *Mod. Pathol.* **2016**, *29*, 582–590. [CrossRef]

25. De Raedt, T.; Beert, E.; Pasmant, E.; Luscan, A.; Brems, H.; Ortonne, N.; Helin, K.; Hornick, J.L.; Mautner, V.; Kehrer-Sawatzki, H.; et al. PRC2 loss amplifies Ras-driven transcription and confers sensitivity to BRD4-based therapies. *Nature* **2014**, *514*, 247–251. [CrossRef]

26. He, S.; Mansour, M.R.; Zimmerman, M.W.; Ki, D.H.; Layden, H.M.; Akahane, K.; Gjini, E.; de Groh, E.D.; Perez-Atayde, A.R.; Zhu, S.; et al. Synergy between loss of NF1 and overexpression of MYCN in neuroblastoma is mediated by the GAP-related domain. *Elife* **2016**, *5*, e14713. [CrossRef]

27. Ki, D.H.; He, S.; Rodig, S.; Look, A.T. Overexpression of PDGFRA cooperates with loss of NF1 and p53 to accelerate the molecular pathogenesis of malignant peripheral nerve sheath tumors. *Oncogene* **2017**, *36*, 1058–1068. [CrossRef]

28. Ki, D.H.; Oppel, F.; Durbin, A.D.; Look, A.T. Mechanisms underlying synergy between DNA topoisomerase I-targeted drugs and mTOR kinase inhibitors in NF1-associated malignant peripheral nerve sheath tumors. *Oncogene* **2019**, *38*, 6585–6598. [CrossRef]

29. Oppel, F.; Tao, T.; Shi, H.; Ross, K.N.; Zimmerman, M.W.; He, S.; Tong, G.; Aster, J.C.; Look, A.T. Loss of atrx cooperates with p53-deficiency to promote the development of sarcomas and other malignancies. *PLoS Genet.* **2019**, *15*, e1008039. [CrossRef]

30. Durbin, A.D.; Ki, D.H.; He, S.; Look, A.T. Malignant Peripheral Nerve Sheath Tumors. *Adv. Exp. Med. Biol.* **2016**, *916*, 495–530.

31. Gregorian, C.; Nakashima, J.; Dry, S.M.; Nghiemphu, P.L.; Smith, K.B.; Ao, Y.; Dang, J.; Lawson, G.; Mellinghoff, I.K.; Mischel, P.S.; et al. PTEN dosage is essential for neurofibroma development and malignant transformation. *Proc. Natl. Acad. Sci. USA* **2009**, *106*, 19479–19484. [CrossRef]

32. Keng, V.W.; Watson, A.L.; Rahrmann, E.P.; Li, H.; Tschida, B.R.; Moriarity, B.S.; Choi, K.; Rizvi, T.A.; Collins, M.H.; Wallace, M.R.; et al. Conditional Inactivation of Pten with EGFR Overexpression in Schwann Cells Models Sporadic MPNST. *Sarcoma* **2012**, *2012*, 620834. [CrossRef]

33. Huijbregts, R.P.; Roth, K.A.; Schmidt, R.E.; Carroll, S.L. Hypertrophic neuropathies and malignant peripheral nerve sheath tumors in transgenic mice overexpressing glial growth factor beta3 in myelinating Schwann cells. *J. Neurosci.* **2003**, *23*, 7269–7280. [CrossRef]

34. Kazmi, S.J.; Byer, S.J.; Eckert, J.M.; Turk, A.N.; Huijbregts, R.P.; Brossier, N.M.; Grizzle, W.E.; Mikhail, F.M.; Roth, K.A.; Carroll, S.L. Transgenic mice overexpressing neuregulin-1 model neurofibroma-malignant peripheral nerve sheath tumor progression and implicate specific chromosomal copy number variations in tumorigenesis. *Am. J. Pathol.* **2013**, *182*, 646–667. [CrossRef]

35. Brosius, S.N.; Turk, A.N.; Byer, S.J.; Longo, J.F.; Kappes, J.C.; Roth, K.A.; Carroll, S.L. Combinatorial therapy with tamoxifen and trifluoperazine effectively inhibits malignant peripheral nerve sheath tumor growth by targeting complementary signaling cascades. *J. Neuropathol. Exp. Neurol.* **2014**, *73*, 1078–1090. [CrossRef]

36. Ling, B.C.; Wu, J.; Miller, S.J.; Monk, K.R.; Shamekh, R.; Rizvi, T.A.; Decourten-Myers, G.; Vogel, K.S.; DeClue, J.E.; Ratner, N. Role for the epidermal growth factor receptor in neurofibromatosis-related peripheral nerve tumorigenesis. *Cancer Cell.* **2005**, *7*, 65–75. [CrossRef]

37. Rahrmann, E.P.; Moriarity, B.S.; Otto, G.M.; Watson, A.L.; Choi, K.; Collins, M.H.; Wallace, M.; Webber, B.R.; Forster, C.L.; Rizzardi, A.E.; et al. Trp53 haploinsufficiency modifies EGFR-driven peripheral nerve sheath tumorigenesis. *Am. J. Pathol.* **2014**, *184*, 2082–2098. [CrossRef]

38. Keng, V.W.; Rahrmann, E.P.; Watson, A.L.; Tschida, B.R.; Moertel, C.L.; Jessen, W.J.; Rizvi, T.A.; Collins, M.H.; Ratner, N.; Largaespada, D.A. PTEN and NF1 Inactivation in Schwann Cells Produces a Severe Phenotype in the Peripheral Nervous System That Promotes the Development and Malignant Progression of Peripheral Nerve Sheath Tumors. *Cancer Res.* **2012**, *72*, 3405–3413. [CrossRef]

39. Mo, W.; Chen, J.; Patel, A.; Zhang, L.; Chau, V.; Li, Y.; Cho, W.; Lim, K.; Xu, J.; Lazar, A.J.; et al. CXCR4/CXCL12 mediate autocrine cell-cycle progression in NF1-associated malignant peripheral nerve sheath tumors. *Cell* **2013**, *152*, 1077–1090. [CrossRef]

40. Johannessen, C.M.; Johnson, B.W.; Williams, S.M.; Chan, A.W.; Reczek, E.E.; Lynch, R.C.; Rioth, M.J.; McClatchey, A.; Ryeom, S.; Cichowski, K. TORC1 is essential for NF1-associated malignancies. *Curr. Biol.* **2008**, *18*, 56–62. [CrossRef]

41. Rhodes, S.D.; He, Y.; Smith, A.; Jiang, L.; Lu, Q.; Mund, J.; Li, X.; Bessler, W.; Qian, S.; Dyer, W.; et al. Cdkn2a (Arf) loss drives NF1-associated atypical neurofibroma and malignant transformation. *Hum. Mol. Genet.* **2019**, *28*, 2752–2762. [CrossRef]

42. Paez-Ribes, M.; Gonzalez-Gualda, E.; Doherty, G.J.; Munoz-Espin, D. Targeting senescent cells in translational medicine. *EMBO Mol. Med.* **2019**, *11*, e10234. [CrossRef]

43. Rahrmann, E.P.; Watson, A.L.; Keng, V.W.; Choi, K.; Moriarity, B.S.; Beckmann, D.A.; Wolf, N.K.; Sarver, A.; Collins, M.H.; Moertel, C.L.; et al. Forward genetic screen for malignant peripheral nerve sheath tumor formation identifies new genes and pathways driving tumorigenesis. *Nat. Genet.* **2013**, *45*, 756–766. [CrossRef]

44. Wu, J.; Keng, V.W.; Patmore, D.M.; Kendall, J.J.; Patel, A.V.; Jousma, E.; Jessen, W.J.; Choi, K.; Tschida, B.R.; Silverstein, K.A.; et al. Insertional Mutagenesis Identifies a STAT3/Arid1b/beta-catenin Pathway Driving Neurofibroma Initiation. *Cell Rep.* **2016**, *14*, 1979–1990. [CrossRef]

45. Laugesen, A.; Hojfeldt, J.W.; Helin, K. Molecular Mechanisms Directing PRC2 Recruitment and H3K27 Methylation. *Mol. Cell* **2019**, *74*, 8–18. [CrossRef]

46. Streubel, G.; Watson, A.; Jammula, S.G.; Scelfo, A.; Fitzpatrick, D.J.; Oliviero, G.; McCole, R.; Conway, E.; Glancy, E.; Negri, G.L.; et al. The H3K36me2 Methyltransferase Nsd1 Demarcates PRC2-Mediated H3K27me2 and H3K27me3 Domains in Embryonic Stem Cells. *Mol. Cell* **2018**, *70*, 371–379 e375. [CrossRef]

47. Vogel, K.S.; Klesse, L.J.; Velasco-Miguel, S.; Meyers, K.; Rushing, E.J.; Parada, L.F. Mouse tumor model for neurofibromatosis type 1. *Science* **1999**, *286*, 2176–2179. [CrossRef]

48. Cichowski, K.; Shih, T.S.; Schmitt, E.; Santiago, S.; Reilly, K.; McLaughlin, M.E.; Bronson, R.T.; Jacks, T. Mouse models of tumor development in neurofibromatosis type 1. *Science* **1999**, *286*, 2172–2176. [CrossRef]

49. Reilly, K.M.; Tuskan, R.G.; Christy, E.; Loisel, D.A.; Ledger, J.; Bronson, R.T.; Smith, C.D.; Tsang, S.; Munroe, D.J.; Jacks, T. Susceptibility to astrocytoma in mice mutant for Nf1 and Trp53 is linked to chromosome 11 and subject to epigenetic effects. *Proc. Natl. Acad. Sci. USA* **2004**, *101*, 13008–13013. [CrossRef]

50. Malone, C.F.; Fromm, J.A.; Maertens, O.; DeRaedt, T.; Ingraham, R.; Cichowski, K. Defining key signaling nodes and therapeutic biomarkers in NF1-mutant cancers. *Cancer Discov.* **2014**, *4*, 1062–1073. [CrossRef]

51. Maertens, O.; McCurrach, M.E.; Braun, B.S.; De Raedt, T.; Epstein, I.; Huang, T.Q.; Lauchle, J.O.; Lee, H.; Wu, J.; Cripe, T.P.; et al. A Collaborative Model for Accelerating the Discovery and Translation of Cancer Therapies. *Cancer Res.* **2017**, *77*, 5706–5711. [CrossRef] [PubMed]

52. Malone, C.F.; Emerson, C.; Ingraham, R.; Barbosa, W.; Guerra, S.; Yoon, H.; Liu, L.L.; Michor, F.; Haigis, M.; Macleod, K.F.; et al. mTOR and HDAC Inhibitors Converge on the TXNIP/Thioredoxin Pathway to Cause Catastrophic Oxidative Stress and Regression of RAS-Driven Tumors. *Cancer Discov.* **2017**, *7*, 1450–1463. [CrossRef] [PubMed]

53. Lock, R.; Ingraham, R.; Maertens, O.; Miller, A.L.; Weledji, N.; Legius, E.; Konicek, B.M.; Yan, S.C.; Graff, J.R.; Cichowski, K. Cotargeting MNK and MEK kinases induces the regression of NF1-mutant cancers. *J. Clin. Invest.* **2016**, *126*, 2181–2190. [CrossRef] [PubMed]

54. Dodd, R.D.; Mito, J.K.; Eward, W.C.; Chitalia, R.; Sachdeva, M.; Ma, Y.; Barretina, J.; Dodd, L.; Kirsch, D.G. NF1 deletion generates multiple subtypes of soft-tissue sarcoma that respond to MEK inhibition. *Mol. Cancer* **2013**, *12*, 1906–1917. [CrossRef]

55. Dodd, R.D.; Lee, C.L.; Overton, T.; Huang, W.; Eward, W.C.; Luo, L.; Ma, Y.; Ingram, D.R.; Torres, K.E.; Cardona, D.M.; et al. NF1(+/−) Hematopoietic Cells Accelerate Malignant Peripheral Nerve Sheath Tumor Development without Altering Chemotherapy Response. *Cancer Res.* **2017**, *77*, 4486–4497. [CrossRef]

56. Huang, J.; Chen, M.; Whitley, M.J.; Kuo, H.C.; Xu, E.S.; Walens, A.; Mowery, Y.M.; Van Mater, D.; Eward, W.C.; Cardona, D.M.; et al. Generation and comparison of CRISPR-Cas9 and Cre-mediated genetically engineered mouse models of sarcoma. *Nat. Commun.* **2017**, *8*, 15999. [CrossRef]

57. Fang, Y.; Elahi, A.; Denley, R.C.; Rao, P.H.; Brennan, M.F.; Jhanwar, S.C. Molecular characterization of permanent cell lines from primary, metastatic and recurrent malignant peripheral nerve sheath tumors (MPNST) with underlying neurofibromatosis-1. *Anticancer Res.* **2009**, *29*, 1255–1262.

58. Sun, D.; Tainsky, M.A.; Haddad, R. Oncogene Mutation Survey in MPNST Cell Lines Enhances the Dominant Role of Hyperactive Ras in NF1 Associated Pro-Survival and Malignancy. *Transl Oncogenomics* **2012**, *5*, 1–7.

59. Castellsague, J.; Gel, B.; Fernandez-Rodriguez, J.; Llatjos, R.; Blanco, I.; Benavente, Y.; Perez-Sidelnikova, D.; Garcia-Del Muro, J.; Vinals, J.M.; Vidal, A.; et al. Comprehensive establishment and characterization of orthoxenograft mouse models of malignant peripheral nerve sheath tumors for personalized medicine. *Embo Mol. Med.* **2015**, *7*, 608–627. [CrossRef]

60. Pollard, K.; Banerjee, J.; Doan, X.; Wang, J.; Guo, X.; Allaway, R.; Langmead, S.; Slobogean, B.; Meyer, C.F.; Loeb, D.M. A clinically and genomically annotated nerve sheath tumor biospecimen repository. *biorxiv* 2020. [CrossRef]

61. Rutkowski, J.L.; Kirk, C.J.; Lerner, M.A.; Tennekoon, G.I. Purification and expansion of human Schwann cells in vitro. *Nat. Med.* **1995**, *1*, 80–83. [CrossRef] [PubMed]

62. Li, H.; Chang, L.J.; Neubauer, D.R.; Muir, D.F.; Wallace, M.R. Immortalization of human normal and NF1 neurofibroma Schwann cells. *Lab. Invest.* **2016**, *96*, 1105–1115. [CrossRef] [PubMed]

63. Watson, A.L.; Rahrmann, E.P.; Moriarity, B.S.; Choi, K.; Conboy, C.B.; Greeley, A.D.; Halfond, A.L.; Anderson, L.K.; Wahl, B.R.; Keng, V.W.; et al. Canonical Wnt/beta-catenin signaling drives human schwann cell transformation, progression, and tumor maintenance. *Cancer Discov.* **2013**, *3*, 674–689. [CrossRef] [PubMed]

64. Carrio, M.; Mazuelas, H.; Richaud-Patin, Y.; Gel, B.; Terribas, E.; Rosas, I.; Jimenez-Delgado, S.; Biayna, J.; Vendredy, L.; Blanco, I.; et al. Reprogramming Captures the Genetic and Tumorigenic Properties of Neurofibromatosis Type 1 Plexiform Neurofibromas. *Stem. Cell Rep.* **2019**, *12*, 411–426. [CrossRef] [PubMed]

65. Dobzhansky, T. Genetics of natural populations; recombination and variability in populations of Drosophila pseudoobscura. *Genetics* **1946**, *31*, 269–290. [PubMed]

66. Hartwell, L.H.; Szankasi, P.; Roberts, C.J.; Murray, A.W.; Friend, S.H. Integrating genetic approaches into the discovery of anticancer drugs. *Science* **1997**, *278*, 1064–1068. [CrossRef] [PubMed]

67. Bryant, H.E.; Schultz, N.; Thomas, H.D.; Parker, K.M.; Flower, D.; Lopez, E.; Kyle, S.; Meuth, M.; Curtin, N.J.; Helleday, T. Specific killing of BRCA2-deficient tumours with inhibitors of poly(ADP-ribose) polymerase. *Nature* **2005**, *434*, 913–917. [CrossRef]

68. Javle, M.; Curtin, N.J. The role of PARP in DNA repair and its therapeutic exploitation. *Br. J. Cancer* **2011**, *105*, 1114–1122. [CrossRef]

69. Huang, A.; Garraway, L.A.; Ashworth, A.; Weber, B. Synthetic lethality as an engine for cancer drug target discovery. *Nat. Rev. Drug Discov.* **2020**, *9*, 23–38. [CrossRef]

70. Semenova, G.; Stepanova, D.S.; Deyev, S.M.; Chernoff, J. Medium throughput biochemical compound screening identifies novel agents for pharmacotherapy of neurofibromatosis type 1. *Biochimie* **2017**, *135*, 1–5. [CrossRef]

71. Ferrer, M.; Gosline, S.J.C.; Stathis, M.; Zhang, X.; Guo, X.; Guha, R.; Ryman, D.A.; Wallace, M.R.; Kasch-Semenza, L.; Hao, H.; et al. Pharmacological and genomic profiling of neurofibromatosis type 1 plexiform neurofibroma-derived schwann cells. *Sci. Data* **2018**, *5*, 180106. [CrossRef] [PubMed]

72. Dombi, E.; Baldwin, A.; Marcus, L.J.; Fisher, M.J.; Weiss, B.; Kim, A.; Whitcomb, P.; Martin, S.; Aschbacher-Smith, L.E.; Rizvi, T.A.; et al. Activity of Selumetinib in Neurofibromatosis Type 1-Related Plexiform Neurofibromas. *N. Engl. J. Med.* **2016**, *375*, 2550–2560. [CrossRef] [PubMed]

73. McCowage, G.B.; Pratilas, C.A.; Hargrave, D.R.; Moertel, C.L.; Whitlock, J.; Fox, E.; Hingorani, P.; Russo, M.W.; Dasgupta, K.; Tseng, L.; et al. Trametinib in pediatric patients with neurofibromatosis type 1 (NF-1)–associated plexiform neurofibroma: A phase I/IIa study. *J. Clin. Oncol.* **2018**, *36*, 10504. [CrossRef]

74. Gross, A.M.; Wolters, P.L.; Dombi, E.; Baldwin, A.; Whitcomb, P.; Fisher, M.J.; Weiss, B.; Kim, A.; Bornhorst, M.; Shah, A.C.; et al. Selumetinib in Children with Inoperable Plexiform Neurofibromas. *N. Engl. J. Med.* **2020**, *382*, 1430–1442. [CrossRef]

75. Daud, A.; Tsai, K. Management of Treatment-Related Adverse Events with Agents Targeting the MAPK Pathway in Patients with Metastatic Melanoma. *Oncologist* **2017**, *22*, 823–833. [CrossRef]

76. Stalnecker, C.A.; Der, C.J. RAS, wanted dead or alive: Advances in targeting RAS mutant cancers. *Sci. Signal.* **2020**, *13*. [CrossRef]

77. Kim, H.A.; Ling, B.; Ratner, N. Nf1-deficient mouse Schwann cells are angiogenic and invasive and can be induced to hyperproliferate: Reversion of some phenotypes by an inhibitor of farnesyl protein transferase. *Mol. Cell Biol.* **1997**, *17*, 862–872. [CrossRef]

78. Dilworth, J.T.; Wojtkowiak, J.W.; Mathieu, P.; Tainsky, M.A.; Reiners, J.J., Jr.; Mattingly, R.R.; Hancock, C.N. Suppression of proliferation of two independent NF1 malignant peripheral nerve sheath tumor cell lines by the pan-ErbB inhibitor CI-1033. *Cancer Biol.* **2008**, *7*, 1938–1946. [CrossRef]

79. Wojtkowiak, J.W.; Fouad, F.; LaLonde, D.T.; Kleinman, M.D.; Gibbs, R.A.; Reiners, J.J., Jr.; Borch, R.F.; Mattingly, R.R. Induction of apoptosis in neurofibromatosis type 1 malignant peripheral nerve sheath tumor cell lines by a combination of novel farnesyl transferase inhibitors and lovastatin. *J. Pharm. Exp.* **2008**, *326*, 1–11. [CrossRef]

80. Widemann, B.C.; Salzer, W.L.; Arceci, R.J.; Blaney, S.M.; Fox, E.; End, D.; Gillespie, A.; Whitcomb, P.; Palumbo, J.S.; Pitney, A.; et al. Phase I trial and pharmacokinetic study of the farnesyltransferase inhibitor tipifarnib in children with refractory solid tumors or neurofibromatosis type I and plexiform neurofibromas. *J. Clin. Oncol.* **2006**, *24*, 507–516. [CrossRef]

81. Widemann, B.C.; Dombi, E.; Gillespie, A.; Wolters, P.L.; Belasco, J.; Goldman, S.; Korf, B.R.; Solomon, J.; Martin, S.; Salzer, W.; et al. Phase 2 randomized, flexible crossover, double-blinded, placebo-controlled trial of the farnesyltransferase inhibitor tipifarnib in children and young adults with neurofibromatosis type 1 and progressive plexiform neurofibromas. *Neuro. Oncol.* **2014**, *16*, 707–718. [CrossRef] [PubMed]

82. Fangusaro, J.; Onar-Thomas, A.; Young Poussaint, T.; Wu, S.; Ligon, A.H.; Lindeman, N.; Banerjee, A.; Packer, R.J.; Kilburn, L.B.; Goldman, S.; et al. Selumetinib in paediatric patients with BRAF-aberrant or neurofibromatosis type 1-associated recurrent, refractory, or progressive low-grade glioma: A multicentre, phase 2 trial. *Lancet. Oncol.* **2019**, *20*, 1011–1022. [CrossRef]

83. Weiss, B.; Widemann, B.C.; Wolters, P.; Dombi, E.; Vinks, A.; Cantor, A.; Perentesis, J.; Schorry, E.; Ullrich, N.; Gutmann, D.H.; et al. Sirolimus for progressive neurofibromatosis type 1-associated plexiform neurofibromas: A neurofibromatosis Clinical Trials Consortium phase II study. *Neuro. Oncol.* **2015**, *17*, 596–603. [CrossRef]

84. Varin, J.; Poulain, L.; Hivelin, M.; Nusbaum, P.; Hubas, A.; Laurendeau, I.; Lantieri, L.; Wolkenstein, P.; Vidaud, M.; Pasmant, E.; et al. Dual mTORC1/2 inhibition induces anti-proliferative effect in NF1-associated plexiform neurofibroma and malignant peripheral nerve sheath tumor cells. *Oncotarget* **2016**, *7*, 35753–35767. [CrossRef] [PubMed]

85. Kim, A.; Lu, Y.; Okuno, S.H.; Reinke, D.; Maertens, O.; Perentesis, J.; Basu, M.; Wolters, P.L.; De Raedt, T.; Chawla, S.; et al. Targeting Refractory Sarcomas and Malignant Peripheral Nerve Sheath Tumors in a Phase I/II Study of Sirolimus in Combination with Ganetespib (SARC023). *Sarcoma* **2020**, *2020*, 5784876. [CrossRef]

86. De Raedt, T.; Walton, Z.; Yecies, J.L.; Li, D.; Chen, Y.; Malone, C.F.; Maertens, O.; Jeong, S.M.; Bronson, R.T.; Lebleu, V.; et al. Exploiting cancer cell vulnerabilities to develop a combination therapy for ras-driven tumors. *Cancer Cell* **2011**, *20*, 400–413. [CrossRef]

87. Widemann, B.C.; Lu, Y.; Reinke, D.; Okuno, S.H.; Meyer, C.F.; Cote, G.M.; Chugh, R.; Milhem, M.M.; Hirbe, A.C.; Kim, A.; et al. Targeting Sporadic and Neurofibromatosis Type 1 (NF1) Related Refractory Malignant Peripheral Nerve Sheath Tumors (MPNST) in a Phase II Study of Everolimus in Combination with Bevacizumab (SARC016). *Sarcoma* **2019**, *2019*, 7656747. [CrossRef]

88. Patwardhan, P.P.; Surriga, O.; Beckman, M.J.; de Stanchina, E.; Dematteo, R.P.; Tap, W.D.; Schwartz, G.K. Sustained inhibition of receptor tyrosine kinases and macrophage depletion by PLX3397 and rapamycin as a potential new approach for the treatment of MPNSTs. *Clin. Cancer Res.* **2014**, *20*, 3146–3158. [CrossRef]

89. Kolberg, M.; Bruun, J.; Murumagi, A.; Mpindi, J.P.; Bergsland, C.H.; Holand, M.; Eilertsen, I.A.; Danielsen, S.A.; Kallioniemi, O.; Lothe, R.A. Drug sensitivity and resistance testing identifies PLK1 inhibitors and gemcitabine as potent drugs for malignant peripheral nerve sheath tumors. *Mol. Oncol.* **2017**, *11*, 1156–1171. [CrossRef]

90. Kohlmeyer, J.L.; Kaemmer, C.A.; Pulliam, C.; Maharjan, C.K.; Moreno Samayoa, A.; Major, H.J.; Cornick, K.E.; Knepper-Adrian, V.; Khanna, R.; Sieren, J.C.; et al. RABL6A is an essential driver of MPNSTs that negatively regulates the RB1 pathway and sensitizes tumor cells to CDK4/6 inhibitors. *Clin. Cancer Res.* 2020. [CrossRef]

91. Wojcik, J.B.; Marchione, D.M.; Sidoli, S.; Djedid, A.; Lisby, A.; Majewski, J.; Garcia, B.A. Epigenomic reordering induced by Polycomb loss drives oncogenesis but leads to therapeutic vulnerabilities in malignant peripheral nerve sheath tumors. *Cancer Res.* **2019**. [CrossRef] [PubMed]

92. Frankowski, K.J.; Wang, C.; Patnaik, S.; Schoenen, F.J.; Southall, N.; Li, D.; Teper, Y.; Sun, W.; Kandela, I.; Hu, D.; et al. Metarrestin, a perinucleolar compartment inhibitor, effectively suppresses metastasis. *Sci. Transl. Med.* **2018**, *10*. [CrossRef] [PubMed]

93. Nair, J.S.; Musi, E.; Schwartz, G.K. Selinexor (KPT-330) Induces Tumor Suppression through Nuclear Sequestration of IkappaB and Downregulation of Survivin. *Clin. Cancer Res.* **2017**, *23*, 4301–4311. [CrossRef] [PubMed]

94. Mahller, Y.Y.; Rangwala, F.; Ratner, N.; Cripe, T.P. Malignant peripheral nerve sheath tumors with high and low Ras-GTP are permissive for oncolytic herpes simplex virus mutants. *Pediatr Blood Cancer.* **2006**, *46*, 745–754. [CrossRef] [PubMed]

95. Farassati, F.; Pan, W.; Yamoutpour, F.; Henke, S.; Piedra, M.; Frahm, S.; Al-Tawil, S.; Mangrum, W.I.; Parada, L.F.; Rabkin, S.D.; et al. Ras signaling influences permissiveness of malignant peripheral nerve sheath tumor cells to oncolytic herpes. *Am. J. Pathol.* **2008**, *173*, 1861–1872. [CrossRef]

96. Antoszczyk, S.; Spyra, M.; Mautner, V.F.; Kurtz, A.; Stemmer-Rachamimov, A.O.; Martuza, R.L.; Rabkin, S.D. Treatment of orthotopic malignant peripheral nerve sheath tumors with oncolytic herpes simplex virus. *Neuro Oncol.* **2014**, *16*, 1057–1066. [CrossRef]

97. Deyle, D.R.; Escobar, D.Z.; Peng, K.W.; Babovic-Vuksanovic, D. Oncolytic measles virus as a novel therapy for malignant peripheral nerve sheath tumors. *Genes* **2015**, *565*, 140–145. [CrossRef]

98. Jackson, J.D.; Markert, J.M.; Li, L.; Carroll, S.L.; Cassady, K.A. STAT1 and NF-kappaB Inhibitors Diminish Basal Interferon-Stimulated Gene Expression and Improve the Productive Infection of Oncolytic HSV in MPNST Cells. *Mol. Cancer Res.* **2016**, *14*, 482–492. [CrossRef]

99. Currier, M.A.; Sprague, L.; Rizvi, T.A.; Nartker, B.; Chen, C.Y.; Wang, P.Y.; Hutzen, B.J.; Franczek, M.R.; Patel, A.V.; Chaney, K.E.; et al. Aurora A kinase inhibition enhances oncolytic herpes virotherapy through cytotoxic synergy and innate cellular immune modulation. *Oncotarget* **2017**, *8*, 17412–17427. [CrossRef]

100. Ghonime, M.G.; Cassady, K.A. Combination Therapy Using Ruxolitinib and Oncolytic HSV Renders Resistant MPNSTs Susceptible to Virotherapy. *Cancer Immunol. Res.* **2018**, *6*, 1499–1510. [CrossRef]

101. Robertson, K.A.; Nalepa, G.; Yang, F.C.; Bowers, D.C.; Ho, C.Y.; Hutchins, G.D.; Croop, J.M.; Vik, T.A.; Denne, S.C.; Parada, L.F.; et al. Imatinib mesylate for plexiform neurofibromas in patients with neurofibromatosis type 1: A phase 2 trial. *Lancet. Oncol.* **2012**, *13*, 1218–1224. [CrossRef]

102. Guo, J.; Grovola, M.R.; Xie, H.; Coggins, G.E.; Duggan, P.; Hasan, R.; Huang, J.; Lin, D.W.; Song, C.; Witek, G.M.; et al. Comprehensive pharmacological profiling of neurofibromatosis cell lines. *Am. J. Cancer Res.* **2017**, *7*, 923–934. [PubMed]

 *genes*

Review

The Role of Polycomb Repressive Complex in Malignant Peripheral Nerve Sheath Tumor

Xiyuan Zhang [1], Béga Murray [1,2], George Mo [1,3] and Jack F. Shern [1,*]

1   Pediatric Oncology Branch, Tumor Evolution and Genomics Section, Center for Cancer Research, National Cancer Institute, National Institutes of Health, Bethesda, MD 20892, USA; xiyuan.zhang@nih.gov (X.Z.); bega.murray@nih.gov (B.M.); george.mo@nih.gov (G.M.)

2   The Patrick G Johnston Centre for Cancer Research, Queen's University Belfast, 97 Lisburn road, Belfast BT9 7AE, UK

3   SUNY Downstate Health Sciences University, Brooklyn, NY 11203, USA

*   Correspondence: john.shern@nih.gov

Received: 21 January 2020; Accepted: 2 March 2020; Published: 9 March 2020

Abstract: Malignant peripheral nerve sheath tumors (MPNSTs) are aggressive soft tissue sarcomas that can arise most frequently in patients with neurofibromatosis type 1 (NF1). Despite an increasing understanding of the molecular mechanisms that underlie these tumors, there remains limited therapeutic options for this aggressive disease. One potentially critical finding is that a significant proportion of MPNSTs exhibit recurrent mutations in the genes *EED* or *SUZ12*, which are key components of the polycomb repressive complex 2 (PRC2). Tumors harboring these genetic lesions lose the marker of transcriptional repression, trimethylation of lysine residue 27 on histone H3 (H3K27me3) and have dysregulated oncogenic signaling. Given the recurrence of PRC2 alterations, intensive research efforts are now underway with a focus on detailing the epigenetic and transcriptomic consequences of PRC2 loss as well as development of novel therapeutic strategies for targeting these lesions. In this review article, we will summarize the recent findings of PRC2 in MPNST tumorigenesis, including highlighting the functions of PRC2 in normal Schwann cell development and nerve injury repair, as well as provide commentary on the potential therapeutic vulnerabilities of a PRC2 deficient tumor cell.

Keywords: neurofibromatosis; malignant peripheral nerve sheath tumor; MPNST; polycomb repressive complex; PRC2

---

1. Introduction

Neurofibromatosis type 1 (NF1) is a common autosomal dominant disorder caused by inactivating mutations in the tumor suppressor gene *NF1* and affects roughly 1/3000 newborns worldwide [1,2]. The gene *NF1* encodes the GTPase-activating protein neurofibromin (also called neurofibromatosis-related protein) that is a negative regulator of the RAS signaling pathway. Both heterozygous and biallelic loss-of-function (LOF) mutations in *NF1* are associated with hyper-activation of RAS signaling and its downstream targets [3–5]. Patients with NF1 are diagnosed when they exhibit two or more of the following symptoms: Six or more café-au-lait macules, two or more neurofibromas or one plexiform neurofibroma (PN), freckling in the axillary or inguinal regions, optic glioma, two or more Lisch nodules, bony dysplasia, or first degree relative with NF1 [6–8]. A life-threatening complication of NF1 is an increased risk of the development of the aggressive and highly metastatic soft tissue sarcoma, malignant peripheral nerve sheath tumor (MPNST) [9]. Patients with NF1 have a risk of developing MPNST that is 1000-fold higher than the general population [10,11]. Currently, there are no effective treatments for MPNST other than complete surgical resection with wide negative margins. There are three types of MPNST: NF1-associated, sporadic, and radiation-related, accounting for 50%, 40%, and 10% of all MPNSTs, respectively [12]. Mutations in *NF1* are found in nearly 90% of MPNSTs and frequently involve biallelic loss of the entire gene [13]. As an important tumor suppressor gene, *NF1* is mutated in 8% of all 10,967 The Cancer Genome Atlas (TCGA) curated samples. Interestingly, mutations in *NF1* are not enriched in its GTPase-activating protein domain; rather, they favor missense or truncating lesions that lead to hyper activated RAS signaling.

Comprehensive genomic and clinical efforts led to the proposal that there are at least three steps required for cellular transformation during the development of MPNST. These steps are outlined in a genetic model for the development of MPNSTs (Figure 1): 1) fifty percent of NF1 patients will suffer from histologically benign PNs that are caused by the biallelic LOF in *NF1* and associated hyperactivation of RAS signaling [14]; 2) atypical neurofibromas (ANFs, here encompassing distinct nodular lesions and atypical neurofibromatous neoplasms of uncertain biologic potential, ANNUBP) arise within PNs and in addition to hyperactivation of RAS, exhibit heterozygous loss of the genomic locus encompassing the gene *CDKN2A* [15,16]; and 3) approximately 8%-13% of NF1 patients will ultimately have their tumors transform into MPNSTs [11,17], where recurrent mutations in *SUZ12* and/or *EED*, two key components of the polycomb repressive complex 2 (PRC2), lead to loss of tri-methylation of histone H3 lysine 27 (H3K27me3) and de-repression of its target genes [18,19]. Undoubtably, this model oversimplifies the genetic progression of the cell towards malignancy and minimizes the contribution of mutations in other genes, such as *TP53*. However, given the high recurrence of PRC2 alterations, this review will focus on summarizing the current understanding of PRC2 loss in MPNST pathogenesis.

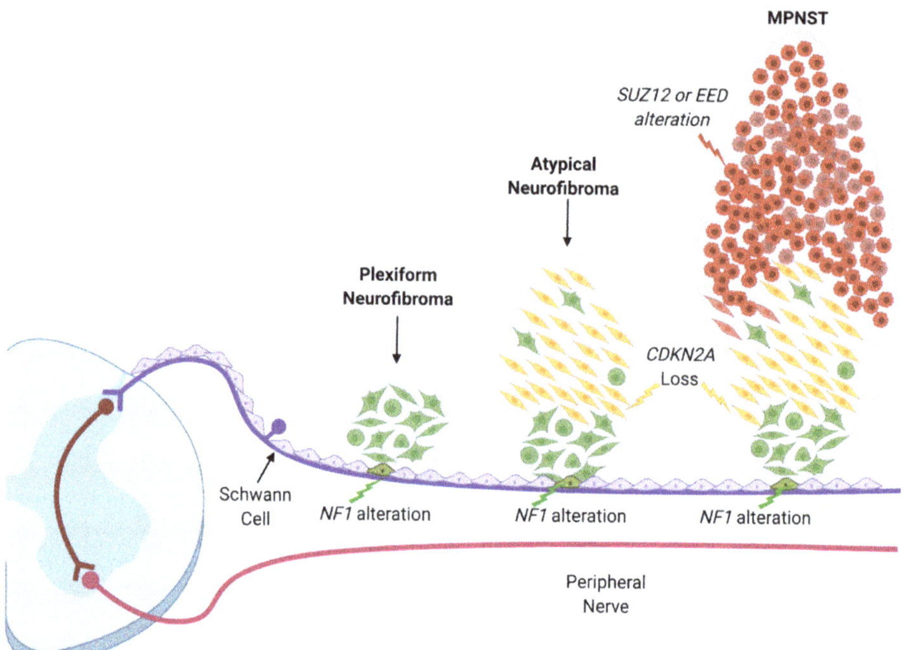

Figure 1. The clinical spectrum and genetic model of nerve tumor development in neurofibromatosis type 1 (NF1). Cells shown here are Schwann cells in the dorsal ganglia root and are affected by the sequential mutations driving the malignant transformation. Green: *NF1* alteration, yellow: *CDKN2A* alteration, and red: PRC2 alteration.

2. Recurrent Mutations in *EED* and *SUZ12* in MPNST

A critical advance in the understanding of the molecular pathogenesis of MPNSTs came from comprehensive genomic analyses of MPNST patient samples through next generation sequencing (NGS). These studies discovered recurrent and frequently mutually exclusive alterations in Embryonic Ectoderm Development *(EED)* and Suppressor of Zeste 12 Protein Homolog (*SUZ12*), two key components of PRC2, in MPNSTs. Three independent groups nearly simultaneously reported their findings of the genetic aberrations using archived patient MPNSTs [18–20]. Lee and colleagues performed whole exome sequencing (WES) of a discovery cohort consisting of 15 MPNSTs and identified five *EED* mutations including four frame-shift and one splice-site alterations, which were associated with loss of heterozygosity, either as a result of deletion of the normal allele or copy-neutral loss and seven *SUZ12* mutations comprised of two homozygous deletions (hom) and six heterozygous loss (het) of one allele [18]. Intriguingly, analysis of WES coupled with whole transcriptome sequencing (RNAseq) of the six MPNSTs with *SUZ12* het loss revealed that two appeared to express the full length of the transcript, with the other 4 exhibiting exonic structural variations (SV). Strikingly, these 4 MPNST samples designated as "het+SV" are all radiation-related, indicating the possibility of local genomic rearrangement caused by previous exposure to radiation in those patients. In the same study, the authors utilized a targeted capture sequencing approach in an additional 37 MPNST samples that were formalin-fixed paraffin-embedded (FFPE) and validated the recurrent mutations in polycomb genes. In another study, Zhang and colleagues performed whole-genome sequencing (WGS) or WES on eight frozen MPNST samples and identified mutations in *EED* or *SUZ12* in 75% of the cases [20]. In their validation cohort of 42 FFPE MPNST specimens, they identified 11 tumors harboring a *SUZ12* mutation (26%). The authors attributed the low fraction of this cohort harboring genetic aberrations in polycomb genes to the partially degraded DNA and false negative rate of the targeted sequencing approach. Finally, DeRaedt et al. used a targeted sequencing approach in a cohort of 51 NF1-associated MPNST samples and discovered 19 samples harboring an *NF1* microdeletion, in which *SUZ12* was frequently co-deleted due to the proximity of these genes in the human genome [19]. More than 50% of the non-microdeletion tumors (32 cases) exhibited inactivating mutations in *SUZ12* or *EED*. Two additional studies verifying the mutation of these polycomb genes were published in 2017. Sohier et al. performed WES in eight NF1-associated MPNSTs and identified *SUZ12* mutations in seven samples and *EED* mutations in two samples [21]. Brohl et al. performed WES on 12 MPNSTs and found genetic alterations in *SUZ12* and/or *EED* in seven samples [13]. The authors also summarized the overall mutational frequency of all five studies that used NGS: *SUZ12* (56.1%) and *EED* (32.5%). Further, MPNST was included in the 2018 soft tissue sarcoma characterization study of TCGA, in which two of the included samples were identified as *SUZ12* mutants in the publicly available data, although no comment was made on these findings in the publication [22]. In summary, a variety of aberrations of the *SUZ12* gene have been identified, including indels, truncating mutations, and missense variants, all of which likely result in aberrant production of this core polycomb protein. Similarly, *EED* is frequently altered through copy number variations in MPNST, as well as through other various LOF mutations including truncating, frameshift, and missense mutations, which lead to abnormal protein production. The *SUZ12* and *EED* mutations identified thus far do not appear to cluster at any known conserved domains within either gene, such as the VEFS binding domain of *SUZ12* or the WD40 protein interaction domain of *EED*, but this may be due to the limited number of MPNST samples that have been characterized for their PRC2 mutant status. Using OncoPrinter and MutationMapper [23,24], we performed meta-analysis of all currently published genome sequencing results of MPNSTs and summarized the accumulated observations of *EED* or *SUZ12* alterations (Figure 2, Table S1).

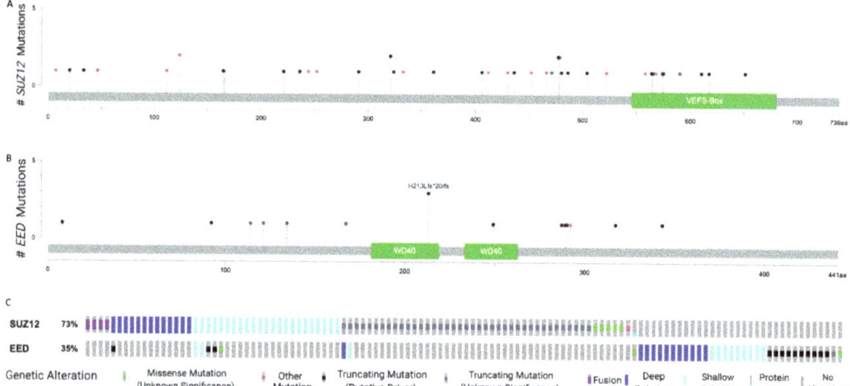

Figure 2. A compiled overview of *SUZ12* and *EED* alterations identified in malignant peripheral nerve sheath tumors (MPNST) published to date, accounting for approximately 75% of sequenced cases. (**A**) A representation of single nucleotide variants (SNVs) discovered in *SUZ12* thus far in MPNST sequencing studies. (**B**) *EED* SNVs identified through sequencing studies. (**C**) An Oncoprint map of the various mutations discovered in both *SUZ12* and *EED* across the MPNST samples sequenced thus far. Figure generated using OncoPrinter and MutationMapper from https://www.cbioportal.org/visualize.

As another core component of PRC2, Enhancer of Zeste Homolog 2 *(EZH2)* has been implicated as an oncogenic driver in a variety of cancers, playing diverse roles in aiding the development and progression of malignancy [25]. In MPNSTs, however, *EZH2* has recurrently been identified intact, despite the high rate of mutation in other key PRC2 components, *SUZ12* and *EED*. Wassef and colleagues revealed that only the combined loss of both EZH1 and EZH2 in immortalized plexiform neurofibroma-derived cells produced deregulatory effects similar to that observed with the sole absence of either EED or SUZ12 [26]. They also noted a lack of independent function of *EZH2* in the presence of either a *SUZ12* or *EED* mutation, and therefore concluded that the functional redundancy of EZH1 and EZH2 was a contributing factor to the lack of mutation in this component of the PRC2 core.

As noted previously, genetic alterations involving PRC2 and its product H3K27me3 have been reported in a variety of cancer types and most of these studies implicate the EZH2 methyltransferase as the driver of oncogenesis. In contrast to the LOF mutations observed in the *SUZ12* and *EED* subunits of the PRC2 complex, *EZH2* is noted to contribute to tumorigenesis through both over-expression as well as gain-of-function (GOF) mutations [27–31]. A prominent example of an oncogenic GOF mutation of this gene is in human B-cell lymphomas. In this tumor type, a mutation in the SET domain of EZH2 (Tyr641) was present in 21.7% of diffuse large B-cell lymphoma and 7.2% of follicular lymphomas of the germinal-center origin [32]. Subsequent biochemical studies showed that mutant Tyr641, through coordination with the wildtype EZH2, exhibited GOF activities by increasing global H3K27me3 levels [27,33]. Additionally, it was noted that this mutation led to a redistribution of the H3K27me3 repressive mark, thereby allowing for transcriptional activation at certain loci and tumorigenesis in rodent B-cell lymphoma and melanoma models [34]. Further GOF mutations have also been identified within this SET domain of EZH2, involving A677G mutant in B-cell lymphoma [35] and A687V mutant in non-Hodgkin lymphoma [36]. In addition to the association of GOF mutations in *EZH2* with oncogenesis, some studies have implicated the overexpression of this gene in cancer development despite an absence of mutations in the coding region, such as in multiple myeloma [37–39], prostate, breast, and endometrial cancers [31,40]. In such cases, the mechanism of EZH2 overexpression could be the result of genomic copy number changes or as a result of epigenomic dysregulation.

Mutations in the PRC2 target histone proteins can also be oncogenic through interference with deposition of the repressive methyl group in the presence of wildtype PRC2. This mechanism is most notable in diffuse intrinsic pontine gliomas and pediatric non-brainstem high-grade gliomas, where mutation in histone H3 yields variant, K27M, that binds and mislocalizes PRC2 in addition to inhibiting its function [41,42]. Interestingly, although these brain tumors exhibit global loss of H3K27me3, several genes retain this repressive mark, and PRC2 itself is required for tumor cell proliferation. This discovery ultimately led to the identification of small molecule inhibitor of EZH2 as potential therapy for this deadly cancer [43].

3. The Biochemical, Epigenetic, and Transcriptomic Consequences of PRC2 Loss in MPNST

Polycomb proteins are important regulators of chromatin structure during early development. PRC2 is a highly conserved multimeric complex that plays a distinct role in the transcriptional regulatory activity of the genome through repressive methylation of H3K27 of target genes, which is required in order to induce transcriptional silencing [44–47]. PRC2 methylates H3K27 to different extents, catalyzing the addition of mono- (H3K27me1), di- (H3K27me2), or tri- methyl groups in cell type specific patterns [48,49]. H3K27me3 is the most well characterized form of methylation at this lysine residue, and occupies around 5%-10% of the genome, while the lesser studied H3K27me2 and H3K27me1 are found at about 50%-70% and 5%-10%, respectively [49–51]. PRC2 is highly mobile, with around 80% of the nuclear-located complex undergoing continuous diffusion throughout it, while the remaining PRC2 is stably bound to chromatin [52]. While stably bound PRC2 is often located at H3K27me3 sites, it is rarely found at the site of dimethyl H3K27 that marks intragenic regions and is suggested to act as a repressor of inappropriate activation of cell-specific enhancers and promoters [25–28]. Unlike the di- and tri- methylated states of H3K27, H3K27me1 does not appear to be involved in transcriptional repression and instead has been found in high abundance at transcribed genes [53]. Prior studies have shown that these global methylation patterns at H3K27 are not regulated by active demethylation at these sites, such as by UTX, Jmjd3, or intracellular demethylases [49]. Target genes that are transcriptionally regulated by PRC2 are essential for embryonic development and cell lineage decisions [48].

Detailed characterization of the biochemical, epigenetic, and transcriptomic consequences of PRC2 loss in the MPNST cell remains a goal of current research efforts. In the 2014 study, Lee and colleagues identified a correlation between PRC2 mutation and H3K27me3 loss in MPNST samples, through a combined use of NGS for PRC2 characterization and immunohistochemistry (IHC) staining, to identify loss of H3K27me3 [18]. The identification of IHC screening for H3K27me3 as a reliable biomarker of MPNST led to additional studies that utilized H3K27me3 IHC as a diagnostic marker for MPNST [54–61]. Wojcik and colleagues used this screen to select samples of both PRC2 mutant and PRC2 wildtype status for use in proteomic analyses [62]. These proteomic analyses revealed that the loss of PRC2 caused global changes in post-translational modifications of histones, including 1) a substantial decrease in the transcriptionally repressive modification H3K27me3, 2) broad distribution of the repressive marker H3K27me2, 3) no compensatory gain of other repressive markers, for instance H3K9me3 or H4K20me3, and 4) significant increase in active chromatin markers, including H3K27 acetylation (H3K27ac) and H3K36me2 [62]. Furthermore, they identified that the loss of H3K27me3 across the genome led to the occupancy of those histone tails solely by H3K36me2. Though this paper assumes the commonly theorized transcriptional activation function of H3K36me2, some literature indicates that the dimethylation of H3K36 might be transcriptionally repressive and in contrast with the activation trimethylation of H3K36. Early investigations regarding the location and function of H3K36 di- and tri- methylation marks suggested both distinct locations and opposing roles of these two marks in the regulation of the *Drosophila* genome [63]. This was further commented on by Turberfield et al., who used genome-wide profiling of H3K36me2 to indicate the widespread deposition of transcriptional mark throughout the genome, but with a notable absence on bodies of highly transcribed genes and CpG island-associated gene promoters [64]. Although methylation of H3K36 is commonly associated with transcriptional activation, it has been shown to participate in other cellular processes, including alternative splicing, DNA replication, as well as transcriptional repression [65]. Interestingly, methylation of H3K36 and H3K27 seems to occur mutually exclusively, suggesting that the elevation of H3K36me2 may be a compensatory mechanism of transcriptional repression in cases of PRC2 and H3K27me3 loss [66].

In addition to the upregulation of recognized PRC2 targets, loss of PRC2 has been implicated in the upregulation of generalized growth and cell division pathways, nucleosome remodeling, and transcriptional activation [18,62]. Somewhat surprisingly, due to the transcriptionally repressive role of PRC2, loss of this complex has also been identified as correlating with the downregulation of certain pathways, including immune-related signaling such as interferon (IFN) signaling and antigen presentation [62]. It is unclear whether these observations are due to alteration of these pathways in the tumor cells or through the reduction of tumor-infiltrating antigen-presenting cells within the tumor. Finally, PRC2 loss was noted to correlate with global DNA hypermethylation at gene promoters and intergenic regions [62]. This DNA hypermethylation was hypothesized by the authors as a potential explanation for the repression of protein expression in the absence of functional PRC2 in MPNSTs. The cooperation of DNA methylation and polycomb complexes in the transcriptional regulation of the genome remains incompletely understood, despite intensive research of the area. It is known that DNA methylation and H3K27me3 mark introduced by PRC2 are typically found in a mutually exclusive pattern across the genome [67–69], and the identification of reduced binding capability of PRC2 to nucleosomes with methylated DNA highlights the possibility of an antagonistic relationship between these two epigenetic marks [70,71]. Further, a study by Cooper et al. noted that a decrease in DNA methylation levels corresponded to a redistribution of the H3K27me3 mark across the genome, indicating a role for DNA methylation in polycomb targeting [72]. The effect of polycomb deposited H3K27me3 on DNA methylation levels are less well known, although recent research indicates a role for PRC2 in maintenance of regions of DNA hypomethylation via TET proteins [73]. It remains to be seen whether the loss of PRC2 leads to the upregulation of DNA methylation in human MPNST samples. Despite the wealth of information provided by these research efforts (as summarized in Figure 3), further investigation of the effects of PRC2 loss on the biochemical, epigenetic, and transcriptomic organization of MPNST is crucial to deciphering the mechanisms of tumor development, metastasis, and discovery of potential treatment options for this aggressive disease.

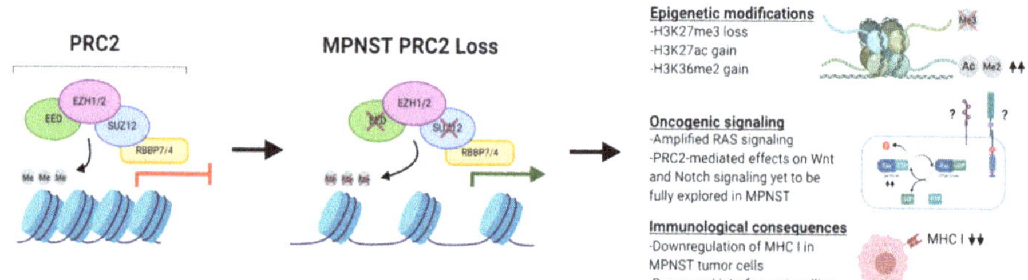

Figure 3. PRC2 structure and consequences of its loss in MPNST. Loss of PRC2 via EED or SUZ12 loss in MPNSTs leads to loss of tri-methylation of histone H3 lysine 27 (H3K27me3) and other potential epigenetic modifications. In addition, PRC2 loss can have a wide variety of consequences on oncogenic signaling and immune surveillance and response.

An unexplored avenue in MPNSTs is the potential relative importance of polycomb repressive complex 1 (PRC1) in a PRC2 deficient cell. PRC1 is a functionally distinct protein complex that plays a critical role in the transcriptional regulation of the genome. PRC1 is responsible for the deposition of mono-ubiquitylation of lysine 119 on histone H2A (H2AK119ub), catalyzed by its E3-ligase subunits, either RING1A or RING1B [74,75]. The mechanism by which PRC1 and PRC2 are recruited to the genome remains an area of debate. It was shown previously that while PRC2 actively methylates H3K27 in target genes and is required in order to induce transcriptional silencing, the Pc subunit of PRC1 also recognizes and binds to this modification, contributing to the transcriptional repression through structural modifications to chromatin, as well as blocking the recruitment of nucleosome remodeling factors such as SWI/SNF [76–79]. In contrast, the H2AK119ub catalyzed by PRC1 can attract bindings by PRC2, therefore affecting deposition of methylation at H3K27 [72,80,81]. Whether the PRC2 loss in MPNST cells affects the PRC1 complex and the ubiquitination of the epigenome remains to be determined but may represent a unique vulnerability and target in this disease.

4. The Role of PRC2 in Schwann Cell Development and Nerve Injury

MPNSTs arise from peripheral nerve branches or fiber sheaths and are thought to be derived from either Schwann cells or pluripotent cells of neural crest origin [82]. In patients with NF1, MPNSTs can arise within the plexiform neurofibromas, and the plexiform neurofibromas grown in the paraspinal region associated with dorsal root ganglia are more likely to go through malignant transformation [83]. In a search for the cells of origin of NF1-associated plexiform neurofibromas, Chen et al. identified a population of GAP43+ PLP+ Schwann cell precursors in the embryonic nerve roots responsible for the neurofibromagenesis [48]. Therefore, understanding the normal development of Schwann cells may be informative for building a model of tumorigenesis of MPNSTs.

Schwann cells are the primary glial cell of the peripheral nervous system and play a variety of functions including nerve impulse conduction [84], maintenance of the nerve microenvironment [85], presentation of antigens [86], and nerve development and regeneration after injury [85,87]. During embryonic development, the development of peripheral nervous system parallels the development of Schwann cells from neural crest cells through a series of phases starting with migration of neural crest cells and differentiation into Schwann cell precursors, which subsequently become immature Schwann cells. These cells can ultimately differentiate into myelinating and non-myelinating Schwann cells of the mature nerves [88–92]. This highly ordered process of Schwann cell development is tightly regulated by a number of signals, including epigenetic and transcriptional regulations (reviewed in detail [93–95]). Notably, in vivo studies showed that disruption of PRC2 or the proper deposit of its product H3K27me3 led to hypermyelination in adult mice [96,97], whilst EZH2 loss in cultured Schwann cells inhibited the myelination process [98]. These inconsistent results may be explained by the differences between in vivo and in vitro systems, and further experiments are needed to resolve these results.

Unlike the ambiguities observed in Schwann cell development, the critical role that PRC2 plays in nerve injury repair is well documented. The Schwann cell injury response involves the reversal of myelin differentiation and downregulation of myelin proteins (reviewed in [99]) and a switch to a repair cell phenotype (reviewed in [87]). In this capacity, the repair Schwann cells can express neurotrophic factors and cytokines that promote neuron survival and axonal regeneration [100,101]. These cytokines recruit macrophages that promote vascularization of distal nerves and assist in the removal of myelin debris that can potentially inhibit axon growth [102,103]. Repair Schwann cells also form tracks known as Bands of Bungner that can guide axon recovery [104].

Nerve injury also induces epigenetic changes in Schwann cells that allows for reprogramming of these cells as they generate the cellular environment required for axon regeneration [105]. PRC2 has been shown to regulate the expression of Schwann cell repair genes and affect nerve injury response via H3K27me3. In this context, PRC2 was found to repress nerve repair genes such as sonic hedgehog (*Shh*), glial-derived neurotrophic factors (*Gdnf*), and brain-derived neurotrophic factors (*Bdnf*) [105]. Nerve injury leads to reversal of PRC2 repression, H3K27 demethylation, and de-repression of these nerve repair genes [106]. Loss of PRC2 repression in an *Eed* conditional knockout mouse model was sufficient to activate these repair genes in uninjured nerves; however, there was no evidence of accelerated nerve injury repair [105]. It is possible that the linkages demonstrating the relationship between nerve injury and PRC2 may yield important clues into the pathogenesis of disease progression of benign neurofibromas to MPNSTs. A critical difference is that PRC2 alteration in nerve injury repair is transient, whereas permanent loss of repression by genetic alteration appears to be needed for malignant transformation. An intriguing result was seen in mice with NF1 deficiencies, where normal mature myelinating Schwann cells exhibited no signs of tumor formation; however, when there was injury to the nerve, neurofibromas developed at those sites [107]. These results and others may indicate that epigenetic programs utilized in the normal process of nerve healing are corrupted by the MPNST cells through the genetic alteration of PRC2.

5. Consequences of PRC2 Loss on Oncogenic Signaling in MPNST

Not surprisingly, the consequences of PRC2 loss on the oncogenic signaling within the MPNST cell has become an intensive area of investigation in the NF1 research community. The current paradigm proposes a combination of H3K27me3 loss and de-repression of PRC2 target genes along with other consequential epigenetic alterations in the chromatin landscape promoting oncogenesis [108]. However, it remains unclear exactly what arethe MPNST specific PRC2 target genes that are de-repressed as the malignant transformation takes place. One strategy to answer this question is through transcriptomic and proteomic profiling of human MPNST samples; comparing the PRC2-negative tumors with the PRC2-wild type ones [18,19,62]. These studies have suggested an amplified oncogenic signaling may be playing a role; however, it remains unknown whether these changes are direct or indirect consequences of PRC2 loss. Furthermore, although global increases in active transcription markers H3K27ac and H3K36me3 were observed in PRC2-loss MPNST samples, it remains unclear how these changes affect the three-dimensional structure of the genome and subsequent transcription. Well-controlled model systems, which allow for interrogation of the epigenetic and transcriptomic landscape of MPNST cells, in the PRC2-deficient and intact states would benefit the field tremendously. The resulting cellular signaling changes are hypothesized to contribute to oncogenesis via cell proliferation and growth, however the exact mechanisms that allow this to occur remain unknown. Here, we summarize what is known about the effects PRC2 loss has on RAS, Wnt, and Notch signaling and speculate on the implications that these findings may have for MPNST pathogenesis.

5.1. PRC2 Loss and RAS Signaling

As mentioned previously, biallelic LOF in *NF1* is observed in all subtypes of MPNSTs [18,20,109] and this disease can therefore be considered a product of hyperactive RAS signaling pathways. Since MPNSTs can arise within a plexiform neurofibroma, treating these benign tumors is considered a valuable preventive strategy. Indeed, the most effective treatment to date for plexiform neurofibromas is to inhibit the RAS pathway by using MEK inhibitors [110]. However, the loss of *NF1* is necessary but not sufficient for the progression of benign neurofibromas into MPNSTs [111]. Additional genetic mutations either through oncogene amplification or deletions in tumor suppressor genes are required for MPNST transformation [112].

There is evidence that PRC2 loss in MPNSTs contributes to the hyperactive RAS signaling through the epigenetic switch from H3K27me3 to H3K27ac. De Raedt and colleagues found that PRC2 loss amplified NF1 loss-mediated RAS activation and signaling. Using gene set enrichment analysis, this group showed that in SUZ12-depleted cells, there was a significant upregulation of RAS signatures. SUZ12 reconstitution in PRC2-deficient MPNST cells confirmed this result, where downregulation of RAS signatures was noted. Because phospho-ERK levels were unaffected by SUZ12 loss or reconstitution, it was speculated that SUZ12 loss amplified RAS signaling via direct chromatin effects [19]. Evidence for this was seen upon treatment of MPNST cell lines with JQ1, a bromodomain inhibitor, where a similar effect on RAS signatures as SUZ12 reconstitution was noted in the PRC2-deficient cells. Furthermore, the combination of JQ1 and a MEK inhibitor PD-901 was found to cause significant tumor regression in a genetically engineered mouse model with *cis* mutations of *Nf1*, *p53*, and *Suz12* compared to JQ1 or PD-901 alone. The effectiveness of JQ1 in treating PRC2-deficient MPNSTs is consistent with the observation that PRC2-loss triggered increased H3K27ac levels, which is a marker of super enhancers [113]. However, it remains unclear how PRC2 loss alters the global super enhancer landscape and whether additional transcriptional regulators might be involved in the process of malignant transformation. Interestingly, a proteomics-based analysis did not observe specific activation of the RAS pathway in human MPNST samples when comparing tumors with and without intact PRC2 [62]. This inconsistency may be explained by the "contamination" caused by tumor microenvironment when using patient samples or the differences in methodology and requires further investigation.

5.2. PRC2 Loss and Wnt Signaling

PRC2 is theorized to suppress Wnt signaling and thereby affect multiple biological processes, such as skeletal muscle differentiation [114], skeletal growth [115], adipogenesis [116], erythropoiesis [117], and intestinal homeostasis [118]. This suppression of Wnt signaling is mediated through a variety of targets within the Wnt pathway, including genes such as Wnt1, Wnt6, Wnt10a, Wnt10b, and Lef1. Therefore, loss of a functional PRC2 repressive complex in MPNST may lead to the upregulation of this signaling pathway, which has previously been identified as a target of oncogenic mutation in many cancer types [119]. Indeed, RNAseq results previously identified enrichment of Wnt signaling in genes significantly upregulated in PRC2-deficient MPNSTs when compared with PRC2-retained samples [18]. Given the active clinical efforts and promising results targeting Wnt signaling, this may represent a tractable therapeutic target in MPNST and supports additional preclinical study and investment.

Activation of the Wnt pathway has previously been described across several different sarcoma types, including osteo-, Ewing, and rhabdomyosarcomas [120–122]. Interestingly, results from an unbiased forward genetic screen highlighted the Wnt signaling pathway as potential driver of oncogenesis in MPNST. In this work, the authors used a *Sleeping Beauty* transposon-based somatic mutagenesis system in mice and found that 17.2% of all genes identified as cooperating with EGFR overexpression were known members of the Wnt/β-catenin pathway [123]. Further, a study by Luscan and colleagues using mRNA expression data and IHC analysis demonstrated altered expression of 20 Wnt genes in MPNST samples compared to benign neurofibromas [124]. These studies provide evidence of Wnt pathway upregulation in MPNST, which could potentially be a direct result of loss of PRC2 in this cancer. The role of PRC2-regulated Wnt signaling has previously been identified in regulating migration and invasion of breast cancer cells, through the regulation of a Wnt signaling pathway inhibitor DKK1 [125], and in multiple myeloma, in which depletion of core PRC2 components EZH1/2 led to overactivation of Wnt signaling [126]. Interestingly, Serresi, Gargiulo, and colleagues have shown that *Eed* deletion cooperated with *Kras* mutant and p53 inactivation to form an invasive mucinous adenocarcinoma [127]. They reported that a chromatin switch between repressive H3K27me3 to its mutually exclusive active mark H3K27ac on the developmental genes of Wnt pathway drove the tumorigenesis. This observation seems to be highly consistent with the genetic alterations reported in MPNSTs. Though the current knowledge of PRC2 regulation of Wnt signaling is limited in the context of carcinogenesis, PRC2 mutant MPNST provides a genetic mechanism and unique model system with which to investigate this interaction further.

5.3. PRC2 Loss and Notch Signaling

Another signaling pathway implicated in MPNST pathogenesis is Notch signaling. The Notch signaling pathway plays a central role in cell differentiation, proliferation, and reprogramming. The Notch family of transmembrane receptors regulate cell fate choices, and aberrant Notch signaling can lead to tumorigenesis in specific cell types such as T-cell lymphomas and pancreatic cancer [128]. While Notch is typically known as a transcriptional activator, several genes have been noted as repressed by Notch activity. The mechanisms as to why this occurs are not fully understood; however, PRC2 may play a role this transcriptional repression.

When Notch receptors are bound and activated, Notch intracellular domains (NICDs) are cleaved and released. NICDs travel to the nucleus and form a ternary complex with the transcriptional coactivator Mastermind (Maml) and DNA-binding transcription factor CSL that can then recruit higher-order transcriptional complexes, resulting in a transcriptional cascade [129–132]. Han and colleagues found that Notch recruited PRC2 in a Lysine Demethylase 1-dependent manner in T-cell lymphomas, and along with the ternary complex forms a stable transcriptional repressor complex. This leads to enrichment of H3K27me3 repression and loss of H3K4me3 activation, contributing to downstream repressive epigenetic changes [133]. In addition, preliminary data showed that Notch activation led to direct EZH2 and SUZ12 transcriptional induction, although no evidence has shown that they are direct Notch-target genes [133]. Intriguingly, there is also interplay between RAS signaling and Notch signaling; in this manner, Notch signaling seems to be downstream of oncogenic RAS, and wildtype Notch1 is needed for oncogenic RAS-mediated neoplastic transformation of human cells in vitro and in vivo [134].

Although a Notch-mediated PRC2 mechanism has yet to be fully explored in MPNST, Notch signaling may contribute to the malignant transformation of MPNSTs from neurofibromas. Li et al. found that in the sNF96.2 MPNST cells, there was active Notch signaling with NICD generation [135]. Transduction of NICD into rat Schwann cells led to loss of Schwann cell differentiation markers and cellular transformation. These transduced cells had elevated levels of phospho-ERK and Cyclins A, D1, and D2 and were capable of growing into tumor masses when injected into rats. Further research into Notch activation in MPNST is warranted, along with its interplay with PRC2 loss and other driver mutations of MPNST formation.

6. Consequences of PRC2 Loss on Tumor Immune Surveillance in MPNST

Given the growing role of immunotherapy in cancer, there is great interest in understanding the effect that PRC2 loss may play in the ability of MPNST cells to evade immune surveillance. In MPNSTs, PRC2 loss downregulated pathways for antigen presentation and IFN signaling [62]. Proteomic studies revealed decreases in major histocompatibility complex class I (MHC I) expression by tumor cells as well as a decreased infiltration of MPNSTs that lost PRC2 by MHC class II-expressing inflammatory cells. These changes were linked to increased H3K36me2 and H3K27ac as a result of H3K27me3 loss. A proposed mechanism for this observation is PRC2 loss contributing to a decreased IFN signaling as well as the loss of MHC expression. Consistently, restoration of a functional PRC2 or depletion of NSD2 (H3K36me2 methyltransferase) in PRC2-deficient MPNST cell lines resulted in increased MHC I expression and restored IFN pathway expression. It remains unclear whether these changes in immune surveillance and the IFN pathway are directly or indirectly caused by the epigenetic switch of PRC2 loss. Future studies using DNA sequencing coupled with chromatin immunoprecipitation would provide additional mechanistic detail to this observation.

Understanding how to therapeutically modulate the PRC2-induced epigenetic changes in MPNST tumor cells and harness the surrounding immune microenvironment remains a goal of immunotherapeutics. Pilot efforts involving treating PRC2-deficient MPNST cell lines with DNA methyltransferase inhibitors (DNMTi) led to halted cell growth and increased cell death that was associated with increased expression of IFN pathway genes. Additionally, both DNMTi and histone deacetylase inhibitor (HDACi) led to increased MHC I expression in MPNST [62]. Regardless of whether the restoration of IFN pathway genes and MHC I expression is due to direct epigenetic changes, the possibility of using drugs that modulate transcriptional activity opens up the exciting therapeutic possibility of restoring tumor immune surveillance and increasing MPNST targetability.

It is important to note, however, that while PRC2 loss may lead to immune evasion in MPNSTs, in many other cancer types, increased PRC2 activity can actually have a similar effect and also lead to immune surveillance escape through decreased MHC I antigen presentation. A recent study found that PRC2 silenced genes associated with MHC I antigen processing such as MHC I heavy chain genes, the transporter associated with antigen processing (TAP), and the immunoproteasome [136]. In addition, PRC2 restricted transcriptional induction of MHC class I in response to cytokine stimulation in MHC class I deficient tumors such as neuroblastoma and small cell lung cancer. EED or EZH1 and EZH2 inhibition restored expression of MHC I antigen processing genes and effective T cell-mediated immunity in MHC I low cancers. EZH2 inhibition also was shown to enhance tumor immunogenicity through increased interferon signaling, production of proinflammatory chemokines CXCL9 and CXCL10, and modulation of immune cell differentiation.

In cancers where increased PRC2 activity leads to evasion of immune surveillance via decreased MHC I antigen presentation, cell lineage likely plays an important role. Namely, these cells appear to harness embryonic and tissue-specific stem cell programs that are typically regulated by PRC2 to mediate immune evasion [137]. A well-studied example of this phenomenon is in human and mouse melanoma, where PRC2 upregulation was found to be promoted by the presence of tumor-infiltrating T cells [138]. In addition, anti-CTLA-4 or IL-2cx immunotherapy led to increased EZH2, subsequent increases in global H3K27me3, and transcriptional silencing of immunogenicity-related genes including MHC I molecules and antigen processing machinery. In this system, EZH2 inactivation via shRNA or an EZH2 small molecule inhibitor upregulated immunogenicity-associated genes post-immunotherapy downregulation, thus demonstrating immunotherapy-induced gene expression changes that are EZH2-dependent. EZH2 inhibition can synergize with anti-melanoma immunotherapy, stimulating CD8$^+$ T cells and suppressing the PD-1/PD-L1 axis.

Thus, the loss or gain of PRC2 depending on the cancer type can both lead to evasion of immune surveillance via decreased antigen presentation by the tumor cells. It seems likely that PRC2 loss in MPNSTs influences the ability of the immune system to recognize these tumors. One possibility is that MPNSTs arise in the context of NF1-loss mediated hyper activated RAS signaling, which causes decreased interferon signaling and antigen presentation [139,140]. Future work dissecting the interactions of the tumor cell and the host immune system using human samples and immune competent animal models will be required to uncover the mechanistic details of these interactions. Given the potency of immunotherapy in controlling other aggressive metastatic tumor types, this work may have a profound therapeutic impact for patients with MPNST.

7. Establishment of Preclinical Modeling of the PRC2 Loss in MPNST as a Pathway to Clinical Translation

Deciphering new vulnerabilities in the MPNST cell that result from PRC2 loss requires the production and characterization of credentialed model systems that faithfully recapitulate human tumors. The most widely used model system in MPNST investigation is that of patient derived tumor cell lines. We summarize here a wide variety of human MPNST cell lines frequently used in preclinical investigations (Table 1). The majority of these cell lines are derived from NF1 patient tumors, and few have been characterized for PRC2 function or SUZ12 expression. This lack of data highlights a potential need for more thorough characterization of these cell lines as we attempt to understand the effects of PRC2 loss in MPNST and its role in oncogenesis. It would also be beneficial to research efforts if *EED* mutant cell lines were identified or developed, which would allow for more complete analyses of the functional consequences of PRC2 complex loss in MPNST. Further, the majority of the commonly used MPNST cell lines were originally obtained from male patient tumors or are of unknown gender origin. This may result in bias of the data obtained from epigenetic research on such MPNST cells, as it has been hypothesized that sex has the potential to affect the epigenetic modification of the nervous system and lead to morphological differences [141,142]. Clarification of the sex of available MPNST cell lines, or the establishment of novel immortalized cell lines from tumors of female patients, may aid the removal of such bias from ongoing MPNST research. Exciting efforts are underway to develop a next generation of model systems, including work by the the NF1 Biospecimen Repository at Johns Hopkins (https://www.hopkinsmedicine.org/kimmel_cancer_center/centers/pediatric_oncology/ research_and_clinical_trials/pratilas/nf1_biospecimen_repository.html). Importantly, these efforts include a fully annotated clinical database and biospecimen bank of NF1-associated MPNST primary tumors, cell lines, and novel patient-derived xenografts (PDXs), which are available on request.

Table 1. A summary of immortalized cell lines used in MPNST research.

Cell Line	Sex	Synonyms	Origin	PRC2 Status	Ref.
T265	/	T265-2c; T265-2C; T265p21	NF1	Loss [62,143]	[144,145]
90-8	/	MPNST 90-8TL; 90-8TL; NF90-8; NF190-8	NF1	Loss [19]	[146]
ST88-3	M	88-3; NF188-3	NF1	Unknown	[147]
ST88-14	M	ST88.14; ST 88-14; ST-8814; ST8814; 88-14; NF188-14	NF1	Loss [18,19]	[147]
sNF02.2	M	sNF02-2	NF1	WT [148]	[149]
sNF10.1	/		NF1	Loss [150]	[150]
sNF94.3	F		NF1	Loss [150]	[151]
sNF96.2	M	SNF96.2; sNF96-2	NF1	Loss [19,148]	[152]
S462	/		NF1	Loss [19,62]	[153]
S462.TY	/	S462-TY; S462TY	NF1	Unknown	[154]
S520	/		NF1	Unknown	[153]
S805	/		NF1	Unknown	[155]
FMS-1	F		NF1	Unknown	[156]
FU-SFT8710	F		NF1	Unknown	[157]
NFS-1	/		NF1	Unknown	[158]
NMS-2	M		NF1	Unknown	[159]
NMS-2PC	M		NF1	Unknown	[159]
MPNST-14	M		NF1	Unknown	[160]
MPNST642	M		NF1	Unknown	[161]
1507.2	/	S1507-2	NF1	Unknown	[153,162]
STS-26T	/	STS26T; STS26	Sporadic	WT [62,143]	[163]
MPNST-724	/	MPNST724	Sporadic	WT [18]	[160]
HS-Sch-2	F		Sporadic	Unknown	[164]
HS-PSS	M		Sporadic	Unknown	[165]
YST-1	F		Sporadic	Unknown	[165]
FU-SFT9817	F		Sporadic	Unknown	[157]
FU-SFT8611	M		Sporadic	Unknown	[157]

F: Derived from a female patient. M: Derived from a male patient. Loss: Normal function of PRC2 loss determined in the indicated reference. WT: Normal function of PRC2 retained in the indicated reference.

A variety of animal models have been used in preclinical investigations of MPNST in an effort to more closely recapitulate the human tumor environment. Murine models are frequently used in this area of research, particularly xenograft or orthograft models involving the engraftment of the human tumor cell lines into mice. Most of the cell lines listed in Table 1 have been utilized in xenograft research of MPNST, including those known to be PRC2 mutant, allowing for biological modelling of tumors possessing this aberration. Another common method of MPNST investigation is the use of PDX models, in which patient tumor is engrafted onto an immunocompromised host, to allow for the investigation of these tumor cells in the context of an in vivo environment. Although there is a growing number of PDX models described [166,167], documentation of PRC2 mutational status has not been routinely commented on.

Genetically engineered mouse models (GEMMs) are another useful tool for the study of cancer development, progression, and therapeutics, but have proved difficult to produce in the case of PRC2 mutant MPNST. Despite the success of GEMM in contributing to the study of NF1 pathogenesis and plexiform development, the development of MPNST models has been a slow and complex process [168]. An effort was made by De Raedt and colleagues to generate MPNST models through the generation of *Nf1*, *p53*, and *Suz12* mutant mice [19]. They generated $Nf1^{+/-}$, $Suz12^{+/-}$ mice in cis, in which the mutant copies of these genes were on a single chromosome, and tumors developed upon spontaneous loss of the wildtype chromosome. This murine model had a high rate of tumor development and decreased survival. They further developed a $Nf1^{+/-}$, $Suz12^{+/-}$, and $p53^{+/-}$ cis model, as the p53 tumor suppressor protein is found to be commonly mutated in MPNST. These mutant models were found to have a high rate of spontaneous tumor development, but the tumors were of a wide histological variety, including histiocytic sarcomas, intestinal adenomas, neurofibromas, hepatocellular carcinomas, as well as MPNSTs. While MPNST development was identified in this study, mice frequently succumbed to other cancerous diseases prior to this tumor formation, indicating a lack of efficiency of this MPNST model. Due to the high rate of tumorigenesis and wide variety of tumors as a result of the combination of *NF1*, *p53*, and *SUZ12* deletion, it is possible that a more effective murine model of MPNST can utilize floxed alleles that conditionally knockout tumor suppressors in the appropriate cells of origin. As mentioned previously, neural crest gives rise to Schwann cell precursors, which subsequently differentiate into immature Schwann cells and then myelinating and non-myelinating Schwann cells after birth [90]. It has been appreciated that the cells of origin that give rise to plexiform neurofibromas and therefore MPNSTs are the Schwann cells in the dorsal ganglia root [83]. Using a genetically engineered *Nf1* floxed, *Cdkn2a/Arf* floxed, and PostnCre mouse model that triggered conditional knockout in the nerve crest derived Schwann cell lineage, Rhodes et al. created one of first models that mimic the human malignant transformation from plexiform to atypical neurofibroma, which eventually developed MPNST with a high penetrance [169]. It is hoped that with the increasingly accurate description of the genetic lesions associated with the tumorigenic formation of human MPNSTs, more MPNST GEMMs will be produced that faithfully recapitulate the genetic lesions and will become available to the researcher community.

8. Conclusions

The advance of NGS enabled the discovery of PRC2 loss in MPNSTs, and the loss of H3K27me3 has become a clinically useful, sensitive, and specific marker for diagnosis. Efforts to understand the consequences of PRC2 loss in MPNST tumorigenesis and to identify novel vulnerabilities in this difficult to treat tumor are areas of intensive focus for both basic and translational researchers. Recent works have discovered that the loss of PRC2 in MPNST likely affects changes in cellular signaling and immune surveillance through alteration of the core epigenetic and transcriptomic landscape in a neuronal specific precursor cell. Further studies will be enabled through a new generation of clinically annotated and genetically profiled patient samples and their derivative MPNST cell lines and PDX models, as well as GEMMs that mimic the clinically observed disease progression from benign plexiform neurofibroma through atypical neurofibroma to MPNST. These and other anticipated advances will hopefully accelerate discovery of mechanistically based strategies for the treatment of this devastating tumor.

Supplementary Materials: The following are available online at http://www.mdpi.com/2073-4425/11/3/287/s1, Table S1: Summary of mutations of *SUZ12* and *EED* in human MPNSTs.

Author Contributions: All authors researched relevant literature, conceived, wrote and edited the manuscript. All authors have read and agreed to the published version of the manuscript.

Acknowledgments: This work was funded by the Center for Cancer Research, Intramural Research Program at the National Cancer Institute. G.M. was supported by the NIH Medical Research Scholars Program, a public-private partnership supported jointly by the NIH and contributions to the Foundation for the NIH from the Doris Duke Charitable Foundation, Genentech, the American Association for Dental Research, the Colgate-Palmolive Company, and other private donors. B.M. was supported by the National Cancer Institute-Queen's University Belfast Graduate Partnership Program, supported by contribution from the Northern Ireland Health and Social Care Research and Development Division. The authors thank Dr. Brigitte Widemann who provided insight and expertise that greatly improved this manuscript.

Conflicts of Interest: The authors declare no conflict of interest.

References

1. Ferner, R.E. Neurofibromatosis 1 and neurofibromatosis 2: A twenty first century perspective. *Lancet Neurol.* **2007**, *6*, 340–351. [CrossRef]
2. Hirbe, A.C.; Gutmann, D.H. Neurofibromatosis type 1: A multidisciplinary approach to care. *Lancet Neurol.* **2014**, *13*, 834–843. [CrossRef]
3. Ratner, N.; Miller, S.J. A RASopathy gene commonly mutated in cancer: The neurofibromatosis type 1 tumour suppressor. *Nat. Publ. Group Nat. Rev. Cancer* **2015**, *15*, 290–301. [CrossRef]
4. Martin, G.A.; Viskochil, D.; Bollag, G.; McCabe, P.C.; Crosier, W.J.; Haubruck, H.; Conroy, L.; Clark, R.; O'Connell, P.; Cawthon, R.M. The GAP-related domain of the neurofibromatosis type 1 gene product interacts with ras p21. *Cell* **1990**, *63*, 843–849. [CrossRef]
5. Karnoub, A.E.; Weinberg, R.A. Ras oncogenes: Split personalities. *Nat. Rev. Mol. Cell Biol.* **2008**, *9*, 517–531. [CrossRef]
6. Jett, K.; Friedman, J.M. Clinical and genetic aspects of neurofibromatosis 1. *Genet. Med.* **2010**, *12*, 1–11. [CrossRef]
7. Ferner, R.E.; Huson, S.M.; Thomas, N.; Moss, C.; Willshaw, H.; Evans, D.G.; Upadhyaya, M.; Towers, R.; Gleeson, M.; Steiger, C.; et al. Guidelines for the diagnosis and management of individuals with neurofibromatosis 1. *J. Med. Genet.* **2007**, *44*, 81–88. [CrossRef]
8. Neurofibromatosis. Conference statement. National Institutes of Health Consensus Development Conference. *Arch. Neurol.* **1988**, *45*, 575–578.
9. Evans, D.G.R.; Salvador, H.; Chang, V.Y.; Erez, A.; Voss, S.D.; Druker, H.; Scott, H.S.; Tabori, U. Cancer and Central Nervous System Tumor Surveillance in Pediatric Neurofibromatosis 2 and Related Disorders. *Clin. Cancer Res.* **2017**, *23*, e54–e61. [CrossRef]
10. Bates, J.E.; Peterson, C.R.; Dhakal, S.; Giampoli, E.J.; Constine, L.S. Malignant peripheral nerve sheath tumors (MPNST): A SEER analysis of incidence across the age spectrum and therapeutic interventions in the pediatric population. *Pediatric Blood Cancer* **2014**, *61*, 1955–1960. [CrossRef]
11. Evans, D.G.R.; Baser, M.E.; McGaughran, J.; Sharif, S.; Howard, E.; Moran, A. Malignant peripheral nerve sheath tumours in neurofibromatosis 1. *J. Med. Genet.* **2002**, *39*, 311–314. [CrossRef]
12. Widemann, B.C. Current status of sporadic and neurofibromatosis type 1-associated malignant peripheral nerve sheath tumors. *Curr. Oncol. Rep.* **2009**, *11*, 322–328. [CrossRef]
13. Brohl, A.S.; Kahen, E.; Yoder, S.J.; Teer, J.K.; Reed, D.R. The genomic landscape of malignant peripheral nerve sheath tumors: Diverse drivers of Ras pathway activation. *Sci. Rep.* **2017**, *7*, 1–5. [CrossRef]
14. Cichowski, K.; Jacks, T. NF1 Tumor Suppressor Gene Function. *Cell* **2001**, *104*, 593–604. [CrossRef]
15. Hirbe, A.C.; Dahiya, S.; Miller, C.A.; Li, T.; Fulton, R.S.; Zhang, X.; McDonald, S.; DeSchryver, K.; Duncavage, E.J.; Walrath, J.; et al. Whole Exome Sequencing Reveals the Order of Genetic Changes during Malignant Transformation and Metastasis in a Single Patient with NF1-plexiform Neurofibroma. *Clin. Cancer Res.* **2015**, *21*, 4201–4211. [CrossRef]
16. Pemov, A.; Hansen, N.F.; Sindiri, S.; Patidar, R.; Higham, C.S.; Dombi, E.; Miettinen, M.M.; Fetsch, P.; Brems, H.; Chandrasekharappa, S.; et al. Low mutation burden and frequent loss of CDKN2A/B and SMARCA2, but not PRC2, define pre-malignant neurofibromatosis type 1-associated atypical neurofibromas. *Neuro Oncol.* **2019**, *21*, 981–992. [CrossRef]
17. Uusitalo, E.; Rantanen, M.; Kallionpää, R.A.; Pöyhönen, M.; Leppävirta, J.; Ylä-Outinen, H.; Riccardi, V.M.; Pukkala, E.; Pitkäniemi, J.; Peltonen, S.; et al. Distinctive Cancer Associations in Patients with Neurofibromatosis Type 1. *J. Clin. Oncol.* **2016**, *34*, 1978–1986. [CrossRef]
18. Lee, W.; Teckie, S.; Wiesner, T.; Ran, L.; Prieto-Granada, C.N.; Lin, M.; Zhu, S.; Cao, Z.; Liang, Y.; Sboner, A.; et al. PRC2 is recurrently inactivated through EED or SUZ12 loss in malignant peripheral nerve sheath tumors. *Nat. Rev. Cancer* **2014**, *46*, 1227–1232. [CrossRef]
19. De Raedt, T.; Beert, E.; Pasmant, E.; Luscan, A.; Brems, H.; Ortonne, N.; Helin, K.; Hornick, J.L.; Mautner, V.; Kehrer-Sawatzki, H.; et al. PRC2 loss amplifies Ras-driven transcription and confers sensitivity to BRD4-based therapies. *Nature* **2016**, *514*, 247–251. [CrossRef]
20. Zhang, M.; Wang, Y.; Jones, S.; Sausen, M.; McMahon, K.; Sharma, R.; Wang, Q.; Belzberg, A.J.; Chaichana, K.; Gallia, G.L.; et al. Somatic mutations of SUZ12 in malignant peripheral nerve sheath tumors. *Nat. Rev. Cancer* **2014**, *46*, 1170–1172. [CrossRef]

21. Sohier, P.; Luscan, A.; Lloyd, A.; Ashelford, K.; Laurendeau, I.; Briand-Suleau, A.; Vidaud, D.; Ortonne, N.; Pasmant, E.; Upadhyaya, M. Confirmation of mutation landscape of NF1-associated malignant peripheral nerve sheath tumors. *Genes Chromosomes Cancer* **2017**, *56*, 421–426. [CrossRef]

22. Cancer Genome Atlas Research Network. Electronic address: Elizabeth.demicco@sinaihealthsystem.ca; Cancer Genome Atlas Research Network Comprehensive and Integrated Genomic Characterization of Adult Soft Tissue Sarcomas. *Cell* **2017**, *171*, 950–965. [CrossRef]

23. Cerami, E.; Gao, J.; Dogrusoz, U.; Gross, B.E.; Sumer, S.O.; Aksoy, B.A.; Jacobsen, A.; Byrne, C.J.; Heuer, M.L.; Larsson, E.; et al. The cBio Cancer Genomics Portal: An Open Platform for Exploring Multidimensional Cancer Genomics Data. *Cancer Discov.* **2012**, *2*, 401–404. [CrossRef]

24. Gao, J.; Aksoy, B.A.; Dogrusoz, U.; Dresdner, G.; Gross, B.; Sumer, S.O.; Sun, Y.; Jacobsen, A.; Sinha, R.; Larsson, E.; et al. Integrative Analysis of Complex Cancer Genomics and Clinical Profiles Using the cBioPortal. *Sci. Signal.* **2013**, *6*, pl1. [CrossRef]

25. Kim, K.H.; Roberts, C.W.M. Targeting EZH2 in cancer. *Nat. Med.* **2016**, *22*, 128–134. [CrossRef]

26. Wassef, M.; Luscan, A.; Aflaki, S.; Zielinski, D.; Jansen, P.W.T.C.; Baymaz, H.I.; Battistella, A.; Kersouani, C.; Servant, N.; Wallace, M.R.; et al. EZH1/2 function mostly within canonical PRC2 and exhibit proliferation-dependent redundancy that shapes mutational signatures in cancer. *Proc. Natl. Acad. Sci. USA* **2019**, *116*, 6075–6080. [CrossRef]

27. Sneeringer, C.J.; Scott, M.P.; Kuntz, K.W.; Knutson, S.K.; Pollock, R.M.; Richon, V.M.; Copeland, R.A. Coordinated activities of wild-type plus mutant EZH2 drive tumor-associated hypertrimethylation of lysine 27 on histone H3 (H3K27) in human B-cell lymphomas. *Proc. Natl. Acad. Sci. USA* **2010**, *107*, 20980–20985. [CrossRef]

28. Li, X.; Gonzalez, M.E.; Toy, K.; Filzen, T.; Merajver, S.D.; Kleer, C.G. Targeted Overexpression of EZH2 in the Mammary Gland Disrupts Ductal Morphogenesis and Causes Epithelial Hyperplasia. *Am. J. Pathol.* **2009**, *175*, 1246–1254. [CrossRef]

29. Karanikolas, B.D.W.; Figueiredo, M.L.; Wu, L. Polycomb Group Protein Enhancer of Zeste 2 Is an Oncogene That Promotes the Neoplastic Transformation of a Benign Prostatic Epithelial Cell Line. *Mol. Cancer Res.* **2009**, *7*, 1456–1465. [CrossRef]

30. Kleer, C.G.; Cao, Q.; Varambally, S.; Shen, R.; Ota, I.; Tomlins, S.A.; Ghosh, D.; Sewalt, R.G.A.B.; Otte, A.P.; Hayes, D.F.; et al. EZH2 is a marker of aggressive breast cancer and promotes neoplastic transformation of breast epithelial cells. *Proc. Natl. Acad. Sci. USA* **2003**, *100*, 11606–11611. [CrossRef]

31. Varambally, S.; Dhanasekaran, S.M.; Zhou, M.; Barrette, T.R.; Kumar-Sinha, C.; Sanda, M.G.; Ghosh, D.; Pienta, K.J.; Sewalt, R.G.A.B.; Otte, A.P.; et al. The polycomb group protein EZH2 is involved in progression of prostate cancer. *Nature* **2002**, *419*, 624–629. [CrossRef]

32. Morin, R.D.; Johnson, N.A.; Severson, T.M.; Mungall, A.J.; An, J.; Goya, R.; Paul, J.E.; Boyle, M.; Woolcock, B.W.; Kuchenbauer, F.; et al. Somatic mutations altering EZH2 (Tyr641) in follicular and diffuse large B-cell lymphomas of germinal-center origin. *Nat. Genet.* **2010**, *42*, 181. [CrossRef]

33. Yap, D.B.; Chu, J.; Berg, T.; Schapira, M.; Cheng, S.W.G.; Moradian, A.; Morin, R.D.; Mungall, A.J.; Meissner, B.; Boyle, M.; et al. Somatic mutations at EZH2 Y641 act dominantly through a mechanism of selectively altered PRC2 catalytic activity, to increase H3K27 trimethylation. *Blood* **2011**, *117*, 2451–2459. [CrossRef]

34. Souroullas, G.P.; Jeck, W.R.; Parker, J.S.; Simon, J.M.; Liu, J.-Y.; Paulk, J.; Xiong, J.; Clark, K.S.; Fedoriw, Y.; Qi, J.; et al. An oncogenic Ezh2 mutation induces tumors through global redistribution of histone 3 lysine 27 trimethylation. *Nat. Med.* **2019**, *22*, 632. [CrossRef]

35. McCabe, M.T.; Graves, A.P.; Ganji, G.; Diaz, E.; Halsey, W.S.; Jiang, Y.; Smitheman, K.N.; Ott, H.M.; Pappalardi, M.B.; Allen, K.E.; et al. Mutation of A677 in histone methyltransferase EZH2 in human B-cell lymphoma promotes hypertrimethylation of histone H3 on lysine 27 (H3K27). *Proc. Natl. Acad. Sci. USA* **2012**, *109*, 2989–2994. [CrossRef]

36. Majer, C.R.; Jin, L.; Scott, M.P.; Knutson, S.K.; Kuntz, K.W.; Keilhack, H.; Smith, J.J.; Moyer, M.P.; Richon, V.M.; Copeland, R.A.; et al. A687V EZH2 is a gain-of-function mutation found in lymphoma patients. *FEBS Lett.* **2012**, *586*, 3448–3451. [CrossRef]

37. Zhan, F.; Hardin, J.; Kordsmeier, B.; Bumm, K.; Zheng, M.; Tian, E.; Sanderson, R.; Yang, Y.; Wilson, C.; Zangari, M.; et al. Global gene expression profiling of multiple myeloma, monoclonal gammopathy of undetermined significance, and normal bone marrow plasma cells. *Blood* **2002**, *99*, 1745–1757. [CrossRef]

38. Croonquist, P.A.; Van Ness, B. The polycomb group protein enhancer of zeste homolog 2 (EZH2) is an oncogene that influences myeloma cell growth and the mutant ras phenotype. *Oncogene* **2005**, *24*, 6269–6280. [CrossRef]

39. Walker, B.A.; Boyle, E.M.; Wardell, C.P.; Murison, A.; Begum, D.B.; Dahir, N.M.; Proszek, P.Z.; Johnson, D.C.; Kaiser, M.F.; Melchor, L.; et al. Mutational Spectrum, Copy Number Changes, and Outcome: Results of a Sequencing Study of Patients With Newly Diagnosed Myeloma. *J. Clin. Oncol.* **2015**, *33*, 3911–3920. [CrossRef]

40. Bachmann, I.M.; Halvorsen, O.J.; Collett, K.; Stefansson, I.M.; Straume, O.; Haukaas, S.A.; Salvesen, H.B.; Otte, A.P.; Akslen, L.A. EZH2 Expression Is Associated With High Proliferation Rate and Aggressive Tumor Subgroups in Cutaneous Melanoma and Cancers of the Endometrium, Prostate, and Breast. *J. Clin. Oncol.* **2006**, *24*, 268–273. [CrossRef]

41. Schwartzentruber, J.; Korshunov, A.; Liu, X.-Y.; Jones, D.T.W.; Pfaff, E.; Jacob, K.; Sturm, D.; Fontebasso, A.M.; Quang, D.-A.K.; Tönjes, M.; et al. Driver mutations in histone H3.3 and chromatin remodelling genes in paediatric glioblastoma. *Nature* **2012**, *482*, 226–231. [CrossRef]

42. The St. Jude Children's Researchs Hospital–Washington University Pediatric Cancer Genome Project. The genomic landscape of diffuse intrinsic pontine glioma and pediatric non-brainstem high-grade glioma. *Nat. Rev. Cancer* **2014**, *46*, 444–450.

43. Mohammad, F.; Weissmann, S.; Leblanc, B.; Pandey, D.P.; Højfeldt, J.W.; Comet, I.; Zheng, C.; Johansen, J.V.; Rapin, N.; Porse, B.T.; et al. EZH2 is a potential therapeutic target for H3K27M-mutant pediatric gliomas. *Nat. Med.* **2017**, *23*, 484. [CrossRef]

44. Margueron, R.; Reinberg, D. The Polycomb complex PRC2 and its mark in life. *Nature* **2011**, *469*, 343–349. [CrossRef]

45. Müller, J.; Verrijzer, P. Biochemical mechanisms of gene regulation by polycomb group protein complexes. *Curr. Opin. Genet. Dev.* **2009**, *19*, 150–158. [CrossRef]

46. Simon, J.A.; Kingston, R.E. Mechanisms of Polycomb gene silencing: Knowns and unknowns. *Nat. Rev. Mol. Cell Biol.* **2009**, *10*, 697–708. [CrossRef]

47. Cao, R.; Zhang, Y. The functions of E(Z)/EZH2-mediated methylation of lysine 27 in histone H3. *Curr. Opin. Genet. Dev.* **2004**, *14*, 155–164. [CrossRef]

48. Bracken, A.P. Genome-wide mapping of Polycomb target genes unravels their roles in cell fate transitions. *Genes Dev.* **2006**, *20*, 1123–1136. [CrossRef]

49. Ferrari, K.J.; Scelfo, A.; Jammula, S.; Cuomo, A.; Barozzi, I.; Stützer, A.; Fischle, W.; Bonaldi, T.; Pasini, D. Polycomb-Dependent H3K27me1 and H3K27me2 Regulate Active Transcription and Enhancer Fidelity. *Mol. Cell* **2014**, *53*, 49–62. [CrossRef]

50. Højfeldt, J.W.; Laugesen, A.; Willumsen, B.M.; Damhofer, H.; Hedehus, L.; Tvardovskiy, A.; Mohammad, F.; Jensen, O.N.; Helin, K. Accurate H3K27 methylation can be established de novo by SUZ12-directed PRC2. *Nat. Struct. Mol. Biol.* **2018**, *25*, 225–232. [CrossRef]

51. Jung, H.R.; Pasini, D.; Helin, K.; Jensen, O.N. Quantitative Mass Spectrometry of Histones H3.2 and H3.3 in Suz12-deficient Mouse Embryonic Stem Cells Reveals Distinct, Dynamic Post-translational Modifications at Lys-27 and Lys-36. *Mol. Cell. Proteom.* **2010**, *9*, 838–850. [CrossRef]

52. Youmans, D.T.; Schmidt, J.C.; Cech, T.R. Live-cell imaging reveals the dynamics of PRC2 and recruitment to chromatin by SUZ12-associated subunits. *Genes Dev.* **2018**, *32*, 794–805. [CrossRef]

53. Barski, A.; Cuddapah, S.; Cui, K.; Roh, T.-Y.; Schones, D.E.; Wang, Z.; Wei, G.; Chepelev, I.; Zhao, K. High-Resolution Profiling of Histone Methylations in the Human Genome. *Cell* **2007**, *129*, 823–837. [CrossRef]

54. Marchione, D.M.; Lisby, A.; Viaene, A.N.; Santi, M.; Nasrallah, M.; Wang, L.-P.; Williams, E.A.; Larque, A.B.; Chebib, I.; Garcia, B.A.; et al. Histone H3K27 dimethyl loss is highly specific for malignant peripheral nerve sheath tumor and distinguishes true PRC2 loss from isolated H3K27 trimethyl loss. *Mod. Pathol.* **2019**, *32*, 1434–1446. [CrossRef]

55. Pekmezci, M.; Cuevas-Ocampo, A.K.; Perry, A.; Horvai, A.E. Significance of H3K27me3 loss in the diagnosis of malignant peripheral nerve sheath tumors. *Mod. Pathol.* **2017**, *30*, 1710–1719. [CrossRef]

56. Asano, N.; Yoshida, A.; Ichikawa, H.; Mori, T.; Nakamura, M.; Kawai, A.; Hiraoka, N. Immunohistochemistry for trimethylated H3K27 in the diagnosis of malignant peripheral nerve sheath tumours. *Histopathology* **2017**, *70*, 385–393. [CrossRef]

57. Le Guellec, S.; Macagno, N.; Velasco, V.; Lamant, L.; Lae, M.; Filleron, T.; Malissen, N.; Cassagnau, E.; Terrier, P.; Chevreau, C.; et al. Loss of H3K27 trimethylation is not suitable for distinguishing malignant peripheral nerve sheath tumor from melanoma: A study of 387 cases including mimicking lesions. *Mod. Pathol.* **2017**, *30*, 1677–1687. [CrossRef]

58. Otsuka, H.; Kohashi, K.; Yoshimoto, M.; Ishihara, S.; Toda, Y.; Yamada, Y.; Yamamoto, H.; Nakashima, Y.; Oda, Y. Immunohistochemical evaluation of H3K27 trimethylation in malignant peripheral nerve sheath tumors. *Pathol. Res. Pract.* **2018**, *214*, 417–425. [CrossRef]

59. Cleven, A.H.G.; Sannaa Al, G.A.; Bruijn, I.B.-D.; Ingram, D.R.; van de Rijn, M.; Rubin, B.P.; de Vries, M.W.; Watson, K.L.; Torres, K.E.; Wang, W.-L.; et al. Loss of H3K27 tri-methylation is a diagnostic marker for malignant peripheral nerve sheath tumors and an indicator for an inferior survival. *Mod. Pathol.* **2016**, *29*, 582–590. [CrossRef]

60. Schaefer, I.-M.; Fletcher, C.D.; Hornick, J.L. Loss of H3K27 trimethylation distinguishes malignant peripheral nerve sheath tumors from histologic mimics. *Mod. Pathol.* **2016**, *29*, 4–13. [CrossRef]

61. Prieto-Granada, C.N.; Wiesner, T.; Messina, J.L.; Jungbluth, A.A.; Chi, P.; Antonescu, C.R. Loss of H3K27me3 Expression Is a Highly Sensitive Marker for Sporadic and Radiation-induced MPNST. *Am. J. Surg. Pathol.* **2016**, *40*, 479–489. [CrossRef]

62. Wojcik, J.B.; Marchione, D.M.; Sidoli, S.; Djedid, A.; Lisby, A.; Majewski, J.; Garcia, B.A. Epigenomic reordering induced by Polycomb loss drives oncogenesis but leads to therapeutic vulnerabilities in malignant peripheral nerve sheath tumors. *Cancer Res.* **2019**, *79*, 3205–3219. [CrossRef]

63. Bell, O.; Wirbelauer, C.; Hild, M.; Scharf, A.N.D.; Schwaiger, M.; MacAlpine, D.M.; Zilbermann, F.; van Leeuwen, F.; Bell, S.P.; Imhof, A.; et al. Localized H3K36 methylation states define histone H4K16 acetylation during transcriptional elongation in Drosophila. *EMBO J.* **2007**, *26*, 4974–4984. [CrossRef]

64. Turberfield, A.H.; Kondo, T.; Nakayama, M.; Koseki, Y.; King, H.W.; Koseki, H.; Klose, R.J. KDM2 proteins constrain transcription from CpG island gene promoters independently of their histone demethylase activity. *Nucleic Acids Res.* **2019**, *47*, 9005–9023. [CrossRef]

65. Wagner, E.J.; Carpenter, P.B. Understanding the language of Lys36 methylation at histone H3. *Nat. Rev. Mol. Cell Biol.* **2012**, *13*, 115–126. [CrossRef]

66. Yuan, W.; Xu, M.; Huang, C.; Liu, N.; Chen, S.; Zhu, B. H3K36 methylation antagonizes PRC2-mediated H3K27 methylation. *J. Biol. Chem.* **2011**, *286*, 7983–7989. [CrossRef]

67. Xie, W.; Schultz, M.D.; Lister, R.; Hou, Z.; Rajagopal, N.; Ray, P.; Whitaker, J.W.; Tian, S.; Hawkins, R.D.; Leung, D.; et al. Epigenomic Analysis of Multilineage Differentiation of Human Embryonic Stem Cells. *Cell* **2013**, *153*, 1134–1148. [CrossRef]

68. Long, H.K.; Sims, D.; Heger, A.; Blackledge, N.P.; Kutter, C.; Wright, M.L.; Grützner, F.; Odom, D.T.; Patient, R.; Ponting, C.P.; et al. Epigenetic conservation at gene regulatory elements revealed by non-methylated DNA profiling in seven vertebrates. *Elife* **2013**, *2*, e00348. [CrossRef]

69. Jeong, M.; Sun, D.; Luo, M.; Huang, Y.; Challen, G.A.; Rodriguez, B.; Zhang, X.; Chavez, L.; Wang, H.; Hannah, R.; et al. Large conserved domains of low DNA methylation maintained by Dnmt3a. *Nat. Rev. Cancer* **2013**, *46*, 17–23. [CrossRef]

70. Bartke, T.; Vermeulen, M.; Xhemalce, B.; Robson, S.C.; Mann, M.; Kouzarides, T. Nucleosome-Interacting Proteins Regulated by DNA and Histone Methylation. *Cell* **2010**, *143*, 470–484. [CrossRef]

71. Wu, H.; Coskun, V.; Tao, J.; Xie, W.; Ge, W.; Yoshikawa, K.; Li, E.; Zhang, Y.; Sun, Y.E. Dnmt3a-Dependent Nonpromoter DNA Methylation Facilitates Transcription of Neurogenic Genes. *Science* **2010**, *329*, 444–448. [CrossRef] [PubMed]

72. Cooper, S.; Dienstbier, M.; Hassan, R.; Schermelleh, L.; Sharif, J.; Blackledge, N.P.; De Marco, V.; Elderkin, S.; Koseki, H.; Klose, R.; et al. Targeting Polycomb to Pericentric Heterochromatin in Embryonic Stem Cells Reveals a Role for H2AK119u1 in PRC2 Recruitment. *Cell Rep.* **2014**, *7*, 1456–1470. [CrossRef] [PubMed]

73. Li, Y.; Zheng, H.; Wang, Q.; Zhou, C.; Wei, L.; Liu, X.; Zhang, W.; Zhang, Y.; Du, Z.; Wang, X.; et al. Genome-wide analyses reveal a role of Polycomb in promoting hypomethylation of DNA methylation valleys. *Genome Biol.* **2018**, *19*, 18. [CrossRef] [PubMed]

74. Wang, H.; Wang, L.; Erdjument-Bromage, H.; Vidal, M.; Tempst, P.; Jones, R.S.; Zhang, Y. Role of histone H2A ubiquitination in Polycomb silencing. *Nature* **2004**, *431*, 873–878. [CrossRef]

75. Cao, R.; Tsukada, Y.-I.; Zhang, Y. Role of Bmi-1 and Ring1A in H2A ubiquitylation and Hox gene silencing. *Mol. Cell* **2005**, *20*, 845–854. [CrossRef]

76. Pengelly, A.R.; Copur, Ö.; Jäckle, H.; Herzig, A.; Müller, J. A Histone Mutant Reproduces the Phenotype Caused by Loss of Histone-Modifying Factor Polycomb. *Science* **2013**, *339*, 698–699. [CrossRef]

77. Francis, N.J.; Kingston, R.E. Mechanisms of transcriptional memory. *Nat. Rev. Mol. Cell Biol.* **2001**, *2*, 409–421. [CrossRef]

78. Cao, R.; Wang, L.; Wang, H.; Xia, L.; Erdjument-Bromage, H.; Tempst, P.; Jones, R.S.; Zhang, Y. Role of Histone H3 Lysine 27 Methylation in Polycomb-Group Silencing. *Science* **2002**, *298*, 1039–1043. [CrossRef]

79. Shao, Z.; Raible, F.; Mollaaghababa, R.; Guyon, J.R.; Wu, C.T.; Bender, W.; Kingston, R.E. Stabilization of chromatin structure by PRC1, a Polycomb complex. *Cell* **1999**, *98*, 37–46. [CrossRef]

80. Kalb, R.; Latwiel, S.; Baymaz, H.I.; Jansen, P.W.T.C.; Müller, C.W.; Vermeulen, M.; Müller, J. Histone H2A monoubiquitination promotes histone H3 methylation in Polycomb repression. *Nat. Struct. Mol. Biol.* **2014**, *21*, 569–571. [CrossRef]

81. Blackledge, N.P.; Farcas, A.M.; Kondo, T.; King, H.W.; McGouran, J.F.; Hanssen, L.L.P.; Ito, S.; Cooper, S.; Kondo, K.; Koseki, Y.; et al. Variant PRC1 complex-dependent H2A ubiquitylation drives PRC2 recruitment and polycomb domain formation. *Cell* **2014**, *157*, 1445–1459. [CrossRef]

82. Gupta, G.; Mammis, A.; Maniker, A. Malignant peripheral nerve sheath tumors. *Neurosurg. Clin. N. Am.* **2008**, *19*, 533–543. [CrossRef]

83. Chen, Z.; Liu, C.; Patel, A.J.; Liao, C.-P.; Wang, Y.; Le, L.Q. Cells of Origin in the Embryonic Nerve Roots for NF1-Associated Plexiform Neurofibroma. *Cancer Cell* **2014**, *26*, 695–706. [CrossRef]

84. Jessen, K.R.; Mirsky, R. The Success and Failure of the Schwann Cell Response to Nerve Injury. *Front. Cell. Neurosci.* **2019**, *13*, 16. [CrossRef]

85. Clements, M.P.; Byrne, E.; Camarillo Guerrero, L.F.; Cattin, A.-L.; Zakka, L.; Ashraf, A.; Burden, J.J.; Khadayate, S.; Lloyd, A.C.; Marguerat, S.; et al. The Wound Microenvironment Reprograms Schwann Cells to Invasive Mesenchymal-like Cells to Drive Peripheral Nerve Regeneration. *Neuron* **2017**, *96*, 98–114. [CrossRef]

86. Wekerle, H.; Schwab, M.; Linington, C.; Meyermann, R. Antigen presentation in the peripheral nervous system: Schwann cells present endogenous myelin autoantigens to lymphocytes. *Eur. J. Immunol.* **1986**, *16*, 1551–1557. [CrossRef]

87. Jessen, K.R.; Mirsky, R. The repair Schwann cell and its function in regenerating nerves. *J. Physiol.* **2016**, *594*, 3521–3531. [CrossRef]

88. Mirsky, R.; Jessen, K.R. Schwann cell development, differentiation and myelination. *Curr. Opin. Neurobiol.* **1996**, *6*, 89–96. [CrossRef]

89. Castelnovo, L.; Bonalume, V.; Melfi, S.; Ballabio, M.; Colleoni, D.; Magnaghi, V. Schwann cell development, maturation and regeneration: A focus on classic and emerging intracellular signaling pathways. *Neural Regen. Res.* **2017**, *12*, 1013.

90. Woodhoo, A.; Sommer, L. Development of the Schwann cell lineage: From the neural crest to the myelinated nerve. *Glia* **2008**, *56*, 1481–1490. [CrossRef]

91. Jessen, K.R.; Mirsky, R. Schwann Cell Precursors; Multipotent Glial Cells in Embryonic Nerves. *Front. Mol. Neurosci.* **2019**, *12*, 366. [CrossRef]

92. Furlan, A.; Adameyko, I. Schwann cell precursor_ a neural crest cell in disguise? *Dev. Biol.* **2018**, *444*, S25–S35. [CrossRef]

93. Stolt, C.C.; Wegner, M. Schwann cells and their transcriptional network: Evolution of key regulators of peripheral myelination. *Brain Res.* **2016**, *1641*, 101–110. [CrossRef]

94. Jacob, C. ScienceDirect Chromatin-remodeling enzymes in control of Schwann cell development, maintenance and plasticity. *Curr. Opin. Neurobiol.* **2017**, *47*, 24–30. [CrossRef]

95. Ma, K.H.; Svaren, J. Epigenetic Control of Schwann Cells. *Neuroscientist* **2018**, *24*, 627–638. [CrossRef]

96. Ness, J.K.; Skiles, A.A.; Yap, E.H.; Fajardo, E.J.; Fiser, A.; Tapinos, N. Nuc-ErbB3 regulates H3K27me3 levels and HMT activity to establish epigenetic repression during peripheral myelination. *Glia* **2016**, *64*, 977–992. [CrossRef]

97. Ma, K.H.; Hung, H.A.; Srinivasan, R.; Xie, H.; Orkin, S.H.; Svaren, J. Regulation of Peripheral Nerve Myelin Maintenance by Gene Repression through Polycomb Repressive Complex 2. *J. Neurosci.* **2015**, *35*, 8640–8652. [CrossRef]

98. Heinen, A.; Tzekova, N.; Graffmann, N.; Torres, K.J.; Uhrberg, M.; Hartung, H.P.; Küry, P. Histone methyltransferase enhancer of zeste homolog 2 regulates Schwann cell differentiation. *Glia* **2012**, *60*, 1696–1708. [CrossRef]

99. Chen, Z.-L.; Yu, W.-M.; Strickland, S. Peripheral Regeneration. *Annu. Rev. Neurosci.* **2007**, *30*, 209–233. [CrossRef]

100. Fontana, X.; Hristova, M.; Da Costa, C.; Patodia, S.; Thei, L.; Makwana, M.; Spencer-Dene, B.; Latouche, M.; Mirsky, R.; Jessen, K.R.; et al. c-Jun in Schwann cells promotes axonal regeneration and motoneuron survival via paracrine signaling. *J. Cell Biol.* **2012**, *198*, 127–141. [CrossRef]

101. Hirota, H.; Kiyama, H.; Kishimoto, T.; Taga, T. Accelerated Nerve Regeneration in Mice by upregulated expression of interleukin (IL) 6 and IL-6 receptor after trauma. *J. Exp. Med.* **1996**, *183*, 2627–2634. [CrossRef]

102. Rotshenker, S. Wallerian degeneration: The innate-immune response to traumatic nerve injury. *J. Neuroinflammation* **2011**, *8*, 109. [CrossRef]

103. Cattin, A.-L.; Burden, J.J.; Van Emmenis, L.; Mackenzie, F.E.; Hoving, J.J.A.; Garcia Calavia, N.; Guo, Y.; McLaughlin, M.; Rosenberg, L.H.; Quereda, V.; et al. Macrophage-Induced Blood Vessels Guide Schwann Cell-Mediated Regeneration of Peripheral Nerves. *Cell* **2015**, *162*, 1127–1139. [CrossRef]

104. Stoll, G.; Müller, H.W. Nerve Injury, Axonal Degeneration and Neural Regeneration: Basic Insights. *Brain Pathol.* **1999**, *9*, 313–325. [CrossRef]

105. Ma, K.H.; Duong, P.; Moran, J.J.; Junaidi, N.; Svaren, J. Polycomb repression regulates Schwann cell proliferation and axon regeneration after nerve injury. *Glia* **2018**, *66*, 2487–2502. [CrossRef]

106. Ma, K.H.; Hung, H.A.; Svaren, J. Epigenomic Regulation of Schwann Cell Reprogramming in Peripheral Nerve Injury. *J. Neurosci.* **2016**, *36*, 9135–9147. [CrossRef]

107. Ribeiro, S.; Napoli, I.; White, I.J.; Parrinello, S.; Flanagan, A.M.; Suter, U.; Parada, L.F.; Lloyd, A.C. Injury signals cooperate with Nf1 loss to relieve the tumor-suppressive environment of adult peripheral nerve. *Cell Rep.* **2013**, *5*, 126–136. [CrossRef]

108. Korfhage, J.; Lombard, D.B. Malignant Peripheral Nerve Sheath Tumors: From Epigenome to Bedside. *Mol. Cancer Res.* **2019**, *17*, 1417–1428. [CrossRef]

109. Bottillo, I.; Ahlquist, T.; Brekke, H.; Danielsen, S.A.; van den Berg, E.; Mertens, F.; Lothe, R.A.; Dallapiccola, B. Germline and somatic NF1 mutations in sporadic and NF1-associated malignant peripheral nerve sheath tumours. *J. Pathol.* **2009**, *217*, 693–701. [CrossRef]

110. Dombi, E.; Baldwin, A.; Marcus, L.J.; Fisher, M.J.; Weiss, B.; Kim, A.; Whitcomb, P.; Martin, S.; Aschbacher-Smith, L.E.; Rizvi, T.A.; et al. Activity of Selumetinib in Neurofibromatosis Type 1–Related Plexiform Neurofibromas. *N. Engl. J. Med.* **2016**, *375*, 2550–2560. [CrossRef]

111. Kim, A.; Stewart, D.R.; Reilly, K.M.; Viskochil, D.; Miettinen, M.M.; Widemann, B.C. Malignant Peripheral Nerve Sheath Tumors State of the Science: Leveraging Clinical and Biological Insights into Effective Therapies. *Sarcoma* **2017**, *2017*, 7429697. [CrossRef]

112. Carroll, S.L. The Challenge of Cancer Genomics in Rare Nervous System Neoplasms: Malignant Peripheral Nerve Sheath Tumors as a Paradigm for Cross-Species Comparative Oncogenomics. *Am. J. Pathol.* **2016**, *186*, 464–477. [CrossRef]

113. Hnisz, D.; Abraham, B.J.; Lee, T.I.; Lau, A.; Saint-André, V.; Sigova, A.A.; Hoke, H.A.; Young, R.A. Super-Enhancers in the Control of Cell Identity and Disease. *Cell* **2013**, *155*, 934–947. [CrossRef]

114. Adhikari, A.; Davie, J. JARID2 and the PRC2 complex regulate skeletal muscle differentiation through regulation of canonical Wnt signaling. *Epigenet. Chromatin* **2018**, *11*, 1–20. [CrossRef]

115. Mirzamohammadi, F.; Papaioannou, G.; Inloes, J.B.; Rankin, E.B.; Xie, H.; Schipani, E.; Orkin, S.H.; Kobayashi, T. Polycomb repressive complex 2 regulates skeletal growth by suppressing Wnt and TGF-β signalling. *Nat. Commun.* **2016**, *7*, 12047. [CrossRef]

116. Wang, L.; Jin, Q.; Lee, J.-E.; Su, I.-H.; Ge, K. Histone H3K27 methyltransferase Ezh2 represses Wnt genes to facilitate adipogenesis. *Proc. Natl. Acad. Sci. USA* **2010**, *107*, 7317–7322. [CrossRef]

117. Rothberg, J.L.M.; Maganti, H.B.; Jrade, H.; Porter, C.J.; Palidwor, G.A.; Cafariello, C.; Battaion, H.L.; Khan, S.T.; Perkins, T.J.; Paulson, R.F.; et al. Mtf2-PRC2 control of canonical Wnt signaling is required for definitive erythropoiesis. *Cell Discov.* **2018**, *4*, 1–16. [CrossRef]

118. Oittinen, M.; Popp, A.; Kurppa, K.; Lindfors, K.; Mäki, M.; Kaikkonen, M.U.; Viiri, K. Polycomb Repressive Complex 2 Enacts Wnt Signaling in Intestinal Homeostasis and Contributes to the Instigation of Stemness in Diseases Entailing Epithelial Hyperplasia or Neoplasia. *Stem Cells* **2017**, *35*, 445–457. [CrossRef]

119. Reya, T.; Clevers, H. Wnt signalling in stem cells and cancer. *Nature* **2005**, *434*, 843–850. [CrossRef]
120. Chen, C.; Zhao, M.; Tian, A.; Zhang, X.; Yao, Z.; Ma, X. Aberrant activation of Wnt/β-catenin signaling drives proliferation of bone sarcoma cells. *Oncotarget* **2015**, *6*, 17570–17583. [CrossRef]
121. Üren, A.; Wolf, V.; Sun, Y.F.; Azari, A.; Rubin, J.S.; Toretsky, J.A. Wnt/Frizzled signaling in Ewing sarcoma. *Pediatric Blood Cancer* **2004**, *43*, 243–249. [CrossRef]
122. Abhinav Adhikari, J.D. Wnt deregulation in rhabdomyosarcoma. *Stem Cell Investig.* **2019**, *6*, 13. [CrossRef]
123. Watson, A.L.; Rahrmann, E.P.; Moriarity, B.S.; Choi, K.; Conboy, C.B.; Greeley, A.D.; Halfond, A.L.; Anderson, L.K.; Wahl, B.R.; Keng, V.W.; et al. Canonical Wnt/β-catenin Signaling Drives Human Schwann Cell Transformation, Progression, and Tumor Maintenance. *Cancer Discov.* **2013**, *3*, 674–689. [CrossRef]
124. Luscan, A.; Shackleford, G.; Masliah-Planchon, J.; Laurendeau, I.; Ortonne, N.; Varin, J.; Lallemand, F.; Leroy, K.; Dumaine, V.; Hivelin, M.; et al. The Activation of the WNT Signaling Pathway Is a Hallmark in Neurofibromatosis Type 1 Tumorigenesis. *Clin. Cancer Res.* **2014**, *20*, 358–371. [CrossRef]
125. Hu, P.; Chu, J.; Wu, Y.; Sun, L.; Lv, X.; Zhu, Y.; Li, J.; Guo, Q.; Gong, C.; Liu, B.; et al. NBAT1 suppresses breast cancer metastasis by regulating DKK1 via PRC2. *Oncotarget* **2015**, *6*, 32410–32425. [CrossRef]
126. Nakagawa, M.; Fujita, S.; Katsumoto, T.; Yamagata, K.; Ogawara, Y.; Hattori, A.; Kagiyama, Y.; Honma, D.; Araki, K.; Inoue, T.; et al. Dual inhibition of enhancer of zeste homolog 1/2 overactivates WNT signaling to deplete cancer stem cells in multiple myeloma. *Cancer Sci.* **2019**, *110*, 194–208. [CrossRef]
127. Serresi, M.; Gargiulo, G.; Proost, N.; Siteur, B.; Cesaroni, M.; Koppens, M.; Xie, H.; Sutherland, K.D.; Hulsman, D.; Citterio, E.; et al. Polycomb Repressive Complex 2 Is a Barrier to KRAS- Driven Inflammation and Epithelial-Mesenchymal Transition in Non-Small-Cell Lung Cancer. *Cancer Cell* **2016**, *29*, 17–31. [CrossRef]
128. Tosello, V.; Ferrando, A.A. The NOTCH signaling pathway: Role in the pathogenesis of T-cell acute lymphoblastic leukemia and implication for therapy. *Ther. Adv. Hematol.* **2013**, *4*, 199–210. [CrossRef]
129. Gordon, W.R.; Arnett, K.L.; Blacklow, S.C. The molecular logic of Notch signaling–A structural and biochemical perspective. *J. Cell Sci.* **2008**, *121*, 3109–3119. [CrossRef]
130. Bray, S.J. Notch signalling: A simple pathway becomes complex. *Nat. Rev. Mol. Cell Biol.* **2006**, *7*, 678–689. [CrossRef]
131. Kopan, R.; Ilagan, M.X.G. The Canonical Notch Signaling Pathway: Unfolding the Activation Mechanism. *Cell* **2009**, *137*, 216–233. [CrossRef]
132. Kovall, R.A. More complicated than it looks: Assembly of Notch pathway transcription complexes. *Oncogene* **2008**, *27*, 5099–5109. [CrossRef]
133. Han, X.; Ranganathan, P.; Tzimas, C.; Weaver, K.L.; Jin, K.; Astudillo, L.; Zhou, W.; Zhu, X.; Li, B.; Robbins, D.J.; et al. Notch Represses Transcription by PRC2 Recruitment to the Ternary Complex. *Mol. Cancer Res.* **2017**, *15*, 1173–1183. [CrossRef]
134. Weijzen, S.; Rizzo, P.; Braid, M.; Vaishnav, R.; Jonkheer, S.M.; Zlobin, A.; Osborne, B.A.; Gottipati, S.; Aster, J.C.; Hahn, W.C.; et al. Activation of Notch-1 signaling maintains the neoplastic phenotype in human Ras-transformed cells. *Nat. Med.* **2002**, *8*, 979–986. [CrossRef]
135. Li, Y.; Rao, P.K.; Wen, R.; Song, Y.; Muir, D.; Wallace, P.; van Horne, S.J.; Tennekoon, G.I.; Kadesch, T. Notch and Schwann cell transformation. *Oncogene* **2004**, *23*, 1146–1152. [CrossRef]
136. Burr, M.L.; Sparbier, C.E.; Chan, K.L.; Chan, Y.-C.; Kersbergen, A.; Lam, E.Y.N.; Azidis-Yates, E.; Vassiliadis, D.; Bell, C.C.; Gilan, O.; et al. An Evolutionarily Conserved Function of Polycomb Silences the MHC Class I Antigen Presentation Pathway and Enables Immune Evasion in Cancer. *Cancer Cell* **2019**, *36*, 385–401. [CrossRef]
137. Agudo, J.; Park, E.S.; Rose, S.A.; Alibo, E.; Sweeney, R.; Dhainaut, M.; Kobayashi, K.S.; Sachidanandam, R.; Baccarini, A.; Merad, M.; et al. Quiescent Tissue Stem Cells Evade Immune Surveillance. *Immunity* **2018**, *48*, 271–285. [CrossRef]
138. Zingg, D.; Arenas-Ramirez, N.; Sahin, D.; Rosalia, R.A.; Antunes, A.T.; Haeusel, J.; Sommer, L.; Boyman, O. The Histone Methyltransferase Ezh2 Controls Mechanisms of Adaptive Resistance to Tumor Immunotherapy. *Cell Rep.* **2017**, *20*, 854–867. [CrossRef]
139. Christian, S.L.; Collier, T.W.; Zu, D.; Licursi, M.; Hough, C.M.; Hirasawa, K. Activated Ras/MEK Inhibits the Antiviral Response of Alpha Interferon by Reducing STAT2 Levels. *J. Virol.* **2009**, *83*, 6717–6726. [CrossRef]
140. AbuSara, N.; Razavi, S.; Derwish, L.; Komatsu, Y.; Licursi, M.; Hirasawa, K. Restoration of IRF1-dependent anticancer effects by MEK inhibition in human cancer cells. *Cancer Lett.* **2015**, *357*, 575–581. [CrossRef]

141. Murray, E.K.; Hien, A.; de Vries, G.J.; Forger, N.G. Epigenetic control of sexual differentiation of the bed nucleus of the stria terminalis. *Endocrinology* **2009**, *150*, 4241–4247. [CrossRef]

142. Speert, D.B.; Konkle, A.T.M.; Zup, S.L.; Schwarz, J.M.; Shiroor, C.; Taylor, M.E.; McCarthy, M.M. Focal adhesion kinase and paxillin: Novel regulators of brain sexual differentiation? *Endocrinology* **2007**, *148*, 3391–3401. [CrossRef]

143. Amirnasr, A.; Verdijk, R.M.; van Kuijk, P.F.; Taal, W.; Sleijfer, S.; Wiemer, E.A.C. Expression and inhibition of BRD4, EZH2 and TOP2A in neurofibromas and malignant peripheral nerve sheath tumors. *PLoS ONE* **2017**, *12*, e0183155. [CrossRef]

144. Fletcher, J.A.; Kozakewich, H.P.; Hoffer, F.A.; Lage, J.M.; Weidner, N.; Tepper, R.; Pinkus, G.S.; Morton, C.C.; Corson, J.M. Diagnostic Relevance of Clonal Cytogenetic Aberrations in Malignant Soft-Tissue Tumors. *N. Engl. J. Med.* **2010**, *324*, 436–443. [CrossRef]

145. Reynolds, J.E.; Fletcher, J.A.; Lytle, C.H.; Nie, L.; Morton, C.C.; Diehl, S.R. Molecular characterization of a 17q11.2 translocation in a malignant schwannoma cell line. *Hum. Genet.* **1992**, *90*, 450–456. [CrossRef]

146. DeClue, J.E.; Papageorge, A.G.; Fletcher, J.A.; Diehl, S.R.; Ratner, N.; Vass, W.C.; Lowy, D.R. Abnormal regulation of mammalian p21ras contributes to malignant tumor growth in von Recklinghausen (type 1) neurofibromatosis. *Cell* **1992**, *69*, 265–273. [CrossRef]

147. Glover, T.W.; Stein, C.K.; Legius, E.; Andersen, L.B.; Brereton, A.; Johnson, S. Molecular and cytogenetic analysis of tumors in von recklinghausen neurofibromatosis. *Genes Chromosom. Cancer* **1991**, *3*, 62–70. [CrossRef]

148. Yang, K.; Guo, W.; Ren, T.; Huang, Y.; Han, Y.; Zhang, H.; Zhang, J. Knockdown of HMGA2 regulates the level of autophagy via interactions between MSI2 and Beclin1 to inhibit NF1-associated malignant peripheral nerve sheath tumour growth. *J. Exp. Clin. Cancer Res.* **2019**, *38*, 1–18. [CrossRef]

149. Li, H.; Zhang, X.; Fishbein, L.; Kweh, F.; Campbell-Thompson, M.; Perrin, G.Q.; Muir, D.; Wallace, M. Analysis of steroid hormone effects on xenografted human NF1 tumor schwann cells. *Cancer Biol. Ther.* **2010**, *10*, 758–764. [CrossRef]

150. Kahen, E.J.; Brohl, A.; Yu, D.; Welch, D.; Cubitt, C.L.; Lee, J.K.; Chen, Y.; Yoder, S.J.; Teer, J.K.; Zhang, Y.O.; et al. Neurofibromin level directs RAS pathway signaling and mediates sensitivity to targeted agents in malignant peripheral nerve sheath tumors. *Oncotarget* **2018**, *9*, 22571–22585. [CrossRef]

151. Perrin, G.Q.; Fishbein, L.; Thomson, S.A.; Thomas, S.L.; Stephens, K.; Garbern, J.Y.; DeVries, G.H.; Yachnis, A.T.; Wallace, M.R.; Muir, D. Plexiform-like neurofibromas develop in the mouse by intraneural xenograft of an NF1 tumor-derived Schwann cell line. *J. Neurosci. Res.* **2007**, *85*, 1347–1357. [CrossRef]

152. Perrin, G.Q.; Li, H.; Fishbein, L.; Thomson, S.A.; Hwang, M.S.; Scarborough, M.T.; Yachnis, A.T.; Wallace, M.R.; Mareci, T.H.; Muir, D. An orthotopic xenograft model of intraneural NF1 MPNST suggests a potential association between steroid hormones and tumor cell proliferation. *Lab. Investig.* **2007**, *87*, 1092–1102. [CrossRef]

153. Frahm, S.; Mautner, V.-F.; Brems, H.; Legius, E.; Debiec-Rychter, M.; Friedrich, R.E.; Knöfel, W.T.; Peiper, M.; Kluwe, L. Genetic and phenotypic characterization of tumor cells derived from malignant peripheral nerve sheath tumors of neurofibromatosis type 1 patients. *Neurobiol. Dis.* **2004**, *16*, 85–91. [CrossRef]

154. Mahller, Y.Y.; Vaikunth, S.S.; Ripberger, M.C.; Baird, W.H.; Saeki, Y.; Cancelas, J.A.; Crombleholme, T.M.; Cripe, T.P. Tissue Inhibitor of Metalloproteinase-3 via Oncolytic Herpesvirus Inhibits Tumor Growth and Vascular Progenitors. *Cancer Res.* **2008**, *68*, 1170–1179. [CrossRef]

155. Mashour, G.A.; Drissel, S.N.; Frahm, S.; Farassati, F.; Martuza, R.L.; Mautner, V.-F.; Kindler-Röhrborn, A.; Kurtz, A. Differential modulation of malignant peripheral nerve sheath tumor growth by omega-3 and omega-6 fatty acids. *Oncogene* **2005**, *24*, 2367–2374. [CrossRef]

156. Hakozaki, M.; Hojo, H.; Sato, M.; Tajino, T.; Yamada, H.; Kikuchi, S.; Abe, M. Establishment and characterization of a novel human malignant peripheral nerve sheath tumor cell line, FMS-1, that overexpresses epidermal growth factor receptor and cyclooxygenase-2. *Virchows Arch.* **2009**, *455*, 517–526. [CrossRef]

157. Aoki, M.; Nabeshima, K.; Nishio, J.; Ishiguro, M.; Fujita, C.; Koga, K.; Hamasaki, M.; Kaneko, Y.; Iwasaki, H. Establishment of three malignant peripheral nerve sheath tumor cell lines, FU-SFT8611, 8710 and 9817: Conventional and molecular cytogenetic characterization. *Int. J. Oncol.* **2006**, *29*, 1421–1428. [CrossRef]

158. Holtkamp, N.; Malzer, E.; Zietsch, J.; Neuro, A.O. EGFR and erbB2 in malignant peripheral nerve sheath tumors and implications for targeted therapy. *Neuro Oncol.* **2008**, *10*, 946–957. [CrossRef]

159. Imaizumi, S.; Motoyama, T.; Ogose, A.; Hotta, T.; Takahashi, H.E. Characterization and chemosensitivity of two human malignant peripheral nerve sheath tumour cell lines derived from a patient with neurofibromatosis type 1. *Virchows Arch.* **1998**, *433*, 435–441. [CrossRef]

160. Subramanian, S.; Thayanithy, V.; West, R.B.; Lee, C.-H.; Beck, A.H.; Zhu, S.; Downs-Kelly, E.; Montgomery, K.; Goldblum, J.R.; Hogendoorn, P.C.; et al. Genome-wide transcriptome analyses reveal p53 inactivation mediated loss of miR-34a expression in malignant peripheral nerve sheath tumours. *J. Pathol.* **2010**, *220*, 58–70. [CrossRef]

161. Lopez, G.; Torres, K.; Liu, J.; Hernandez, B.; Young, E.; Belousov, R.; Bolshakov, S.; Lazar, A.J.; Slopis, J.M.; McCutcheon, I.E.; et al. Autophagic Survival in Resistance to Histone Deacetylase Inhibitors: Novel Strategies to Treat Malignant Peripheral Nerve Sheath Tumors. *Cancer Res.* **2011**, *71*, 185–196. [CrossRef]

162. Spyra, M.; Kluwe, L.; Hagel, C.; Nguyen, R.; Panse, J.; Kurtz, A.; Mautner, V.-F.; Rabkin, S.D.; Demestre, M. Cancer Stem Cell-Like Cells Derived from Malignant Peripheral Nerve Sheath Tumors. *PLoS ONE* **2011**, *6*, e21099. [CrossRef]

163. Badache, A.; De Vries, G.H. Neurofibrosarcoma-derived Schwann cells overexpress platelet-derived growth factor (PDGF) receptors and are induced to proliferate by PDGF BB. *J. Cell. Physiol.* **1998**, *177*, 334–342. [CrossRef]

164. Sonobe, H.; Takeuchi, T.; Furihata, M.; Taguchi, T.; Kawai, A.; Ohjimi, Y.; Iwasaki, H.; Kaneko, Y.; Ohtsuki, Y. A new human malignant peripheral nerve sheath tumour-cell line, HS-sch-2, harbouring p53 point mutation. *Int. J. Oncol.* **2000**, *17*, 347–352. [CrossRef]

165. Kolberg, M.; Bruun, J.; Murumägi, A.; Mpindi, J.P.; Bergsland, C.H.; Høland, M.; Eilertsen, I.A.; Danielsen, S.A.; Kallioniemi, O.; Lothe, R.A. Drug sensitivity and resistance testing identifies PLK1 inhibitors and gemcitabine as potent drugs for malignant peripheral nerve sheath tumors. *Mol. Oncol.* **2017**, *11*, 1156–1171. [CrossRef]

166. Schoffski, P.; Van Renterghem, B.; Cornillie, J.; Wang, Y.; Gebreyohannes, Y.K.; Lee, C.-J.; Wellens, J.; Vanleeuw, U.; Nysen, M.; Hompes, D.; et al. XenoSarc: A comprehensive platform of patient-derived xenograft (PDX) models of soft tissue sarcoma (STS) for early drug testing. *J. Glob. Oncol.* **2019**, *5*, 37. [CrossRef]

167. Castellsague, J.; Gel, B.; Fernandez-Rodriguez, J.; Llatjos, R.; Blanco, I.; Benavente, Y.; Perez-Sidelnikova, D.; Garcia-del Muro, J.; Vinals, J.M.; Vidal, A.; et al. Comprehensive establishment and characterization of orthoxenograft mouse models of malignant peripheral nerve sheath tumors for personalized medicine. *EMBO Mol. Med.* **2015**, *7*, 608–627. [CrossRef]

168. Brossier, N.M.; Carroll, S.L. Genetically engineered mouse models shed new light on the pathogenesis of neurofibromatosis type I-related neoplasms of the peripheral nervous system. *Brain Res. Bull.* **2012**, *88*, 58–71. [CrossRef]

169. Rhodes, S.D.; He, Y.; Smith, A.; Jiang, L.; Lu, Q.; Mund, J.; Li, X.; Bessler, W.; Qian, S.; Dyer, W.; et al. Cdkn2a (Arf) loss drives NF1-associated atypical neurofibroma and malignant transformation. *Hum. Mol. Genet.* **2019**, *28*, 2752–2762. [CrossRef]

MDPI
St. Alban-Anlage 66
4052 Basel
Switzerland
Tel. +41 61 683 77 34
Fax +41 61 302 89 18
www.mdpi.com

Genes Editorial Office
E-mail: genes@mdpi.com
www.mdpi.com/journal/genes